Lecture Notes in Mathematics 1693

Editors:
J.-M. Morel, Cachan
F. Takens, Groningen
B. Teissier, Paris

Stephen Simons

From Hahn-Banach
to Monotonicity

2nd, expanded edition

 Springer

Stephen Simons
Department of Mathematics
University of California
Santa Barbara, CA 93105-3080
USA
simons@math.ucsb.edu
http://www.math.ucsb.edu/~simons

1st edition 1998 LNM 1693: Minimax and Monotonicity

ISBN 978-1-4020-6918-5   e-ISBN 978-1-4020-6919-2

DOI 10.1007/978-1-4020-6919-2

Lecture Notes in Mathematics ISSN print edition: 0075-8434
                             ISSN electronic edition: 1617-9692

Library of Congress Control Number: 2007942159

Mathematics Subject Classification (2000): 46A22, 49J35, 47N10, 49J52, 47H05

Cover design: *WMXDesign GmbH*

Printed on acid-free paper

9 8 7 6 5 4 3 2 1

springer.com

*For Jacqueline,
whose support and patience
are unbounded.*

Preface

A more accurate title for these notes would be: "The Hahn–Banach–Lagrange theorem, Convex analysis, Symmetrically self–dual spaces, Fitzpatrick functions and monotone multifunctions".

The Hahn–Banach–Lagrange theorem is a version of the Hahn–Banach theorem that is admirably suited to applications to the theory of monotone multifunctions, but it turns out that it also leads to extremely short proofs of the standard existence theorems of functional analysis, a minimax theorem, a Lagrange multiplier theorem for constrained convex optimization problems, and the Fenchel duality theorem of convex analysis.

Another feature of the Hahn–Banach–Lagrange theorem is that it can be used to transform problems on the existence of continuous linear functionals into problems on the existence of a single real constant, and then obtain a sharp lower bound on the norm of the linear functional satisfying the required condition. This is the case with both the Lagrange multiplier theorem and the Fenchel duality theorem applications mentioned above.

A multifunction from a Banach space into the subsets of its dual can, of course, be identified with a subset of the product of the space with its dual. Simon Fitzpatrick defined a convex function on this product corresponding with any such multifunction. So part of these notes is devoted to the rather special convex analysis for the product of a Banach space with its dual.

The product of a Banach space with its dual is a special case of a "symmetrically self–dual space". The advantage of going to this slightly higher level of abstraction is not only that it leads to more general results but, more to the point, it cuts the length of each proof approximately in half which, in turn, gives a much greater insight into the nature of the processes involved. Monotone multifunctions then correspond to subsets of the symmetrically self–dual space that are "positive" with respect to a certain quadratic form.

We investigate a particular kind of convex function on a symmetrically self–dual space, which we call a "BC–function". Since the Fitzpatrick function of a maximally monotone multifunction is always a BC–function, these BC–functions turn out to be very successful for obtaining results on maximally monotone multifunctions on reflexive spaces.

The situation for nonreflexive spaces is more challenging. Here, it turns out that we must consider two symmetrically self–dual spaces, and we call the corresponding convex functions "$\widetilde{\mathrm{BC}}$–functions". In this case, a number of different subclasses of the maximally monotone multifunctions have been introduced over the years — we give particular attention to those that are "of type (ED)". These have the great virtue that all the common maximally monotone multifunctions are of type (ED), and maximally monotone multifunctions of type (ED) have nearly all the properties that one could desire. In order to study the maximally monotone multifunctions of type (ED), we have to introduce a weird topology on the bidual which has a number of very nice properties, despite that fact that it is not normally compatible with its vector space structure.

These notes are somewhere between a sequel to and a new edition of [99]. As in [99], the essential idea is to reduce questions on monotone multifunctions to questions on convex functions. In [99], this was achieved using a "big convexification" of the graph of the multifunction and the "minimax technique" for proving the existence of linear functionals satisfying certain conditions. The "big convexification" is a very abstract concept, and the analysis is quite heavy in computation. The Fitzpatrick function gives another, more concrete, way of associating a convex functions with a monotone multifunction. The problem is that many of the questions on convex functions that one obtains require an analysis of the special properties of convex functions on the product of a Banach space with its dual, which is exactly what we do in these notes. It is also worth noting that the minimax theorem is hardly used here.

We envision that these notes could be used for four different possible courses/seminars:

• An introductory course in functional analysis which would, at the same time, touch on minimax theorems and give a grounding in convex Lagrange multiplier theory and the main theorems in convex analysis.
• A course in which results on monotonicity on general Banach spaces are established using symmetrically self–dual spaces and Fitzpatrick functions.
• A course in which results on monotonicity on reflexive Banach spaces are established using symmetrically self–dual spaces and Fitzpatrick functions.
• A seminar in which the the more technical properties of maximal monotonicity on general Banach spaces that have been established since 1997 are discussed.

We give more details of these four possible uses at the end of the introduction.

I would like to express my sincerest thanks to Heinz Bausckhe, Patrick Combettes, Michael Crandall, Carl de Boor, Radu Ioan Boţ, Juan Enrique Martínez-Legaz, Xianfu Wang and Constantin Zălinescu for reading preliminary versions of parts of these notes, making a number of excellent suggestions and, of course, finding a number of errors.

Of course, despite all the excellent efforts of the people mentioned above, these notes doubtless still contain errors and ambiguities, and also doubtless have other stylistic shortcomings. At any rate, I hope that there are not too many of these. Those that do exist are entirely my fault.

Stephen Simons
September 23, 2007
Santa Barbara
California

Table of Contents

Introduction 1

I The Hahn-Banach-Lagrange theorem and some consequences

1 The Hahn–Banach–Lagrange theorem 15

2 Applications to functional analysis 23

3 A minimax theorem 24

4 The dual and bidual of a normed space 25

5 Excess, duality gap, and minimax criteria for weak
   compactness 28

6 Sharp Lagrange multiplier and KKT results 32

II Fenchel duality

7 A sharp version of the Fenchel Duality theorem 41

8 Fenchel duality with respect to a bilinear form —
   locally convex spaces 44

9 Some properties of $\frac{1}{2}\|\cdot\|^2$ 49

10 The conjugate of a sum in the locally convex case 51

11 Fenchel duality vs the conjugate of a sum 54

12 The restricted biconjugate and Fenchel–Moreau points ... 58

13 Surrounding sets and the dom lemma 60

14 The ⊖-theorem 62

15 The Attouch–Brezis theorem 65

16 A bivariate Attouch–Brezis theorem 67

III Multifunctions, SSD spaces, monotonicity and Fitzpatrick functions

17 Multifunctions, monotonicity and maximality 71

18 Subdifferentials are maximally monotone 74

19 SSD spaces, q–positive sets and BC–functions 79

20 Maximally q–positive sets in SSD spaces 86

21 SSDB spaces 88

22 The SSD space $E \times E^*$ 93

23 Fitzpatrick functions and fitzpatrifications 99

24 The maximal monotonicity of a sum 103

IV Monotone multifunctions on general Banach spaces

25 Monotone multifunctions with bounded range 107

26 A general local boundedness theorem 108

27 The six set theorem and the nine set theorem 108

28 $D(S_\varphi)$ and various hulls 111

V Monotone multifunctions on reflexive Banach spaces

29 Criteria for maximality, and Rockafellar's surjectivity
   theorem 117

30 Surjectivity and an abstract Hammerstein theorem 123

31 The Brezis–Haraux condition 125

32 Bootstrapping the sum theorem 128

33 The $>$ six set and the $>$ nine set theorems for pairs
   of multifunctions 130

34 The Brezis–Crandall–Pazy condition 132

VI Special maximally monotone multifunctions

35 The norm–dual of the space $E \times E^*$ and $\widetilde{BC}$–functions ... 139

36 Subclasses of the maximally monotone multifunctions ... 147

37 First application of Theorem 35.8: type (D) implies
type (FP) 153

38 $\mathcal{T}_{CLB}(E^{**})$, $\mathcal{T}_{CLBN}(B^*)$ and type (ED) 154

39 Second application of Theorem 35.8: type (ED) implies
type (FPV)... 157

40 Final applications of Theorem 35.8: type (ED) implies
strong... 158

41 Strong maximality and coercivity 159

42 Type (ED) implies type (ANA) and type (BR)........... 161

43 The closure of the range 167

44 The sum problem and the closure of the domain 170

45 The biconjugate of a maximum and $\mathcal{T}_{CLB}(E^{**})$ 172

46 Maximally monotone multifunctions with convex graph .. 180

47 Possibly discontinuous positive linear operators 183

48 Subtler properties of subdifferentials 188

49 Saddle functions and type (ED) 192

VII The sum problem for general Banach spaces

50 Introductory comments 197

51 Voisei's theorem 197

52 Sums with normality maps 198

53 A theorem of Verona–Verona 199

VIII Open problems   203

IX Glossary of classes of multifunctions   205

X A selection of results   207

References 233
Subject index 239
Symbol index 243

Introduction

These notes fall into three distinct parts. In Chapter I, we discuss the "Hahn–Banach–Lagrange theorem", a new version of the Hahn–Banach theorem, which gives very efficient proofs of the main existence theorems in functional analysis, optimization theory, minimax theory and convex analysis. In Chapter II, we zero in on the applications to convex analysis. In the remaining five chapters, we show how the results of the first two chapters can be used to obtain a large number of results on monotone multifunctions, many of which have not yet appeared in print.

Chapter I: The main result in Chapter I is the "Hahn-Banach-Lagrange" theorem, which first appeared in [103]. We prove this result in Theorem 1.11, discuss the classical functional analytic applications in Section 2 (namely the "Sandwich theorem" in Corollary 2.1, the "extension form of the Hahn–Banach theorem" in Corollary 2.2, and the "one dimensional form of the Hahn–Banach theorem" in Corollary 2.4) and give an application to a classical minimax theorem in Section 3. In Section 4, we introduce the results from classical Banach space theory that we shall need. In Section 5, we prove, among other things, a minimax criterion for a subset of a Banach space to be weakly compact using the concepts of "excess" and "duality gap". The contents of this section first appeared in [102].

In Section 6, we give a necessary and sufficient condition for the existence of Lagrange multipliers for constrained convex optimization problems (generalizing the classical sufficient "Slater condition"), with a sharp lower bound on the norm of the multiplier. We also prove a similar result for Karush–Kuhn–Tucker problems for functions with convex Gâteaux derivatives. Some of the results on Lagrange multipliers first appeared in [104]. In the flowchart below, we show the dependencies of the sections in Chapter I. We note, in particular, that Section 6 does not depend on Sections 2–5.

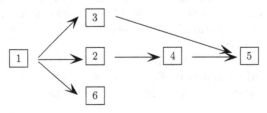

Chapter II: As explained above, Chapter II is about convex analysis. We start our discussion in Section 7 by using the Hahn-Banach-Lagrange theorem to obtain a necessary *and sufficient* condition for the Fenchel duality theorem to hold for two convex functions on a normed space, with a sharp lower bound on the norm of the functional obtained. (Incidentally, this approach avoids the aggravating problem of the "vertical hyperplane" that so destroys the elegance of the usual approach through the Eidelheit separation theorem.) This sharp version of the Fenchel duality theorem is in Theorem 7.4, and it is explained in Remark 7.6 how the lower bound obtained is of a very geometric character.

While the concept of Fenchel conjugate is introduced in Section 7 with reference to a convex function on a normed space, in fact this causes no end of confusion when dealing with monotone multifunctions on a nonreflexive Banach space. The way out of this problem (as has been observed by many authors) is to define Fenchel conjugates with respect to a dual pair of spaces. This is what we do in Section 8, and it enables a painless transition to the locally convex case. As we will see in Section 22, this is exactly what we need for our discussion of monotone multifunctions on a nonreflexive Banach space. We present a necessary and sufficient condition for the Fenchel duality theorem to be true in this sense in Theorem 8.1, and in Theorem 8.4 we present a unifying sufficient condition that implies the results that are used in practice, the versions due to Rockafellar and Attouch–Brezis. Theorem 8.4 uses the binary operation $\ominus$ defined in Notation 8.3.

In Section 9, we return to the normed case and give some results of a more numerical character, in which we explore the properties of the function $\frac{1}{2}\|\cdot\|^2$. These results will enable us to give a precise expression for the minimum norm of the resolvent of a maximally monotone multifunction on a reflexive Banach space in Theorem 29.5.

We bootstrap Theorem 8.4 in Section 10, and obtain sufficient conditions for the "inf–convolution" formula for the conjugate of a sum to hold, and give as application in Corollary 10.4 a consequence that will be applied in Theorem 21.10 to the existence of autoconjugates in SSDB spaces. This bootstrapping operation exhibits the well known fact that results on the conjugate of a sum are very close to the Fenchel duality theorem. However, these concepts are not interchangeable, and in Section 11 we give examples which should serve to distinguish them (giving examples of the failure of "stability" in duality).

In Section 12, we introduce the concepts of the *biconjugate* of a convex function, and the *Fenchel–Moreau points* of a convex function on a locally convex space. We deduce the Fenchel–Moreau formula in Corollary 12.4 in the case where the function is lower semicontinuous. Some of these results first appeared in [103].

We collect together in Sections 13 and 14 various results on convex functions that depend ultimately on Baire's theorem. The "dom lemma", Lemma 13.3, is a generalization to convex functions of the classical uniform boundedness (Banach Steinhaus) theorem (see Remark 13.6) and the "⊖-theorem", Theorem 14.2 (which uses the operation ⊖ already mentioned) is a generalization to convex functions of the classical open mapping theorem (see Remark 14.4). Both of these results will be applied later on to obtain results on monotonicity. We can think of the dom lemma and the ⊖-theorem as "quantitative" results, since their main purpose is to provide numerical bounds. Associated with them are two "qualitative" results, the "dom corollary", Corollary 13.5, and the "⊖-corollary", Corollary 14.3, from which the numerics have been removed. The ⊖-corollary will also be of use to us later on. In Remark 14.5, we give a brief discussion of convex Borel sets and functions.

In Theorem 15.1, we show how the ⊖-theorem leads to the Attouch–Brezis version of the Fenchel duality theorem, which we will use (via the local transversality theorem, Theorem 21.12, and Theorem 30.1) to prove various surjectivity results, including an abstract Hammerstein theorem; and in Theorem 16.4 we obtain a bivariate version of the Attouch-Brezis theorem, which we will use in Theorem 24.1 (via Lemma 22.9) and in Theorem 35.8 (a result that is fundamental for the understanding of maximally monotone multifunctions on nonreflexive Banach spaces). This bivariate version of the Attouch–Brezis theorem first appeared in [109].

Chapter III: In Chapter III, we will discuss the basic result on monotonicity. Section 17 starts off with a conventional discussion of multifunctions, monotonicity and maximal monotonicity. Remark 17.1 is a bridge in which we show that if E is a Banach space then there is a vector space B and a quadratic form q on B such that if $S\colon E \rightrightarrows E^*$ is a multifunction then there is a subset A of B such that S is monotone if and only if $b, c \in A \implies q(b-c) \geq 0$. Actually $B = E \times E^*$ and $A = G(S)$, but this paradigm leads to a strict generalization of monotonicity, in which the proofs are much more concise.

In Section 18, we digress a little from the general theory in order to give a short proof of Rockafellar's fundamental result that the subdifferential of a proper convex lower semicontinuous function on a Banach space is maximally monotone. In Theorem 18.1 and Theorem 18.2, we give the formula for the subdifferential of the sum of convex functions under two different hypotheses, in Corollary 18.5 and Theorem 18.6, we show how to deduce the Brøndsted–Rockafellar theorem from Ekeland's variational principle and the Hahn–Banach–Lagrange theorem, and then we come finally to our proof of the maximal monotonicity of subdifferentials in Theorem 18.7, which is based on the very elegant one found recently by M. Marques Alves and B. F. Svaiter in [60]. We also give in Corollary 18.3 and Theorem 18.10 two results about normal cones that will be useful later on. Readers who are familiar with the formula for the subdifferential of the sum of convex functions and the Brøndsted–Rockafellar theorem should be able to understand this section without having to read any of the previous sections.

We return to our development of the general theory in Sections 19–21. In Section 19, we introduce the concept of a SSD (symmetrically self–dual) space, a nonzero real vector space with a symmetric bilinear form which separates points. This bilinear form defines a quadratic form, q, in the obvious way. In general, this quadratic form is not positive, but we isolate certain subsets of a SSD space that we will call "q–positive". Appropriate convex functions on the SSD space define q–positive sets. We zero in on a subclass of the convex functions on a SSD space which we call "BC–functions". Critical to this enterprise is the self–dual property, because the conjugate of a convex function has the same domain of definition as the original convex function. Lemma 19.12 contains an unexpected result on BC–functions, but the most important result on BC–functions is undoubtedly the transversality theorem, Theorem 19.16, which leads (via Theorem 21.4) to generalizations of Rockafellar's classical surjectivity theorem for maximally monotone multifunctions on a reflexive Banach space (see Theorem 29.5) together with a sharp lower bound on the norm of solutions in terms of the Fitzpatrick function (see Theorem 29.6), and to sufficient conditions for the sum of maximally monotone multifunctions on a reflexive Banach space to be maximally monotone (see Theorem 24.1). Section 19 concludes with a discussion of how every q–positive set, A, gives rise to a convex function, Φ_A (this construction is an abstraction of the construction of the "Fitzpatrick function" that we will consider in Section 23). In Section 20, we introduce maximally q–positive sets, and show that the convex function determined by a maximally q–positive set is a BC–function.

In Section 21, we introduce the SSDB spaces, which are SSD spaces with an appropriate Banach norm. Roughly speaking, the additional structure that SSDB spaces possess over SSD spaces is ultimately what accounts for the fact that maximally monotone multifunctions on reflexive Banach spaces are much more tractable than maximally monotone multifunctions on general Banach spaces. That is not to say that the SSD space determined by a nonreflexive Banach space does not have a norm structure, the problem is that this norm structure is not "appropriate". Apart from Theorem 21.4, which we have already mentioned, the other important results in this section are Theorem 21.10 on the existence of autoconjugates, Theorem 21.11, which gives a formula for a maximally q–positive superset of a given nonempty q–positive set, and the local transversality theorem, Theorem 21.12, which leads ultimately to a number of surjectivity results, including an abstract Hammerstein theorem in Section 30.

We start considering in earnest the special SSD space $E \times E^*$ (where E is a nonzero Banach space) in Section 22. We first prove some preliminary results which depend ultimately on Rockafellar's version of the Fenchel duality theorem introduced in Corollary 8.6. It is important to realize that, despite the fact that E is a Banach space, we need Corollary 8.6 for *locally convex spaces* since the topology we are using for this result is the topology

$\mathcal{T}_{\parallel\ \parallel}(E) \times w(E^*, E)$ on $E \times E^*$. Theorem 22.5 has a precise description of the projection on E of the domain of the conjugate of a proper convex function on $E \times E^*$ in terms of a related convex function on E. In Theorem 22.8, we establish the equality of six sets determined by certain proper convex functions on $E \times E^*$, and in Lemma 22.9 we prove a result which will be critical for our treatment of sum theorems for maximally monotone multifunctions in Theorem 24.1.

In Section 23, we show how the concepts introduced in Sections 19–21 specialize to the situation considered in Section 22. The q–positive sets introduced in Section 19 then become the graphs of monotone multifunctions, the maximally q–positive sets introduced in Section 20 then become the graphs of maximally monotone multifunctions, and the function Φ_A determined by a q–positive set A introduced in Section 19 becomes the Fitzpatrick function, φ_S, determined by a monotone multifunction S. The Fitzpatrick function was originally introduced in [42] in 1988, but lay dormant until it was rediscovered by Martínez-Legaz and Théra in [63] in 2001.

This is an appropriate place for us to make a comparison between the analysis presented in these notes with the analysis presented in [99]. In both cases, the essential idea is to reduce questions on monotone multifunctions to questions on convex functions. In [99], this was achieved using a "big convexification" of the graph of the multifunction and the "minimax technique" for proving the existence of linear functionals satisfying certain conditions. This technique is very successful for working back from conjectures, and finding conditions under which they hold. On the other hand, the "big convexification" is a very abstract concept, and the analysis is quite heavy in computation. Now the Fitzpatrick function gives another way of associating a convex functions with a monotone multifunction, and this can also be used to reduce questions on monotone multifunctions to questions on convex functions. The problem is that many of the questions on convex functions that one obtains require an analysis of the special properties of convex functions on $E \times E^*$. This is exactly the analysis that we perform in Section 22, and later on in Section 35. As already explained, the SSD spaces introduced in Sections 19–21 give us a strict generalization of monotonicity. More to the point, the fact that the notation is more concise enables us to get a much better grasp of the underlying structures. A good example of this is Theorem 35.8, a relatively simple result with far–reaching applications to the classification of maximally monotone multifunctions on nonreflexive spaces. Another example is provided by Section 46 on maximally monotone multifunctions with convex graph.

We now return to our discussion of Section 23. We also introduce the "fitzpatrification", S_φ, of a monotone multifunction, S. This is a multifunction with convex graph which is normally much larger than the graph of S. S_φ is, in general, not monotone, but it is very useful since its use shortens the statements of many results considerably. The final result of this section,

Lemma 23.9, will be used in our discussion of the sum problem in Theorem 24.1 and the Brezis–Haraux condition in Theorem 31.4.

In Section 24, we give sufficient conditions for the sum of maximally monotone multifunctions to be maximally monotone. These results will be extended in the reflexive case in Section 32, and we will discuss the nonreflexive case in Chapter VII.

Chapter IV: In Chapter IV, we use results from Sections 4, 12, 18 and 23 to establish a number of results on monotone multifunctions on general Banach spaces. Section 25 is devoted to the single result that a maximally monotone multifunction with bounded range has full domain, and in Section 26, we prove a local boundedness theorem for any (not necessarily maximally) monotone multifunction on a Banach space. Specifically, we prove that a *monotone multifunction, S, is locally bounded at any point surrounded by* $D(S_\varphi)$.

In Section 27, we prove the "six set theorem", Theorem 27.1, that if S is maximally monotone then the six sets $\operatorname{int} D(S)$, $\operatorname{int}(\operatorname{co} D(S))$, $\operatorname{int} D(S_\varphi)$, $\operatorname{sur} D(S)$, $\operatorname{sur}(\operatorname{co} D(S))$ and $\operatorname{sur} D(S_\varphi)$ coincide, and its consequence, the "nine set theorem", Theorem 27.3, that, if $\operatorname{sur} D(S_\varphi) \neq \emptyset$, then the nine sets $\overline{D(S)}$, $\overline{\operatorname{co} D(S)}$, $\overline{D(S_\varphi)}$, $\operatorname{int} D(S)$, $\operatorname{int}(\operatorname{co} D(S))$, $\operatorname{int} D(S_\varphi)$, $\operatorname{sur} D(S)$, $\operatorname{sur}(\operatorname{co} D(S))$ and $\operatorname{sur} D(S_\varphi)$ coincide. ("Sur" is defined in Section 13.) The six set theorem and the nine set theorem not only extend the results of Rockafellar that $\operatorname{int} D(S)$ is convex and that, if $\operatorname{int}(\operatorname{co} D(S)) \neq \emptyset$ then $\overline{D(S)}$ is convex, but also answer in the affirmative a question raised by Phelps, namely whether an absorbing point of $D(S)$ is necessarily an interior point. In Theorem 27.5 and Theorem 27.6, we give sufficient conditions that $\overline{D(S)} = \overline{D(S_\varphi)}$ and $\overline{R(S)} = \overline{R(S_\varphi)}$ — these conditions do not have any interiority hypotheses.

Section 28 contains the results that if S is maximally monotone then the closed linear hull of $D(S_\varphi)$ is identical with the closed linear hull of $D(S)$, and the closed affine hull of $D(S_\varphi)$ is identical with the closed affine hull of $D(S)$. The arguments here are quite simple, which is in stark contrast with the similar question for closed convex hulls. This section also contains some results for pairs of multifunctions, which will be used in our analysis of bootstrapped sum theorems for reflexive spaces in Section 32. We also give some results on the "restriction" of a monotone multifunction to a closed subspace. The results in this section depend ultimately on the result of Lemma 20.4 on q–positive sets that are "flattened" by certain elements of a SSD space.

Chapter V: Chapter V is concerned with maximally monotone multifunctions on reflexive Banach spaces. In Section 29, we use the theory of SSDB spaces developed in Section 21 to obtain various criteria for a monotone multifunctions on a reflexive Banach space to be maximally monotone. We deduce in Theorem 29.5 and Theorem 29.6 Rockafellar's surjectivity theorem, together with a sharp lower bound on the norm of solutions in terms of the Fitzpatrick function. Theorem 29.8 contains an expression for a maximally monotone extension of a given nontrivial monotone multifunction on a reflexive space, and Theorem 29.9 gives Torralba's analog in the context of maximally monotone multifunctions of the Brøndsted–Rockafellar theorem for convex lower semicontinuous functions.

In Section 30, we discuss more subtle surjectivity results. The main result here is Theorem 30.1, a general existence theorem for BC–functions, which has as applications (in Theorem 30.2) a nontrivial generalization of Theorem 29.5, and (in Theorem 30.4) an abstract Hammerstein theorem.

Section 31 is devoted to the Brezis–Haraux condition for $R(S + T)$ to be close to $R(S) + R(T)$. In fact, we show in Theorem 31.4(c) that stronger results are true under the condition $R(S) + R(T) \subset R(S_\varphi + T_\varphi)$, and then in Corollary 31.6, that $R(S) + R(T) \subset R(S_\varphi + T_\varphi)$ under the original Brezis–Haraux hypotheses.

The final three sections of Chapter V are concerned with various sufficient conditions for the sum of maximally monotone multifunctions on a reflexive Banach space to be maximally monotone, together with certain related identities. In Section 32, we use the results of Section 28 to bootstrap the result of Theorem 24.1(a), obtaining the "sandwiched closed subspace theorem", Theorem 32.2, which first appeared in [109], and unifies sufficient conditions that have been established by various authors for the sum theorem to hold. In Section 33, we again use Theorem 24.1(a), this time to establish that several of the sets appearing in the above mentioned sufficient conditions are, in fact, identical. Finally, in Section 34, we use the theory of BC–functions to obtain various generalization of the Brezis–Crandall–Pazy "perturbation" result on the maximal monotonicity of the sum.

Chapter VI: In Chapter VI, we return to the discussion of monotonicity on possibly nonreflexive Banach spaces that we initiated in Chapters III and IV. Many of the nice results that we established in Chapter V either fail in this context, or the situation is not clear. The precise problem can be traced to the difference between SSD spaces and SSDB spaces.

Section 35 is a continuation of Section 22, but now we use the topology $\mathcal{T}_{\| \|}(E \times E^*)$ instead of the topology $\mathcal{T}_{\| \|}(E) \times w(E^*, E)$ on $E \times E^*$. In this case, we are led to consider two SSD spaces, $E \times E^*$ and $(E \times E^*)^* = E^{**} \times E^*$. We start off Section 35 with some examples, and then give, in Lemma 35.4, Lemma 35.5, Lemma 35.6 and Lemma 35.7, analogs to our present situation of Lemma 19.13, Theorem 19.16, Theorem 21.4(b) and Lemma 22.9. These results are all combined to obtain the main result of this section, Theorem 35.8, which will ultimately be applied in Sections 37, 39, 40 and 41.

Many subclasses of the class of maximally monotone multifunctions have been introduced, the basic idea being to define subclasses for which some of the properties of maximally monotone multifunctions on reflexive spaces continue to hold.

In Section 36, we introduce those that are "of type (D)", "of type (NI)", "of type (FP)", "of type (FPV)", "strongly maximally monotone", "of type

(ANA)", and "of type (BR)". The oldest of them, the maximally monotone multifunctions of "type (D)", were introduced by Gossez in 1971, while the others are much more recent. In addition to giving the definitions of these subclasses, we also discuss a number of related open problems. There is an eighth subclass of the class of maximally monotone multifunctions which has a very interesting theory, those that are "of type (ED)". The definition of these requires more preliminary work, and so it will be postponed until Section 38. All of these subclasses share the property that it is hard to find a maximally monotone multifunctions that does *not* belong to the subclass. We now know that there are various inclusions between the subclasses. One of these will be the subject of Section 37, where we will prove that every maximally monotone multifunction of type (D) is automatically of type (FP). The result of Section 37 first appeared in [101].

It is true (and was realized by Gossez) that it is advantageous to replace the topology $w(E^{**}, E^*) \times \mathcal{T}_{\| \|}(E^*)$ on $E^{**} \times E^*$ in the definition of "type (D)" by a stronger one. In Section 38, we will define $\mathcal{T}_{C\mathcal{LBN}}(E^{**} \times E^*)$, which is such a replacement, and produces a subclass of the maximally monotone multifunctions that has a number of extremely attractive properties. $\mathcal{T}_{C\mathcal{LBN}}(E^{**} \times E^*)$ is defined in terms of a topology, $\mathcal{T}_{C\mathcal{LB}}(E^{**})$, on E^{**} which lies between the weak* topology $w(E^{**}, E^*)$ and the norm topology $\mathcal{T}_{\| \|}(E^{**})$. Despite the fact that $\mathcal{T}_{C\mathcal{LB}}(E^{**})$ has a number of pleasant properties, for reasons explained in the beginning of Section 38, $\left(E^{**}, \mathcal{T}_{C\mathcal{LB}}(E^{**})\right)$ will normally fail to be a topological vector space. The corresponding class of maximally monotone multifunctions, those that are of "type (ED)", will be introduced in Definiton 38.3. We now give a graphic that serves to show the central position occupied by maximally monotone multifunctions of type (ED), and will provide a roadmap to some of the results of the following sections. We assume that E is a nonzero Banach space and $S\colon E \rightrightarrows E^*$ is maximally monotone.

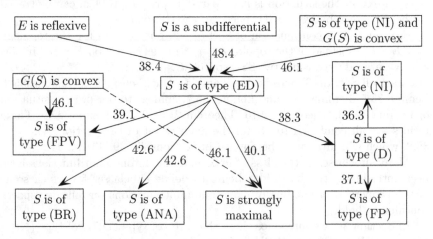

In Sections 39–42, we use the results of Section 35 to prove that maximally monotone multifunctions of type (ED) are always of type (FPV), strongly maximal, of type (ANA) and of type (BR), and also explore the connection between these multifunctions and coercivity. In particular, we deduce in Corollary 41.4 the classical result that a coercive maximally monotone multifunction on a reflexive space is surjective. The analysis of these four sections is based on results that first appeared in [101] and [107]. It is worth pointing out that we do not know of a maximally monotone multifunction of type (D) which is *not* of type (ED).

In [44] and [45], it was proved that if S is maximally monotone of type (FP), or a certain condition involving *approximate resolvents* holds, then $\overline{R(S)}$ is convex. In Section 43, we use Theorem 27.6 to show that, under either of these hypotheses, in fact $\overline{R(S)} = \overline{R(S_\varphi)}$. In Section 44, we consider the fascinating question (already referred to in Problem 31.3) whether the maximal monotonicity of S implies the convexity of $\overline{D(S)}$. In order to put this question into context, we must now discuss *Rockafellar's sum problem*. This is the following: if S, T are maximally monotone and $D(S) \cap \mathrm{int}\, D(T) \neq \emptyset$ then is $S + T$ maximally monotone? A solution to this problem has been announced recently, but the jury is still out on this. It follows from Theorem 44.1 that if the solution to this problem is positive then *every maximally monotone multifunction is of type (FPV)*. Furthermore, it would then follow from Theorem 44.2 that, *for every maximally monotone multifunction* S, $\overline{D(S)} = \overline{\mathrm{co}\, D(S)} = \overline{D(S_\varphi)}$ so, in particular, *for every maximally monotone multifunction* S, $\overline{D(S)}$ *is convex*.

In Section 45 we prove that, under certain circumstances, the biconjugate of the pointwise maximum of a finite number of functions is the maximum of their biconjugates. (See Corolllary 45.5.) What is curious is that we can establish this result without having a simple explicit formula for the *conjugate* of this pointwise maximum. This result will be applied in Lemma 45.9 to obtain the fundamental property that $\widehat{E}$ is dense in $\big(E^{**}, \mathcal{T}_{\mathcal{CLB}}(E^{**})\big)$, and also a stronger result too complicated to discuss here. Lemma 45.9 will be applied in our work on maximally monotone multifunctions with convex graph in Section 46, and in our proof that subdifferentials are maximally monotone of type (ED) in Section 48. Theorem 45.12 gives an unexpected characterization of the closure of certain convex subsets of $E^{**} \times E^*$ with respect to $\mathcal{T}_{\mathcal{CLBN}}(E^{**} \times E^*)$ — this will also be used in Section 46.

As we have already observed, in Section 46, we discuss maximally monotone multifunctions with convex graph. This is an important subclass of the maximally monotone multifunctions since it includes all affine maximally monotone operators, and all maximally monotone multifunctions whose inverse is an affine function. We prove in Theorem 46.1 that any such multifunction is always strongly maximal and of type (FPV) and, further, if such a multifunction is of type (NI) then it is of type (ED) and, in Theorem 46.3, we give a sufficient condition of "Attouch–Brezis" type for the sum of two such maximally monotone multifunctions to be maximally monotone. In Section 47, we apply the results of Section 46 to possibly discontinuous linear operators, explaining the connections with known results. In addition, Theorem 47.1 contains a necessary and sufficient condition for a positive linear operator to be maximally monotone and we prove in Theorem 47.7 that every *continuous* positive linear operator is of type (ANA).

In Section 48, we first prove in Theorem 48.1 that if $f \in \mathcal{PCLSC}(E)$ then $\iota(G(\partial f))$ is dense in $G^{-1}(\partial f^*)$ in $\mathcal{T}_{\mathcal{CLBN}}(E^{**} \times E^*)$, from which we deduce in Theorem 48.4 that subdifferentials are of type (ED), of type (FP), of type (FPV), strongly maximally monotone, of type (ANA) and of type (BR). We also deduce in Corollary 48.8 a result that is approximately a considerable generalization of the Brøndsted–Rockafellar theorem. We emphasize that the results in Section 48 depend on Lemma 45.9(a), which depends ultimately on the formula for the biconjugate of a maximum that we established in Theorem 45.3.

In Section 49, we prove that the "subdifferential" of a closed saddle–function on the product of a Banach space and a reflexive Banach space is maximally monotone of type (ED).

Chapter VII: In this chapter, we give various sufficient conditions for the sum of maximally monotone multifunctions on a general Banach space to be maximally monotone. Andrew Eberhard and Jonathan Borwein have announced the following result: *if E is a nonzero Banach space, $S, T: E \rightrightarrows E^*$ are maximally monotone and $D(S) \cap \operatorname{int} D(T) \neq \emptyset$ then $S + T$ is maximally monotone.* While their paper is not in definitive form, in Section 50, we discuss the far–reaching implication of such a result. In Section 51, we use results from Section 24 and Section 28 to prove a mild generalization of Voisei's theorem on the maximal monotonicity of the sum of two maximally monotone multifunctions S and T when $D(S)$ and $D(T)$ are closed and convex. In Section 52, we give sufficient conditions for $S + N_C$ to be maximally monotone, where $S: E \rightrightarrows E^*$ is monotone and N_C is the normality map of a nonempty closed convex set and, in Section 53, we give a short proof of the result of Verona–Verona that $S + T$ is maximally monotone when S is a subdifferential and T is maximally monotone with full domain.

Chapters VIII–X: In Chapter VIII, we collect together some of the open problems that have appeared in the body of the text, in Chapter IX, we provide a glossary of the definitions of the various classes of monotone multifunctions that we have introduced in the body of the text and, in Chapter X, we give a selection of the results proved in the body of the text.

Epilog

We will shortly provide four flowcharts giving various possible uses for this volume. It is worth noting how important Section 23 is in the second, third and fourth of these. This simple idea due to Simon Fitzpatrick has had an enormous impact on the theory of monotonicity. Simon's death in 2004 at the untimely age of 51 was a tremendous loss to mathematics. Some idea of the scope of Simon's work can be obtained from the memorial volume [43], and there is a short history of his life by Borwein et al. in [19].

The first of the promised charts shows the flow of logic for Chapters I and II. This material could be used for an introductory course in functional analysis which would, at the same time, touch on minimax theorems and give a grounding in convex Lagrange multiplier theory and the main theorems in convex analysis.

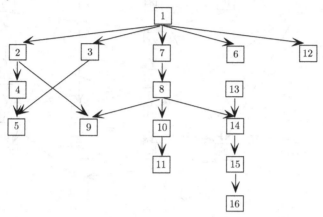

The next chart shows the flow of logic in Chapters III, IV and VII, starting from the appropriate sections in Chapters I and II. This material could be used as a basis for a course in which results on monotonicity on general Banach spaces are established using SSD spaces and Fitzpatrick functions.

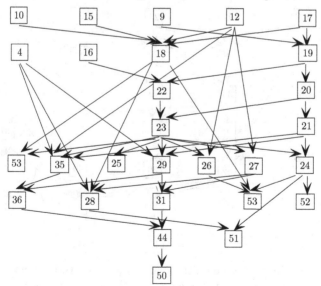

The third chart shows the flow of logic in Chapter V, starting from the appropriate sections in Chapters I and II. This material could be used as a basis for a course in which results on monotonicity on reflexive Banach spaces are established using SSD spaces and Fitzpatrick functions.

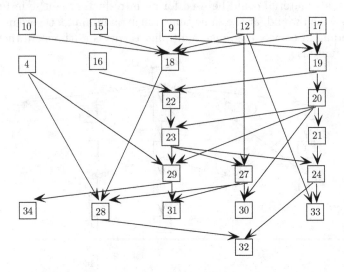

The final chart shows the flow of logic in Chapter VI, starting from the appropriate sections in Chapters I, II, III, and IV. This contains an exposition of the more technical properties of maximal monotonicity on general Banach spaces that have been established since 1997.

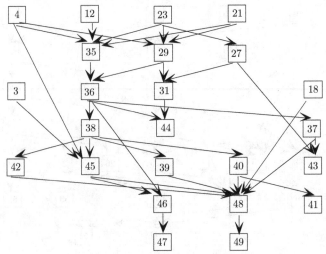

These notes in no way claim to be an exhaustive study of convex analysis or monotonicity. We refer the reader to Zălinescu, [119], for such a study of convex analysis, and to Rockafellar and Wets, [83], for such a study of monotonicity in finite dimensional spaces.

We do not discuss the theory of *cyclical monotonicity* introduced by Rockafellar in [80]. We refer the reader to Bartz, Bauschke, Borwein, Reich and Wang, [5], for recent developments in this direction using a sequence of Fitzpatrick functions. We also do not discuss the *Asplund decomposition* of a maximally monotone multifunction into the sum of a sub-differential and an acyclic maximally monotone multifunction, or of *monotone variational inequalities*. We refer the reader to Borwein, [17, Theorem 3.4, p. 571] and [17, Section 5.3, pp. 580–581] for a discussion of these subjects.

Another topic that we do not discuss is the theory of monotonicity in Hilbert spaces. The connection between Fitzpatrick functions and the Kirszbraun-Valentine extension theorem in Hilbert spaces was explored by Reich–Simons in [73] and Borwein–Zhu in [24, Theorem 5.1.33, p. 179], and further results on this topic have been obtained by Bauschke in [7]. There is also the forthcoming book [10] by Bauschke and Combettes, in which convex analysis and monotonicity are discussed specifically in the Hilbert space setting.

We also do not discuss the semicontinuity of multifunctions, uscos and cuscos. These are treated in [24, Sections 5.1.4–5, pp. 173–177].

I The Hahn-Banach-Lagrange theorem and some consequences

1 The Hahn–Banach–Lagrange theorem

The main topic of this section is the Hahn-Banach-Lagrange theorem, Theorem 1.11. This will be extended in Theorem 1.13, where we outline a generalization in which the functions are defined on an abstract set rather than on a convex subset of a vector space.

Theorem 1.11 is proved using the technique of the "auxiliary sublinear functional". Most of the work for it is actually done in the technical Lemma 1.5.

In Lemma 1.2, we give a relatively little–known proof of the "sublinear form" of the Hahn–Banach theorem, using an argument which goes back to [88] (in which there were also applications to Choquet theory) and König, [53]. See Remark 1.3 for a discussion of why we use this particular method of proof. Here is the notation that we will need. Let E be a nonzero vector space. (All vector spaces in these notes will be *real*.) For the moment we do not need any additional structure for E. A *sublinear functional on E* is a map $P: E \to \mathbb{R}$ such that

$$P \text{ is } \textit{subadditive:} \quad x_1, x_2 \in E \implies P(x_1 + x_2) \le P(x_1) + P(x_2)$$

and

$$P \text{ is } \textit{positively homogeneous:} \quad x \in E \text{ and } \lambda > 0 \implies P(\lambda x) = \lambda P(x).$$

It follows from this that $P(0) = P(2 \times 0) = 2 \times P(0)$, hence $P(0) = 0$. Consequently, the definition of positive homogeneity implies that:

$$x \in E \text{ and } \lambda \ge 0 \implies P(\lambda x) = \lambda P(x).$$

We exclude the $\lambda = 0$ case from the *definition* in order to reduce the amount of work necessary in Lemmas 1.1, 1.2 and 1.5. We explain in Remark 1.4, why we cannot allow P to take the value $+\infty$.

We recall that a *seminorm* on E is a subadditive map $P: E \to [0, \infty[$ such that $x \in E$ and $\lambda \in \mathbb{R} \implies P(\lambda x) = |\lambda| P(x)$.

Note that a norm or a seminorm is an example of a sublinear functional. So also are linear functionals. Since nonzero linear functionals are never positive, and norms and seminorms are, by definition, always positive, there are plenty of examples of sublinear functionals that are not norms or seminorms or linear. See the paper [55] by König for some of the subtler properties of sublinear functionals.

Our proof of Lemma 1.2 depends on the following simple lemma, which will be considerably generalized in Lemma 1.5. If P and Q are sublinear functionals on E, we write $Q \leq P$ if, for all $x \in E$, $Q(x) \leq P(x)$.

Lemma 1.1. *Let E be a nonzero vector space and P: $E \to \mathbb{R}$ be sublinear. Let $y \in E$. For all $x \in E$, let*

$$P_y(x) := \inf_{\lambda > 0} \left[P(x + \lambda y) - \lambda P(y) \right]. \tag{1.1}$$

Then P_y: $E \to \mathbb{R}$, P_y is sublinear, $P_y \leq P$ and $P_y(-y) \leq -P(y)$.

Proof. If $x \in E$ and $\lambda > 0$ then $\lambda P(y) = P(\lambda y) \leq P(x + \lambda y) + P(-x)$, hence

$$P(x + \lambda y) - \lambda P(y) \geq -P(-x).$$

Taking the infimum over $\lambda > 0$, $P_y(x) \geq -P(-x) > -\infty$. Thus P_y: $E \to \mathbb{R}$. We now prove that P_y is subadditive. To this end, let $x_1, x_2 \in E$. Let $\lambda_1, \lambda_2 > 0$ be arbitrary. Then

$$\left[P(x_1 + \lambda_1 y) - \lambda_1 P(y) \right] + \left[P(x_2 + \lambda_2 y) - \lambda_2 P(y) \right]$$
$$\geq P\big(x_1 + x_2 + (\lambda_1 + \lambda_2)y\big) - (\lambda_1 + \lambda_2)P(y)$$
$$\geq P_y(x_1 + x_2).$$

Taking the infimum over λ_1 and λ_2, $P_y(x_1) + P_y(x_2) \geq P_y(x_1 + x_2)$. Thus P_y is subadditive. Next, let $x \in E$ and $\mu > 0$. Then

$$P_y(\mu x) = \inf_{\lambda > 0} \left[P(\mu x + \lambda y) - \lambda P(y) \right]$$
$$= \mu \inf_{\lambda > 0} \left[P(x + (\lambda/\mu)y) - (\lambda/\mu)P(y) \right]$$
$$= \mu \inf_{\nu > 0} \left[P(x + \nu y) - \nu P(y) \right] = \mu P_y(x).$$

Thus P_y is positively homogeneous and, consequently, sublinear. If x is an arbitrary element of E then (taking $\lambda = 1$)

$$P_y(x) \leq P(x + y) - P(y) \leq P(x),$$

thus $P_y \leq P$. Similarly,

$$P_y(-y) \leq P(-y + y) - P(y) = -P(y). \qquad \square$$

We now prove the classical Hahn–Banach theorem for sublinear functionals, which is undoubtedly one of the most important results in functional analysis.

Lemma 1.2. *Let E be a nonzero vector space and P: $E \to \mathbb{R}$ be sublinear. Then there exists a linear functional L on E such that $L \leq P$.*

Proof. Let $\mathcal{Q}$ stand for the set of all sublinear functionals Q on E such that $Q \leq P$. We first prove that every nonempty totally ordered subset, $\mathcal{T}$, of $\mathcal{Q}$ has a lower bound in $\mathcal{Q}$. To this end, let $Q(x) := \inf\{T(x) \colon T \in \mathcal{T}\}$. If $x \in E$ and $T \in \mathcal{T}$ then, since T is subadditive, $T(x) \geq -T(-x) \geq -P(-x)$. Taking the infimum over $T \in \mathcal{T}$, $Q(x) \geq -P(-x) > -\infty$. Thus $Q \colon E \to \mathbb{R}$. We now prove that Q is subadditive. To this end, let $x_1, x_2 \in E$. Let $T_1, T_2 \in \mathcal{T}$ be arbitrary. If $T_1 \geq T_2$ then

$$T_1(x_1) + T_2(x_2) \geq T_2(x_1) + T_2(x_2) \geq T_2(x_1 + x_2) \geq Q(x_1 + x_2),$$

while if $T_2 \geq T_1$ then

$$T_1(x_1) + T_2(x_2) \geq T_1(x_1) + T_1(x_2) \geq T_1(x_1 + x_2) \geq Q(x_1 + x_2).$$

So, in either case, $T_1(x_1) + T_2(x_2) \geq Q(x_1 + x_2)$. Taking the infimum over T_1 and T_2 gives $Q(x_1) + Q(x_2) \geq Q(x_1 + x_2)$. Thus Q is subadditive. It is easy to see that Q is positively homogeneous, and so Q is sublinear. Obviously, $Q \in \mathcal{Q}$.

From Zorn's lemma, there exists a minimal element, L, of $\mathcal{Q}$. Now let $y \in E$. In the notation of Lemma 1.1, $L_y \colon E \to \mathbb{R}$ is sublinear, $L_y \leq L$ (from which $L \in \mathcal{Q}$) and $L_y(-y) \leq -L(y)$. Since L is minimal in $\mathcal{Q}$, in fact $L_y = L$, and so $L(-y) \leq -L(y)$. On the other hand, since L is subadditive, $L(-y) \geq -L(y)$. Combining together these two inequalities, $L(-y) = -L(y)$, so we can "pull minus signs" through L. If now $x \in E$ and $\lambda < 0$ then

$$L(\lambda x) = L\big(-(-\lambda)x\big) = -L\big((-\lambda)x\big) = -(-\lambda)L(x) = \lambda L(x), \qquad (1.2)$$

and so L is homogeneous. If $x_1, x_2 \in E$ then, since the subadditivity of L gives $L(-x_1 - x_2) \leq L(-x_1) + L(-x_2)$, we have from three applications of (1.2) that

$$
\begin{aligned}
L(x_1 + x_2) = L\big(-(-x_1 - x_2)\big) &= -L(-x_1 - x_2) \\
&\geq -L(-x_1) - L(-x_2) = L(x_1) + L(x_2) \geq L(x_1 + x_2),
\end{aligned}
$$

and so $L(x_1 + x_2) = L(x_1) + L(x_2)$. Thus L is linear. $\qquad\square$

Remark 1.3. See Kelley–Namioka, [52, 3.4, p. 21] for a proof of Lemma 1.2 using cones, and Rudin, [84, Theorem 3.2, pp. 56–57] for a proof of Lemma 1.2 using an "extension by subspaces" argument. Since we will give Corollary 2.2 as (ultimately) a consequence of Lemma 1.2, it would seem better to avoid the "extension by subspaces" proof in order to dispel any suspicion of circularity. We have chosen to use the "minimal sublinear functional" argument, since it is most in tune with the other analysis in this section.

Remark 1.4. An *extended sublinear functional on E* is a subadditive and positively homogeneous map $P \colon E \to \,]-\infty, \infty]$ such that $P(0) = 0$. We first give an example (taken from [87]) of an extended sublinear functional for which the analog of Lemma 1.2 fails. Let E be the vector space of finitely nonzero sequences $\{x_n\}_{n\geq 1}$. If $m \geq 1$, let

$$E_m := \left\{ \{x_n\}_{n\geq 1} \in E \colon x_1, \ldots, x_{m-1} \leq 0,\ x_m < 0,\ x_{m+1} = x_{m+2} = \cdots = 0 \right\}.$$

Define $P\colon E \to\,]-\infty, \infty]$ by

$$P\big(\{x_n\}_{n\geq 1}\big) := \begin{cases} 0, & \text{if } \{x_n\}_{n\geq 1} = 0; \\ m(x_1 + \cdots + x_m), & \text{if } m \geq 1 \text{ and } \{x_n\}_{n\geq 1} \in E_m; \\ \infty, & \text{otherwise.} \end{cases}$$

P is an extended sublinear functional on E. For all $m \geq 1$, let $e^{(m)}$ be the mth basic vector of E. If $\varepsilon > 0$ then $-e^{(1)} - \varepsilon e^{(m)} \in E_m$ and so $P\big(-e^{(1)} - \varepsilon e^{(m)}\big) = -m(1+\varepsilon)$. Suppose now that L were a linear functional on E such that $L \leq P$. Then, for all $m \geq 1$ and $\varepsilon > 0$, we would have:

$$-L\big(e^{(1)}\big) - \varepsilon L\big(e^{(m)}\big) = L\big(-e^{(1)} - \varepsilon e^{(m)}\big) \leq -m(1 + \varepsilon).$$

If we let $\varepsilon \to 0$ in this, we would obtain:

$$m \geq 1 \quad \Longrightarrow \quad -L\big(e^{(1)}\big) \leq -m,$$

which is manifestly impossible since $L\big(e^{(1)}\big) \in \mathbb{R}$. On the other hand, the analog of Lemma 1.2 for extended sublinear functionals is true if E is finite–dimensional (see [87, Corollary 3, p. 115]).

The next main result that we establish on our way to Theorem 1.11 is the *Mazur–Orlicz theorem*, Lemma 1.6. The Mazur-Orlicz theorem is not nearly as well known as it deserves to be — we refer the reader to the paper [54] by König for a number of applications of it to other fields of analysis. Splitting the proof of Theorem 1.11, as we have done here, enables us to avoid a considerable amount of computation compared to the original method used in [103]. The use of the Mazur–Orlicz theorem to prove the Hahn–Banach–Lagrange theorem was suggested to the author by Michael Crandall. We first prove a preliminary lemma, which subsumes Lemma 1.1.

Lemma 1.5. *Let E be a nonzero vector space and $P\colon E \to \mathbb{R}$ be sublinear. Let D be a nonempty convex subset of a vector space and $\beta := \inf_D P \in \mathbb{R}$. For all $x \in E$, let*

$$Q(x) := \inf_{d \in D,\ \lambda > 0} \big[P(x + \lambda d) - \lambda\beta\big]. \tag{1.3}$$

Then $Q\colon E \to \mathbb{R}$, Q is sublinear, $Q \leq P$ and, for all $d \in D$, $-Q(-d) \geq \beta$.

Proof. If $x \in E$, $d \in D$ and $\lambda > 0$ then

$$P(x + \lambda d) - \lambda\beta \geq -P(-x) + \lambda P(d) - \lambda\beta \geq -P(-x) > -\infty.$$

Taking the infimum over $d \in D$ and $\lambda > 0$, $Q(x) \geq -P(-x) > -\infty$. Thus $Q\colon E \to \mathbb{R}$. It is easy to check that Q is positively homogeneous, so to prove that Q is sublinear it remains to show that Q is subadditive. To this end, let $x_1, x_2 \in E$. Let $d_1, d_2 \in D$ and $\lambda_1, \lambda_2 > 0$ be arbitrary. Write $x := x_1 + x_2$, $\lambda := \lambda_1 + \lambda_2$ and $d := (\lambda_1/\lambda)d_1 + (\lambda_2/\lambda)d_2$. Then

$$\begin{aligned}
\big[P(x_1 + \lambda_1 d_1) - \lambda_1\beta\big] + \big[P(x_2 + \lambda_2 d_2) - \lambda_2\beta\big] &\geq P(x + \lambda_1 d_1 + \lambda_2 d_2) - \lambda\beta \\
&= P(x + \lambda d) - \lambda\beta \\
&\geq Q(x) = Q(x_1 + x_2).
\end{aligned}$$

Taking the infimum over d_1, d_2, λ_1 and λ_2 gives $Q(x_1) + Q(x_2) \geq Q(x_1 + x_2)$. Thus Q is subadditive, and consequently, sublinear. Fix $d \in D$. Let x be an arbitrary element of E. Then, for all $\lambda > 0$, $Q(x) \leq P(x) + \lambda\big[P(d) - \beta\big]$. Letting $\lambda \to 0$, $Q(x) \leq P(x)$. Thus $Q \leq P$. Finally, let d be an arbitrary element of D. Then, taking $\lambda = 1$ in (1.3),

$$Q(-d) \leq P(-d + d) - \beta = -\beta,$$

hence $-Q(-d) \geq \beta$, which completes the proof of Lemma 1.5.   □

We now come to the Mazur–Orlicz theorem:

Lemma 1.6. *Let E be a nonzero vector space, $P: E \to \mathbb{R}$ be sublinear and D be a nonempty convex subset of E. Then there exists a linear functional L on E such that $L \leq P$ and*

$$\inf_D L = \inf_D P.$$

Proof. Let $\beta := \inf_D P$. If $\beta = -\infty$, the result is immediate from Lemma 1.2 (take any linear functional L on E such that $L \leq P$). So we can suppose that $\beta \in \mathbb{R}$. Define the auxiliary sublinear functional, Q, as in Lemma 1.5. From Lemma 1.2, there exists a linear functional L on E such that $L \leq Q$. Since $Q \leq P$, $L \leq P$, as required. Let $d \in D$. Then

$$L(d) = -L(-d) \geq -Q(-d) \geq \beta.$$

Taking the infimum over $d \in D$,

$$\inf_D L \geq \beta = \inf_D P.$$

On the other hand, since $L \leq P$,

$$\inf_D L \leq \inf_D P.$$ □

Remark 1.7. It is worth pointing out that the definition of the auxiliary sublinear functional used to prove Lemma 1.6 is "forced" in the sense that if L is linear, $L \leq P$ and $\beta = \inf_D P = \inf_D L \in \mathbb{R}$ then, as the reader can easily verify, $L \leq Q$.

We shall see in Corollary 2.4 that any sublinear functional is the pointwise supremum of the linear functionals that it dominates. Thus if $\inf_D P \in \mathbb{R}$ then Q is the pointwise supremum of the linear functionals L such that $L \leq P$ and $\inf_D L = \inf_D P$.

Let X be a nonempty convex subset of a vector space, and $f: X \to \mathbb{R}$. We say that f is *convex* if

$$x, y \in X \text{ and } \lambda \in \,]0, 1[\quad \Longrightarrow \quad f(\lambda x + (1 - \lambda)y) \leq \lambda f(x) + (1 - \lambda)f(y).$$

We say that f is *concave* if

$$x, y \in X \text{ and } \lambda \in \,]0, 1[\quad \Longrightarrow \quad f(\lambda x + (1 - \lambda)y) \geq \lambda f(x) + (1 - \lambda)f(y).$$

Definition 1.8. If X is a nonempty set and $f\colon X \to \,]-\infty, \infty]$, we write $\operatorname{dom} f := \{x \in X\colon f(x) \in \mathbb{R}\}$. The set $\operatorname{dom} f$ is the *effective domain* of f. We say that f is *proper* if $\operatorname{dom} f \neq \emptyset$. Let C be a nonempty convex subset of a vector space and $\mathcal{PC}(C)$ stand for the set of all proper functions $k\colon C \to \,]-\infty, \infty]$ such that $\operatorname{dom} k$ is convex, and the restriction of k to $\operatorname{dom} k$ is convex in the sense already defined. Equivalently, instead of looking at the restriction of k to $\operatorname{dom} k$, we can say

$$x, y \in C \text{ and } \lambda \in \,]0, 1[\quad \Longrightarrow \quad k(\lambda x + (1 - \lambda)y) \leq \lambda k(x) + (1 - \lambda)k(y),$$

provided that we interpret $\infty + \infty$ to be ∞, and $\lambda \times \infty$ to be ∞ for $\lambda > 0$.

This extension of the definition of convex function is motivated by constrained optimization — if X is a nonempty subset of a set C, $f: X \to \mathbb{R}$ and we are trying to find a minimum of f over X, we can extend the definition of f to be $+\infty$ on $C \backslash X$, and thereby produce a function defined over C. Clearly a minimum of the extended function over C is identical with a minimum of the original function over X. So we frequently assume that our functions are defined on C, but take values in $\,]-\infty, \infty]$.

Now let X and Z be nonempty convex subsets of vector spaces and $f: X \to Z$. We say that f is *affine* if

$$x, y \in X \text{ and } \lambda \in \,]0, 1[\quad \Longrightarrow \quad f(\lambda x + (1 - \lambda)y) = \lambda f(x) + (1 - \lambda)f(y).$$

Definition 1.9. Let E be a nonzero vector space and $P\colon E \to \mathbb{R}$ be sublinear. Define the vector ordering "$\leq_P$" on E by declaring that $y \leq_P z$ if $P(y - z) \leq 0$. Let C be a nonempty convex subset of a vector space and $j\colon C \to E$. We say that j is *P–convex* if

$$x_1, x_2 \in C, \mu_1, \mu_2 > 0 \text{ and } \mu_1 + \mu_2 = 1$$
$$\Longrightarrow \quad j(\mu_1 x_1 + \mu_2 x_2) \leq_P \mu_1 j(x_1) + \mu_2 j(x_2).$$

An affine function is clearly P–convex. Apart from the applications in Lemma 3.1 and Theorem 6.4(a), all the P–convex functions that we consider will, in fact, be affine.

Remark 1.10. "P–convex" can mean different things under different circumstances. Consider the special case when $E = \mathbb{R}$. If $P(y) := |y|$, $P(y) := y$, $P(y) := -y$ or $P(y) := 0$, respectively, then "P–convex" means "affine", "convex", "concave" or "arbitrary", respectively.

We now come to the Hahn–Banach–Lagrange theorem, which will be used implicitly in nearly all the sections of these notes. In fact, we will use it explicitly in Corollary 2.1, Corollary 2.2, Lemma 3.1, Theorem 6.4, Example 7.1, Example 7.2, Theorem 8.1, Lemma 9.2, Theorem 12.2, Corollary 18.5, Theorem 46.1 and Lemma 47.2.

Theorem 1.11. *Let E be a nonzero vector space and $P\colon E \to \mathbb{R}$ be sublinear. Let C be a nonempty convex subset of a vector space, $k \in \mathcal{PC}(C)$ and $j\colon C \to E$ be P–convex. Then there exists a linear functional L on E such that $L \le P$ and*

$$\inf_C \left[L \circ j + k\right] = \inf_C \left[P \circ j + k\right]. \tag{1.4}$$

In fact, if we write $\mathcal{L}$ for the set of linear functionals L on E such that $L \le P$, then we have

$$\inf_C \left[P \circ j + k\right] = \max_{L \in \mathcal{L}} \inf_C \left[L \circ j + k\right]. \tag{1.5}$$

Proof. Define $Q\colon E \times \mathbb{R} \to \mathbb{R}$ by

$$Q(y, \lambda) := P(y) + \lambda \qquad \left((y, \lambda) \in E \times \mathbb{R}\right).$$

Then, as the reader can easily verify, Q is sublinear. Let

$$
\begin{aligned}
D &:= \bigcup_{x \in C} \left\{(y, \lambda) \in E \times \mathbb{R}\colon P\big(j(x) - y\big) \le 0,\ k(x) \le \lambda\right\} \\
&= \bigcup_{x \in C} \left\{(y, \lambda) \in E \times \mathbb{R}\colon j(x) \le_P y,\ k(x) \le \lambda\right\}.
\end{aligned}
$$

D is a convex subset of $E \times \mathbb{R}$. The Mazur–Orlicz theorem, Lemma 1.6, with E replaced by $E \times \mathbb{R}$ and P by Q, now gives a linear functional M on $E \times \mathbb{R}$ such that

$$M \le Q \text{ on } E \times \mathbb{R} \quad \text{and} \quad \inf_D M = \inf_D Q.$$

Since $M \le Q$ on $E \times \mathbb{R}$, there exists a linear functional L on E such that

$$L \le P \text{ on } E \quad \text{and} \quad (y, \lambda) \in E \times \mathbb{R} \Longrightarrow M(y, \lambda) = L(y) + \lambda.$$

We now derive (1.4) since

$$\inf_D M = \inf_C \left[L \circ j + k\right] \quad \text{and} \quad \inf_D Q = \inf_C \left[P \circ j + k\right],$$

which follows easily from the fact that, for all $x \in C$ and $y \in E$,

$$P\big(j(x) - y\big) \le 0 \Longrightarrow P \circ j(x) \le P\big(j(x) - y\big) + P(y) \Longrightarrow P \circ j(x) \le P(y)$$

and

$$P\big(j(x) - y\big) \le 0 \Longrightarrow L\big(j(x) - y\big) \le 0 \iff L \circ j(x) \le L(y).$$

Finally, (1.5) follows immediately from (1.4). $\qquad\square$

The following generalization of the Mazur–Orlicz theorem is due to König. The proof is similar to that of Lemma 1.6, but somewhat more technical.

Lemma 1.12. *Let E be a nonzero vector space, $P: E \to \mathbb{R}$ be sublinear and D be a nonempty subset of E such that*

for all $d_1, d_2 \in D$, there exists $d \in D$ such that $P\left(d - \frac{1}{2}d_1 - \frac{1}{2}d_2\right) \le 0$.

Then there exists a linear functional L on E such that $L \le P$ and

$$\inf_D L = \inf_D P.$$

Proof. See König, [54, Basic Theorem, p. 583].   □

We now give a generalization of Theorem 1.11, in which we dispense with the structure of C as a convex set:

Theorem 1.13. *Let E be a nonzero vector space, $P: E \to \mathbb{R}$ be sublinear, X be a nonempty set, $k: X \to \,]-\infty, \infty]$ be proper and $j: X \to E$. Suppose that, for all $x_1, x_2 \in \mathrm{dom}\, k$, there exists $u \in \mathrm{dom}\, k$ such that*

$$j(u) \le_P \tfrac{1}{2}j(x_1) + \tfrac{1}{2}j(x_2) \quad \text{and} \quad k(u) \le \tfrac{1}{2}k(x_1) + \tfrac{1}{2}k(x_2). \qquad (1.6)$$

Then there exists a linear functional L on E such that $L \le P$ and

$$\inf_X \big[L \circ j + k\big] = \inf_X \big[P \circ j + k\big].$$

In fact, if we write $\mathcal{L}$ for the set of linear functionals L on E such that $L \le P$, then we have

$$\inf_X \big[P \circ j + k\big] = \max_{L \in \mathcal{L}} \inf_X \big[L \circ j + k\big].$$

Proof. Let $D := \bigcup_{x \in X} \big\{(y, \lambda) \in E \times \mathbb{R}: \; j(x) \le_P y, \; k(x) \le \lambda\big\}$, and define the sublinear functional $Q: E \times \mathbb{R} \to \mathbb{R}$ by $Q(y, \lambda) := P(y) + \lambda$. Let $d_1, d_2 \in D$. Then, for $i = 1, 2$, $d_i = (y_i, \lambda_i)$ where, for some $x_i \in \mathrm{dom}\, k$, $j(x_i) \le_P y_i$ and $k(x_i) \le \lambda_i$. Choose $u \in \mathrm{dom}\, k$ as in (1.6). Since

$$j(u) \le_P \tfrac{1}{2}j(x_1) + \tfrac{1}{2}j(x_2) \le_P \tfrac{1}{2}y_1 + \tfrac{1}{2}y_2$$

and

$$k(u) \le \tfrac{1}{2}k(x_1) + \tfrac{1}{2}k(x_2) \le \tfrac{1}{2}\lambda_1 + \tfrac{1}{2}\lambda_2,$$

it follows that

$$\left(\tfrac{1}{2}y_1 + \tfrac{1}{2}y_2, \tfrac{1}{2}\lambda_1 + \tfrac{1}{2}\lambda_2\right) \in D.$$

The Mazur–Orlicz–König theorem, Lemma 1.12, now gives us a linear functional M on $E \times \mathbb{R}$ such that $M \le Q$ on $E \times \mathbb{R}$ and $\inf_D M = \inf_D Q$. The rest of the proof continues exactly as in Theorem 1.11.   □

The hypothesis (1.6) is somewhat similar to that of the minimax theorem in König, [53, p. 486] and our [90, Théorème 4, pp. 2-3].

2 Applications to functional analysis

Corollary 2.1 is the *sandwich theorem* (see [53, Theorem 1.7, p. 112]). It follows immediately from Theorem 1.11 with $C := E$ and $j(x) := x$.

Corollary 2.1. *Let E be a nonzero vector space, $P: E \to \mathbb{R}$ be sublinear, $k \in \mathcal{PC}(E)$ and $-k \leq P$ on E. Then there exists a linear functional L on E such that $-k \leq L \leq P$ on E.*

Corollary 2.2 is the *extension form of the Hahn–Banach theorem*. It follows immediately from Theorem 1.11 with $C := F$, $j(x) := x$ and $k := -M$. It can also be deduced from Corollary 2.1 (see [53, Corollary 1.8, p. 112]).

Corollary 2.2. *Let E be a nonzero vector space, F be a linear subspace of E, $P: E \to \mathbb{R}$ be sublinear, $M: F \to \mathbb{R}$ be linear and $M \leq P$ on F. Then there exists a linear functional L on E such that $L \leq P$ on E and $L|_F = M$.*

Remark 2.3. The analog of Corollary 2.2 for extended sublinear functionals fails, even for $E = \mathbb{R}^2$. The following example is taken from [87]: define $P: \mathbb{R}^2 \to \,]-\infty, \infty]$ by

$$P(x_1, x_2) := \begin{cases} 0, & \text{if } x_1 \leq 0 \text{ and } x_2 = 0; \\ x_1, & \text{if } x_2 < 0; \\ \infty, & \text{otherwise.} \end{cases}$$

P is an extended sublinear functional on $\mathbb{R}^2$. Let $F := \mathbb{R} \times \{0\} \subset \mathbb{R}^2$ and $M := 0$ on F. If L is a linear functional on $\mathbb{R}^2$ and $L \leq P$ then, for all $\varepsilon > 0$,

$$-L(1,0) - \varepsilon L(0,1) = L(-1, -\varepsilon) \leq P(-1, -\varepsilon) = -1.$$

Letting $\varepsilon \to 0$ yields $-L(1,0) \leq -1$, thus $L|_F \neq M$.

Our next result, the "one–dimensional Hahn–Banach theorem" (which can also be deduced from Corollary 2.2, see Rudin, [84, Theorem 3.2, pp. 56–57] (exercise!)) follows immediately from the Mazur–Orlicz theorem, Lemma 1.6 by taking $D := \{x\}$.

Corollary 2.4. *Let P be a sublinear functional on E and $x \in E$. Then there exists a linear functional L on E such that*

$$L \leq P \text{ on } E \quad \text{and} \quad L(x) = P(x).$$

3 A minimax theorem

As a general matter of notation: if λ, $\mu \in \mathbb{R}$, we write $\lambda \vee \mu$ for the maximum value of λ and μ, and $\lambda \wedge \mu$ for the minimum value of λ and μ. The result contained in Lemma 3.1 can also be deduced from Fan–Glicksberg–Hoffman, [41, Theorem 1, p. 618], after some simple transformations. We note the total absence of topological hypotheses in Lemma 3.1 — this will be important for us later. Lemma 3.1 will not only be used in Theorem 3.2, but also twice in Theorem 5.1 and Lemma 45.1. In one of these uses, it is a substitute for the theory of locally convex spaces and the "bipolar theorem".

Lemma 3.1. *Let C be a nonempty convex subset of a vector space and $f_1, \ldots, f_m$ be convex real functions on C. Then there exist $\lambda_1, \ldots, \lambda_m \geq 0$ such that $\lambda_1 + \cdots + \lambda_m = 1$ and*

$$\inf_C [f_1 \vee \cdots \vee f_m] = \inf_C [\lambda_1 f_1 + \cdots + \lambda_m f_m].$$

In fact, if $\mathcal{L} := \big\{ (\lambda_1, \ldots, \lambda_m) \in \mathbb{R}^m \colon \lambda_1, \ldots, \lambda_m \geq 0, \ \lambda_1 + \cdots + \lambda_m = 1 \big\}$, then

$$\inf_C [f_1 \vee \cdots \vee f_m] = \max_{(\lambda_1, \ldots, \lambda_m) \in \mathcal{L}} \inf_C [\lambda_1 f_1 + \cdots + \lambda_m f_m].$$

Proof. Let $E := \mathbb{R}^m$. Define the sublinear functional $P \colon E \to \mathbb{R}$ by

$$P(\mu_1, \ldots, \mu_m) := \mu_1 \vee \cdots \vee \mu_m,$$

the P–convex function $j \colon C \to E$ by $j := \big(f_1(\cdot), \ldots, f_m(\cdot) \big)$ and the convex function $k \colon C \to \mathbb{R}$ by $k := 0$. The result now follows from Theorem 1.11. $\square$

Let X, Y be nonempty sets, and $h \colon X \times Y \to \mathbb{R}$. We shall write $\inf_X \sup_Y h$ for $\inf_{x \in X} \sup_{y \in Y} h(x, y)$, and use a corresponding shorthand for other similar expressions. It is easily seen that (exercise!) the inequality

$$\sup_Y \inf_X h \leq \inf_X \sup_Y h \tag{3.1}$$

is always satisfied. This inequality can be strict, take for instance $X = Y = \{0, 1\}$ and $h(x, y) = 0$ if $x \neq y$ and $h(x, y) = 1$ if $x = y$. A *minimax theorem* is a theorem that gives conditions under which

$$\sup_Y \inf_X h = \inf_X \sup_Y h.$$

One final remark on notation: we shall say that h is "convex on X" if, for all $y \in Y$, $h(\cdot, y)$ is convex, and use a corresponding shorthand for other similar expressions.

There are many different minimax theorems (see our survey [95]). In Theorem 3.2, we shall use Lemma 3.1 to give a simple proof of a minimax theorem that follows from a result of Fan (see [40]). (See also the paper [53] by König and our paper [90] for simple generalizations of Fan's result.) It is important to note that the set X has no topological structure. It is worth mentioning that Lemma 3.1 can be deduced from Theorem 3.2 (exercise!). Theorem 3.2 will be used explicitly in Theorem 46.1.

Theorem 3.2. *Let X be a nonempty convex subset of a vector space, Y be a nonempty convex subset of a vector space and Y also be a compact Hausdorff topological space. Let $h\colon X \times Y \to \mathbb{R}$ be convex on X, and concave and upper semicontinuous on Y. Then*

$$\inf_X \max_Y h = \max_Y \inf_X h.$$

Proof. We can write "max" instead of "sup" because h is upper semicontinuous on Y and Y is compact. Let $\alpha := \inf_X \max_Y h$. Let $x_1, \ldots, x_m \in X$. Then, from Lemma 3.1 with $C := Y$ and $f_i := -h(x_i, \cdot)$, there exist $\lambda_1, \ldots \lambda_m \geq 0$ such that $\lambda_1 + \cdots + \lambda_m = 1$ and

$$\max_{y \in Y} \big[h(x_1, y) \wedge \cdots \wedge h(x_m, y) \big] = \max_{y \in Y} \big[\lambda_1 h(x_1, y) + \cdots + \lambda_m h(x_m, y) \big].$$

Since h is convex on X,

$$\max_{y \in Y} \big[h(x_1, y) \wedge \cdots \wedge h(x_m, y) \big] \geq \max_{y \in Y} h(\lambda_1 x_1 + \cdots + \lambda_m x_m, y) \geq \alpha.$$

Thus

$$\{ y \in Y \colon h(x_1, y) \geq \alpha \} \cap \cdots \cap \{ y \in Y \colon h(x_m, y) \geq \alpha \} \neq \emptyset.$$

Since Y is compact Hausdorff, and h is upper semicontinuous on Y, from the "finite intersection property" of the closed sets in Y,

$$\bigcap_{x \in X} \{ y \in Y \colon h(x, y) \geq \alpha \} \neq \emptyset.$$

Hence $\max_{y \in Y} \inf_{x \in X} h(x, y) \geq \alpha$, that is to say,

$$\max_Y \inf_X h \geq \inf_X \max_Y h.$$

The result of Theorem 3.2 now follows from (3.1). $\qquad\square$

4 The dual and bidual of a normed space

If E is a normed space, E^* will stand for the *dual space* of E, the set of continuous linear functionals on E. If L is a linear functional on E then L is continuous if, and only if,

$$\|L\| := \sup_{x \in E,\ \|x\| \leq 1} L(x) < \infty.$$

If $x^* \in E^*$, we will write $\langle x, x^* \rangle$ for the value of x^* at x. The reason for this notation is that we sometimes want to consider the elements of E as functions on E^* and "$\langle x, \cdot \rangle$" is easier to read than "$\cdot(x)$". However, as we shall see later, despite the more symmetric nature of this notation, the situation is not totally symmetric. Anyhow, with this notation,

$$\text{for all } x^* \in E^*, \quad \|x^*\| = \sup_{x \in E,\ \|x\| \leq 1} \langle x, x^* \rangle.$$

Note that "$\| \ \|$" does double duty both as the original norm of E, and as the dual norm that we have defined for E^*. However, no confusion will arise from this. Now the norm of E defines a topology on E. However, it is frequently useful to consider the *weak topology*, $w(E, E^*)$, which is defined as the smallest topology on E with respect to which all the functions $\langle \cdot, x^* \rangle$ $(x^* \in E^*)$ are continuous. We now introduce two other important concepts. The first is the *weak* topology*, $w(E^*, E)$, which is defined as the smallest topology on E^* with respect to which all the functions $\langle x, \cdot \rangle$ $(x \in E)$ are continuous. The main result about $w(E^*, E)$ is the celebrated Banach–Alaoglu theorem, which we shall use in Corollary 5.4 and Theorem 41.1:

Theorem 4.1. *Let E be a normed space and $M \geq 0$. Then the set $\{x^* \in E^*\colon \|x^*\| \leq M\}$ is $w(E^*, E)$–compact.*

Proof. See Kelley–Namioka, [52, 17.4, p. 155]. □

In fact there is a stronger result that we shall use in Theorem 5.1(c), namely (exercise!):

Theorem 4.2. *Let E be a normed space and $P\colon E \to \mathbb{R}$ be a sublinear functional dominated by some scalar multiple of the norm of E. Then*

$$\{x^* \in E^*\colon x^* \leq P \text{ on } E\} \text{ is } w(E^*, E)\text{–compact.}$$

One is tempted to hope that the result "dual" to Theorem 4.1 is true, that is to say: *if $M \geq 0$ then $\{x \in E\colon \|x\| \leq M\}$ is $w(E, E^*)$–compact.* We shall see below that this is not true in general.

The second important concept that we need at this time is that of the *bidual*, E^{**}, of E, which is defined to be the dual of the normed space E^*. Any element of E "gives" an element of E^{**}. More precisely, we can define a linear operator $\hat{\ }\colon E \to E^{**}$ such that

$$x \in E \text{ and } x^* \in E^* \implies \langle x^*, \hat{x} \rangle = \langle x, x^* \rangle.$$

In some senses, the map $\hat{\ }$ gives a very tight connection between E and E^{**}. For instance, $\hat{\ }$ is an "isometry", that is to say, for all $x \in E$, $\|\hat{x}\| = \|x\|$ (see Holmes, [51, §16.D, pp. 122–123]). It follows from this that if B is a norm–closed subset of a Banach space E then the image $\widehat{B}$ of B is a norm–closed subset of E^{**}. Also, $\hat{\ }$ is a homeomorphism of E onto $\widehat{E}$ with respect to $w(E, E^*)$ and $w(E^{**}, E^*)$ (exercise!). The tightest connection would be that $\widehat{E} = E^{**}$. We say that E is *reflexive* when this happens. (See [51, §16.F, pp. 125–127]). Some of the common Banach spaces are reflexive (for instance, Hilbert spaces and the spaces ℓ^p, $(1 < p < \infty)$), and some are not (for instance, ℓ^1, ℓ^∞, c_0 and $C[0,1]$.) The connection with the concepts introduced in the preceding paragraph is contained in the next result:

Theorem 4.3. *Let E be a Banach space. Then E is reflexive if, and only if, $\{x \in E\colon \|x\| \leq 1\}$ is $w(E, E^*)$–compact.*

Proof. See [51, §16.F, pp. 126–127]. ☐

We now show the connection between the Mazur–Orlicz theorem and another well known result in functional analysis, one of the *separation theorems* (see Rudin, [84, Theorem 3.4, pp. 58–59] for a more general result).

Theorem 4.4. *Let C be a nonempty convex subset of a normed space E and $x \in E \setminus \overline{C}$. Then there exists $z^* \in E^*$ such that $\sup_C z^* < \langle x, z^* \rangle$.*

Proof. From the Mazur–Orlicz theorem, Lemma 1.6, with $P := \|\cdot\|$ and $D := x - C$, there exists a linear function L on E such that

$$L \le \|\cdot\| \text{ on } E \quad \text{and} \quad \inf_D L = \inf_D \|\cdot\|.$$

Since $L \le \|\cdot\|$ on E, $L \in E^*$ (and in fact $\|L\| \le 1$). Furthermore, $\inf_D \|\cdot\|$ is the (strictly positive) distance from x to C, and $\inf_D L = L(x) - \sup_C L$. Now let $z^* := L$. ☐

We now give two corollaries of Theorem 4.4. We leave the proofs of these as exercises. We shall use Corollary 4.5 (see also [84, Theorem 3.12, pp. 64–65]) in Corollary 5.3, and Corollary 4.6 (see also [84, Theorem 3.5, p. 59]) in Theorem 47.1.

Corollary 4.5. *If F is a nonempty closed convex subset of a normed space E then F is $w(E, E^*)$–closed.*

Corollary 4.6. *If D is a subspace of a normed space E and*

$$z^* \in E^* \text{ and } \langle y, z^* \rangle = 0 \text{ for all } y \in D \quad \Longrightarrow \quad z^* = 0,$$

then D is dense in E.

Remark 4.7. One can easily deduce the following standard "separation theorem" (which we will not need) from Theorem 4.4, with $D := F - C$. If F is a nonempty closed convex subset of a normed space E, C is a nonempty $w(E, E^*)$–compact convex subset of E and $F \cap C = \emptyset$ then there exists $z^* \in E^*$ such that $\sup_F z^* < \inf_C z^*$. (See Rudin, [84, Theorem 3.4, pp. 58–59].)

We conclude this section by mentioning James's theorem, one of the most beautiful results in functional analysis: *if C is a nonempty bounded closed convex subset of a Banach space E then C is $w(E, E^*)$–compact if, and only if, for all $x^* \in E^*$, there exists $x \in C$ such that $\langle x, x^* \rangle = \max_C x^*$.* James's theorem is not easy — we refer the reader to Pryce, [71], or Ruiz Galán–Simons, [85], for a proof. We shall use the following special case, which follows from Theorem 4.3, in the construction of a couple of examples.

Theorem 4.8. *A Banach space E is reflexive if, and only if, for all $x^* \in E^*$, there exists $x \in E$ such that $\|x\| \le 1$ and $\langle x, x^* \rangle = \|x^*\|$.*

5 Excess, duality gap, and minimax criteria for weak compactness

In this section, we will explore the connection between two concepts from metric space and optimization theory, "excess" and "duality gap", and minimax criteria for weak* and weak compactness. The main result on excess and duality gap appears in Theorem 5.1(b), and it leads rapidly to Theorem 5.1(c), a minimax criterion for certain subsets of a dual space to be weak* compact. Corollary 5.4 contains an application of Theorem 5.1(d) to reflexive spaces. The initial results in this section first appeared in [102].

One of the side–effects of the Banach–Alaoglu theorem is that weak* compactness is much less interesting than weak compactness, and so we show in Theorem 5.6 how to adapt Theorem 5.1 to obtain results on weak compactness. It is interesting and significant that the bidual is not mentioned in the statement of Theorem 5.6. Theorem 5.6, which first appeared in [90], was originally obtained in connection with James's theorem.

If (X, ρ) is a metric space and Y is a nonempty subset of X, we define the function $D_Y \colon X \to [0, \infty[$ by $D_Y(x) := \inf_{y \in Y} \rho(x, y)$ $(x \in X)$. If now $Y \subset Z \subset X$, then the *excess* of Z over Y is defined by $e(Z, Y) := \sup_{z \in Z} D_Y(z)$. The concept of excess has been extensively used in metric space theory and, in particular, in epigraphical analysis. See Beer, [15, §1.5, pp. 28–33] for more information on this concept.

If $X \neq \emptyset$, $Y \neq \emptyset$ and $f \colon X \times Y \to \mathbb{R}$, the *duality gap* of f over $X \times Y$ is defined by

$$\mathrm{dgap}_f(X, Y) := \inf_X \sup_Y f - \sup_Y \inf_X f \geq 0.$$

The concept of duality gap has been extensively studied in optimization theory. However, we should caution the reader that some authors use the phrase "duality gap" for the interval $\left[\sup_Y \inf_X f, \inf_X \sup_Y f \right]$. Since we will use the concept of duality gap only with reference to the canonical bilinear form associated with a normed space and its dual, we will suppress the subscript. In other words, as far as we are concerned,

$$\mathrm{dgap}(X, Y) := \inf_{x \in X} \sup_{y^* \in Y} \langle x, y^* \rangle - \sup_{y^* \in Y} \inf_{x \in X} \langle x, y^* \rangle \geq 0.$$

The sublinear functional P defined in Theorem 5.1 below is known as the *support functional* of Y. Lemma 3.1 is used twice in Theorem 5.1. Its use in Theorem 5.1(b) is a substitute for the theory of locally convex spaces and the "bipolar theorem".

Theorem 5.1. *Let F be a nonzero normed space with dual F^* and unit ball F_1. Let Y be a nonempty bounded convex subset of F^*, and define the sublinear functional P on F by $P := \sup\langle \cdot, Y \rangle$. Write $\widetilde{Y}$ for the set of linear* functionals on F dominated by P on F. Finally, let $\mathcal{CS}$ stand for the set of all nonempty convex subsets of F_1, and $\mathcal{CCS}$ stand for the set of all nonempty closed convex subsets of F_1.

(a) $Y \subset \widetilde{Y} \subset F^*$. Further,

$$e(\widetilde{Y}, Y) = \sup_{X \in \mathcal{CS}} \mathrm{dgap}(X, Y) \tag{5.1}$$

$$= \sup_{X \in \mathcal{CCS}} \mathrm{dgap}(X, Y). \tag{5.2}$$

(b) Let $\ddot{Y}$ be the $w(F^*, F)$–closure of Y. Then $\widetilde{Y} = \ddot{Y}$.

(c) Let $\overline{Y}$ be the norm–closure of Y. Then the conditions (5.3)–(5.5) are equivalent.

$$\overline{Y} \text{ is } w(F^*, F)\text{–compact}. \tag{5.3}$$

$$\text{For all } X \in \mathcal{CS}, \quad \inf_{x \in X} \sup_{y^* \in Y} \langle x, y^* \rangle = \sup_{y^* \in Y} \inf_{x \in X} \langle x, y^* \rangle. \tag{5.4}$$

$$\text{For all } X \in \mathcal{CCS}, \quad \inf_{x \in X} \sup_{y^* \in Y} \langle x, y^* \rangle = \sup_{y^* \in Y} \inf_{x \in X} \langle x, y^* \rangle. \tag{5.5}$$

Proof. (a) It is obvious that $Y \subset \widetilde{Y}$. Further, since Y is bounded, P is dominated by some scalar multiple of the norm of F, which implies that $\widetilde{Y} \subset F^*$. Now let X be an arbitrary element of $\mathcal{CS}$. Then, from the Mazur–Orlicz theorem, Lemma 1.6, there exists $\widetilde{y} \in \widetilde{Y}$ such that $\inf_X \widetilde{y} = \inf_X P$. Let y^* be an arbitrary element of Y and x be an arbitrary element of X. Since $x \in F_1$, $\langle x, y^* \rangle + \|\widetilde{y} - y^*\| \geq \langle x, y^* \rangle + \langle x, \widetilde{y} - y^* \rangle = \langle x, \widetilde{y} \rangle$. Taking the infimum over $x \in X$ and using the choice of $\widetilde{y}$ and the definition of P, we derive that

$$\inf_{x \in X} \langle x, y^* \rangle + \|\widetilde{y} - y^*\| \geq \inf_{x \in X} \langle x, \widetilde{y} \rangle = \inf_X P = \inf_{x \in X} \sup_{z^* \in Y} \langle x, z^* \rangle,$$

from which $\|\widetilde{y} - y^*\| \geq \inf_{x \in X} \sup_{z^* \in Y} \langle x, z^* \rangle - \inf_{x \in X} \langle x, y^* \rangle$. Taking the infimum over $y^* \in Y$,

$$D_Y(\widetilde{y}) \geq \inf_{x \in X} \sup_{z^* \in Y} \langle x, z^* \rangle - \sup_{y^* \in Y} \inf_{x \in X} \langle x, y^* \rangle = \mathrm{dgap}(X, Y).$$

Thus we have established that $e(\widetilde{Y}, Y) \geq \mathrm{dgap}(X, Y)$, which gives the inequality "$\geq$" in (5.1). The inequality "$\geq$" in (5.2) is now immediate, so it only remains for (a) to prove that

$$e(\widetilde{Y}, Y) \leq \sup_{X \in \mathcal{CCS}} \mathrm{dgap}(X, Y). \tag{5.6}$$

There is so much "loss" in the the proof of the inequality "$\geq$" in (5.1) above that it seems unlikely that (5.6) holds. However, it does, and here is a proof. Let $\widetilde{y}$ be an arbitrary element of $\widetilde{Y}$, and $\varepsilon > 0$. Since $\langle F_1, \widetilde{y} \rangle$ is bounded and $\widetilde{y}$ is continuous and linear on F, we can write $F_1 = X_1 \cup \cdots \cup X_m$, where, for all $i \in \{1, \ldots, m\}$, $X_i \in \mathcal{CCS}$, and $\sup_{X_i} \widetilde{y} \leq \inf_{X_i} \widetilde{y} + \varepsilon$. Thus, using the definition of $\widetilde{Y}$, $\widetilde{y} \leq P$ on F and so

$$\sup_{X_i} \widetilde{y} \leq \inf_{X_i} P + \varepsilon = \inf_{x \in X_i} \sup_{y^* \in Y} \langle x, y^* \rangle + \varepsilon. \tag{5.7}$$

Further, from the definition of "supremum", for all $i \in \{1, \ldots, m\}$, there exists $y_i^* \in Y$ such that $\inf_{X_i} y_i^* \geq \sup_{y^* \in Y} \inf_{x \in X_i} \langle x, y^* \rangle - \varepsilon$. Combining this with (5.7) and using the fact that $X_i \in \mathcal{CCS}$,

$$\sup_{X_i} (\widetilde{y} - y_i^*) \leq \sup_{X_i} \widetilde{y} - \inf_{X_i} y_i^* \leq \mathrm{dgap}(X_i, Y) + 2\varepsilon \leq \alpha + 2\varepsilon,$$

where $\alpha := \sup_{X \in \mathcal{CCS}} \mathrm{dgap}(X, Y)$. Since $F_1 = X_1 \cup \ldots \cup X_m$,

$$\sup_{F_1} \left[(\widetilde{y} - y_1^*) \wedge \cdots \wedge (\widetilde{y} - y_m^*) \right] \leq \alpha + 2\varepsilon.$$

From Lemma 3.1 with $C = F_1$ and $f_i := y_i^* - \widetilde{y}$, there exist $\lambda_1, \ldots, \lambda_m \geq 0$ such that $\lambda_1 + \ldots + \lambda_m = 1$ and

$$\sup_{F_1} \left[\lambda_1 (\widetilde{y} - y_1^*) + \cdots + \lambda_m (\widetilde{y} - y_m^*) \right] \leq \alpha + 2\varepsilon.$$

Setting $y^* := \lambda_1 y_1^* + \cdots + \lambda_m y_m^* \in Y$, we have $\sup_{F_1} (\widetilde{y} - y^*) \leq \alpha + 2\varepsilon$, i.e., $\|\widetilde{y} - y^*\| \leq \alpha + 2\varepsilon$. Thus we have proved that $D_Y(\widetilde{y}) \leq \alpha + 2\varepsilon$. Since this construction can be carried out for all $\widetilde{y} \in \widetilde{Y}$, and ε can be made arbitrarily small, $e(\widetilde{Y}, Y) \leq \alpha$. This completes the proof of (5.6), and hence of (a).

(b) $\widetilde{Y}$ is obviously $w(F^*, F)$–closed, and so it follows from (a) that $\widetilde{Y} \supset \ddot{Y}$. Conversely, let $\widetilde{y}$ be an arbitrary element of $\widetilde{Y}$, $m \geq 1$, $x_1, \ldots, x_m \in F$ and $\varepsilon > 0$. From Lemma 3.1 with $C = Y$ and $f_i := \langle x_i, \widetilde{y} - \cdot \rangle$, there exist $\lambda_1, \ldots, \lambda_m \geq 0$ such that $\lambda_1 + \cdots + \lambda_m = 1$ and $\inf_Y \left[f_1 \vee \cdots \vee f_m \right] = \inf_Y \left[\lambda_1 f_1 + \cdots + \lambda_m f_m \right]$. Let $x = \lambda_1 x_1 + \cdots + \lambda_m x_m \in F$. We have

$$\inf_Y \left[\lambda_1 f_1 + \cdots + \lambda_m f_m \right] = \inf_{y^* \in Y} \langle x, \widetilde{y} - y^* \rangle = \langle x, \widetilde{y} \rangle - P(x) \leq 0,$$

and so $\inf_Y \left[f_1 \vee \cdots \vee f_m \right] \leq 0$, from which there exists $y^* \in Y$ such that, for all $i = 1, \ldots m$, $\langle x_i, \widetilde{y} - y^* \rangle \leq \varepsilon$. This implies that $\widetilde{y} \in \ddot{Y}$. Thus $\widetilde{Y} \subset \ddot{Y}$, which completes the proof of (b).

(c)((5.3)$\Longrightarrow$(5.4)) If (5.3) is satisfied then $\overline{Y}$ is $w(F^*, F)$–closed and (b) implies that $\widetilde{Y} \subset \overline{Y}$. Thus $e(\widetilde{Y}, Y) \leq e(\overline{Y}, Y) = 0$, and (5.4) follows from (a).

((5.4) $\Longrightarrow$ (5.5)) This is immediate.

((5.5) $\Longrightarrow$ (5.3)) It is easy to see that, in any case, $\widetilde{Y} \supset \overline{Y}$. If now (5.5) is satisfied then (a) gives $e(\widetilde{Y}, Y) = 0$, and so $\widetilde{Y} \subset \overline{Y}$. Thus $\overline{Y} = \widetilde{Y}$, and (5.3) follows from Theorem 4.2. $\qquad \square$

Remark 5.2. Of course, by performing an appropriate scaling, we can obtain two further equivalent conditions in Theorem 5.1(c) by replacing $\mathcal{CS}$ in (5.4) by the set of all nonempty bounded convex subsets of F (instead of F_1), and replacing $\mathcal{CCS}$ in (5.5) by the set of all nonempty bounded closed convex subsets of F.

Corollary 5.3. *Let F be a nonzero normed space, X be a nonempty bounded convex subset of F and Y be a nonempty convex subset of F^* such that the $w(F^*, F^{**})$–closure of Y is $w(F^*, F)$–compact. Then*

$$\inf_{x \in X} \sup_{y^* \in Y} \langle x, y^* \rangle = \sup_{y^* \in Y} \inf_{x \in X} \langle x, y^* \rangle. \tag{5.8}$$

Proof. Since Y is convex, Corollary 4.5 implies that its norm–closure is identical with its $w(F^*, F^{**})$–closure. The result now follows from Theorem 5.1(c), enhanced as in Remark 5.2. □

The author is grateful to Jonathan Borwein for pointing out to him that the following result can be proved in other ways.

Corollary 5.4. *Let F be a nonzero reflexive Banach space, X be a nonempty bounded convex subset of F and Y be a nonempty bounded convex subset of F^*. Then (5.8) holds.*

Proof. This follows from the Banach–Alaoglu theorem, Theorem 4.1, and Corollary 5.3, since the reflexivity of F implies that $w(F^*, F^{**}) = w(F^*, F)$. □

Remark 5.5. Let F be a nonzero reflexive Banach space. It follows from Theorem 3.2 that if X is a (possibly unbounded) nonempty convex subset of F and Y is a nonempty bounded closed convex subset of F^* then (5.8) holds. It is tempting to surmise that (5.8) holds if X is a nonempty closed convex subset of F and Y is a nonempty bounded convex subset of F^* (we are transferring the "closedness" from Y to X – another way of looking at this is an attempted "interpolation" between Theorem 3.2 and Corollary 5.4). However, this is false, as can be seen from the following simple example. Let $F := \mathbb{R}$, $X := [0, \infty[$ and $Y :=]-1, 0[$. Then $\inf_{x \in X} \sup_{y^* \in Y} xy^* = 0$ and $\sup_{y^* \in Y} \inf_{x \in X} xy^* = -\infty$.

We now show how the above results can be parlayed into results on weak compactness. We point out again, as we have already done in the introduction, the significant fact that the bidual of E is *not* mentioned in the statement of Theorem 5.6 below.

Theorem 5.6. *Let E be a nonzero normed space with dual E^*, $\mathcal{CS}$ be the set of all nonempty convex subsets of the unit ball of E^* and $\mathcal{CCS}$ be the set of all nonempty norm–closed convex subsets of the unit ball of E^*. Let Z be a nonempty bounded norm–complete convex subset of E. Then the conditions (5.9)–(5.11) are equivalent.*

$$Z \text{ is } w(E, E^*)\text{–compact.} \tag{5.9}$$

$$\text{For all } X \in \mathcal{CS}, \quad \inf_{x^* \in X} \sup_{z \in Z} \langle z, x^* \rangle = \sup_{z \in Z} \inf_{x^* \in X} \langle z, x^* \rangle. \tag{5.10}$$

$$\text{For all } X \in \mathcal{CCS}, \quad \inf_{x^* \in X} \sup_{z \in Z} \langle z, x^* \rangle = \sup_{z \in Z} \inf_{x^* \in X} \langle z, x^* \rangle. \tag{5.11}$$

Proof. Let $Y = \widehat{Z} \subset E^{**}$. Since the canonical map is a $w(E, E^*)$–$w(E^{**}, E^*)$ homeomorphism from Z onto Y, (5.9) is equivalent to

$$Y \text{ is } w(E^{**}, E^*)\text{–compact.} \tag{5.12}$$

However, the canonical map is also a norm–isometry from Z onto Y, hence Y is complete and so $Y = \overline{Y}$, the norm–closure of Y in E^{**}. Thus (5.12) is equivalent to

$$\overline{Y} \text{ is } w(E^{**}, E^*)\text{–compact.}$$

The result now follows from Theorem 5.1(c), with $F := E^*$. □

Remark 5.7. As was the case with Theorem 5.1(c), by performing an appropriate scaling, we can obtain two further equivalent conditions in Theorem 5.6 by replacing CS in (5.10) by the set of all nonempty bounded convex subsets of E^*, and replacing CCS in (5.11) by the set of all nonempty bounded closed convex subsets of E^*.

We conclude this section with a useful consequence of Theorem 5.1(b) (exercise!), which can also be proved using the theory of locally convex spaces and the "bipolar theorem".

Theorem 5.8. *Let E be a nonzero normed space with dual E^*, C be a nonempty $w(E^*, E)$–compact convex subset of E^*, and define the sublinear functional P on E by $P := \max\langle\cdot, C\rangle$. Let $x^* \in E^*$ and $x^* \leq P$ on E. Then $x^* \in C$.*

6 Sharp Lagrange multiplier and KKT results

This is the first of two sections in which we show how Theorem 1.11 can be used to convert problems on the existence of continuous linear functionals on a normed space to problems on the existence of a single real constant, and obtain sharp lower bound on the norm of the linear functional satisfying the required condition.

In this section, we consider the existence of Lagrange multipliers (see Definition 6.3) for the infinite–dimensional constrained convex optimization problem described in (6.2) below. The main result is Theorem 6.4(a), which gives a *necessary and sufficient* condition for the existence of a Lagrange multiplier with, in Theorem 6.4(c), a sharp lower bound on its norm. We also show in Theorem 6.6(b) how Theorem 6.4 implies the classical *sufficient* "Slater condition" result, with an upper bound on the norm as a bonus. We then show in Theorem 6.9 how the results of Theorem 6.4 and Theorem 6.6 lead to a short proof of a Karush–Kuhn–Tucker theorem for minimization problems for functions with convex Gâteaux derivatives. This section concludes (in Remark 6.10) with a sketch of a generalization of our Lagrange multiplier results to the case where the functions are defined on an abstract set rather than a convex subset of a vector space.

Let $(E, \|\cdot\|)$ be a nonzero normed space with dual E^* and $\preceq$ be a partial ordering on E compatible with its vector space structure. Let N be the negative cone $\{z \in E\colon z \preceq 0\}$, and $D_N, D_{E\setminus N}\colon E \to [0, \infty[$ be defined by

$$D_N := \mathrm{dist}\,(\cdot, N) = \inf_{z \in N} \|\cdot - z\|$$

and

$$D_{E\setminus N} := \mathrm{dist}\,(\cdot, E \setminus N) = \inf_{z \in E\setminus N} \|\cdot - z\|.$$

The following lemma ties up the situation here with the concepts introduced in Section 1.

Lemma 6.1. *Let $M \geq 0$, and define $P: E \to [0, \infty[$ by $P := MD_N$. Then P is sublinear, $P = 0$ on N and $P \leq M\|\cdot\|$ on E.*

Proof. Let $y_1, y_2 \in E$ and $z_1, z_2 \in N$ be arbitrary. Then, since $z_1 + z_2 \in N$,

$$M\|y_1 - z_1\| + M\|y_2 - z_2\| \geq M\|y_1 + y_2 - (z_1 + z_2)\| \geq P(y_1 + y_2).$$

Take the infimum over z_1 and z_2, $P(y_1) + P(y_2) \geq P(y_1 + y_2)$. Thus P is subadditive. It is easy to see that P is positively homogeneous, so P is sublinear. It is obvious that $P = 0$ on N. $\square$

Up to Theorem 6.6, we suppose that C is a nonempty convex subset of a vector space and $j: C \to E$ is convex with respect to $\preceq$, that is to say

$$w, x \in C, \text{ and } \lambda \in]0, 1[\implies j(\lambda w + (1 - \lambda)x) \preceq \lambda j(w) + (1 - \lambda)j(x). \quad (6.1)$$

Lemma 6.2. *Under the preceding conditions, j is P–convex.*

Proof. This is clear from Lemma 6.1 since, if $y, z \in E$, then

$$y \preceq z \iff y - z \in N \implies P(y - z) = 0 \implies y \leq_P z. \quad \square$$

Now let $k \in \mathcal{PC}(C)$, and $\mu \in \mathbb{R}$ be the *constrained infimum*

$$\mu = \inf_{j^{-1}N} k = \inf\{k(x): x \in C, \ j(x) \preceq 0\}. \quad (6.2)$$

Definition 6.3. A *Lagrange multiplier* for the infimization problem (6.2) is an element z^* of E^* such that z^* is $\preceq$–positive, that is to say,

$$z^* \leq 0 \text{ on } N, \quad (6.3)$$

and

$$\inf_{x \in C}\left[\langle j(x), z^*\rangle + k(x)\right] = \mu. \quad (6.4)$$

Theorem 6.4. (a) *There exists a Lagrange multiplier if, and only if*

$$\text{there exists } M \geq 0 \text{ such that} \quad MD_N \circ j + k \geq \mu \text{ on } C. \quad (6.5)$$

In this case, there exists a Lagrange multiplier z^* such that $\|z^*\| \leq M$.

If $k \geq \mu$ on C then we can take $M = 0$ above, and 0 is a Lagrange multiplier. In order to exclude this trivial case, we shall now suppose that $\inf_C k < \mu$. Let

$$W := \{w \in C: k(w) < \mu\} \neq \emptyset.$$

(b) *If z^* is a Lagrange multiplier then $j(W) \subset E \setminus \overline{N}$ and*

$$0 < \sup_W \frac{\mu - k}{D_N \circ j} \leq \|z^*\| < \infty.$$

(c) *If $j(W) \subset E \setminus \overline{N}$ and $0 < M := \sup_W \dfrac{\mu - k}{D_N \circ j} < \infty$ then*

$$\min\{\|z^*\|: z^* \text{ is a Lagrange multiplier}\} = M.$$

Proof. (a) Our proof is based on the (disarmingly simple) observation that $z^* \in E^*$ if, and only if, $z^* = L$, where L is a linear functional on E for which there exists $M \geq 0$ such that $L \leq M\| \cdot \|$ on E. So there exists a Lagrange multiplier if, and only if, there exists $M \geq 0$ for which there exists a linear functional L on E such that

$$L \leq M\| \cdot \| \text{ on } E, \quad L \leq 0 \text{ on } N \tag{6.6}$$

and

$$\inf{}_C \left[L \circ j + k \right] = \mu. \tag{6.7}$$

Now suppose that there exists a Lagrange multiplier, and let M and L be as in (6.6). Let $P := MD_N$. Let y be an arbitrary element of E and z be an arbitrary element of N. Then, using (6.6),

$$L(y) \leq L(y) - L(z) = L(y - z) \leq M\|y - z\|$$

and, taking the infimum over $z \in N$, $L(y) \leq MD_N(y)$. Thus we have proved that

$$L \leq P \text{ on } E. \tag{6.8}$$

Clearly, (6.7) and (6.8) imply that

$$\inf{}_C \left[P \circ j + k \right] \geq \mu, \tag{6.9}$$

and so (6.5) is satisfied. The converse of this is also true: suppose that M is as in (6.5). Then, writing $P := MD_N$ again, (6.9) is satisfied. From Lemmas 6.1 and 6.2, P is sublinear and j is P–convex, and so Theorem 1.11 provides a linear functional L on E satisfying (6.8) such that

$$\inf{}_C \left[L \circ j + k \right] \geq \mu. \tag{6.10}$$

Lemma 6.1 and (6.8) now give us (6.6), from which

$$x \in j^{-1}N \Longrightarrow j(x) \in N \Longrightarrow L \circ j(x) \leq 0,$$

and (6.2) now gives

$$\inf{}_C \left[L \circ j + k \right] \leq \inf{}_{j^{-1}N} \left[L \circ j + k \right] \leq \inf{}_{j^{-1}N} k = \mu.$$

(6.7) now follows by combining this with (6.10), and so L is a Lagrange multiplier. Furthermore, (6.6) also implies that $\|L\| \leq M$.

(b) We can go through the proof of (a)($\Longrightarrow$) with $M := \|z^*\|$, and so (6.9) implies that $\|z^*\|D_N \circ j \geq \mu - k$ on C, from which $\|z^*\|D_N \circ j \geq \mu - k > 0$ on W. (b) is immediate from this and the fact that $D_N(y) > 0 \Longleftrightarrow y \notin \overline{N}$.

(c) The definition of M clearly implies that $MD_N \circ j + k \geq \mu$ on W. On the other hand, $MD_N \circ j + k \geq 0 + \mu = \mu$ on $C \setminus W$. Thus (6.9) is true, and (a)($\Longleftarrow$) gives a Lagrange multiplier z^* such that $\|z^*\| \leq M$. (c) follows by combining this with (b). $\qquad \square$

The following simple lemma will enable us to use Theorem 6.4 to obtain a generalization of the classical Slater condition for the existence of Lagrange multipliers. We now write

$$V := \{v \in \operatorname{dom} k \colon\ j(v) \in \operatorname{int} N\} \subset C.$$

As above, let $W := \{w \in C \colon\ k(w) < \mu\} \neq \emptyset.$

Lemma 6.5. *Let $w \in W$ and $v \in V$. Then*

$$D_N \circ j(w)\big(k(v) - \mu\big) \geq D_{E \setminus N} \circ j(v)\big(\mu - k(w)\big) > 0. \qquad (6.11)$$

Consequently, $j(w) \in E \setminus \overline{N}$ *and* $k(v) > \mu$. *If $V \neq \emptyset$ then*

$$0 < \sup_W \frac{\mu - k}{D_N \circ j} \leq \inf_V \frac{k - \mu}{D_{E \setminus N} \circ j}. \qquad (6.12)$$

Proof. Since $k(w) < \mu$ and $j(v) \in N$, (6.2) implies that $j(w) \notin N$ and $\mu \leq k(v) < \infty$. Furthermore, since $j(v) \in \operatorname{int} N$, $D_{E \setminus N} \circ j(v) > 0$. Let $z \in N$, $0 < \eta < D_{E \setminus N} \circ j(v)$, $\alpha := \|j(w) - z\| > 0$ and $\lambda := \eta/(\alpha + \eta) \in\,]0, 1[$. Now

$$\left\| \frac{\lambda}{1 - \lambda}\big(j(w) - z\big) \right\| = \left\| \frac{\eta}{\alpha}\big(j(w) - z\big) \right\| = \eta < D_{E \setminus N} \circ j(v)$$

and so $\dfrac{\lambda}{1 - \lambda}\big(j(w) - z\big) + j(v) \in N$, from which $\lambda j(w) + (1 - \lambda)j(v) \preceq \lambda z \preceq 0$. (6.1) now gives $j\big(\lambda w + (1 - \lambda)v\big) \preceq 0$, and it follows from the convexity of k and (6.2) that

$$\lambda k(w) + (1 - \lambda)k(v) \geq k\big(\lambda w + (1 - \lambda)v\big) \geq \mu,$$

from which $(1 - \lambda)\big(k(v) - \mu\big) \geq \lambda\big(\mu - k(w)\big)$, i.e., $\alpha\big(k(v) - \mu\big) \geq \eta\big(\mu - k(w)\big)$. We now obtain (6.11) by letting $\eta \to D_{E \setminus N} \circ j(v)$ and then taking the infimum over $z \in N$. From (6.11),

$$0 < \frac{\mu - k(w)}{D_N \circ j(w)} \leq \frac{k(v) - \mu}{D_{E \setminus N} \circ j(v)},$$

and (6.12) now follows immediately. $\qquad\square$

The classical *Slater condition*, which is a sufficient condition for there to exist a Lagrange multiplier, is that $V \neq \emptyset$. This will be strengthened in Theorem 6.6(b) below. (See [59, Theorem 8.3.1, pp. 217–218] for an exposition of Slater's result — [59] also contains applications to control theory.) Theorem 6.6(a) will be used in our analysis of Karush–Kuhn–Tucker theorems in Theorem 6.9.

Theorem 6.6. *Suppose that* $V = \{v \in \operatorname{dom} k \colon\ j(v) \in \operatorname{int} N\} \neq \emptyset$. *Then:*
(a) $\mu = \inf_V k$.
(b) *There exists a Lagrange multiplier z^* such that* $\|z^*\| \leq \inf_V \dfrac{k - \mu}{D_{E \setminus N} \circ j}$.

Proof. (a) Since $\mu \in \mathbb{R}$, it follows from (6.2) that

$$\mu = \inf \left\{ k(x) \colon x \in C, \; j(x) \preceq 0 \right\} = \inf \left\{ k(x) \colon x \in \operatorname{dom} k, \; j(x) \preceq 0 \right\}. \quad (6.13)$$

Now let v be a fixed element of V. Let $x \in \operatorname{dom} k$ and $j(x) \preceq 0$. Let $\lambda \in \,]0,1[$. Then $\lambda v + (1 - \lambda)x \in \operatorname{dom} k$ and, from (6.1),

$$j\big(\lambda v + (1 - \lambda)x\big) \in \lambda j(v) + (1 - \lambda)j(x) + N \subset \lambda \operatorname{int} N + (1 - \lambda)N + N.$$

Kelley–Namioka, [52, 13.1(i), pp. 110–111] (for instance), now implies that $j\big(\lambda v + (1 - \lambda)x\big) \in \operatorname{int} N$, from which $\lambda v + (1 - \lambda)x \in V$. Consequently,

$$\inf_V k \leq k\big(\lambda v + (1 - \lambda)x\big) \leq \lambda k(v) + (1 - \lambda)k(x).$$

Letting $\lambda \to 0$, $\inf_V k \leq k(x)$, and it now follows from (6.13) by taking the infimum over x that $\inf_V k \leq \mu$. On the other hand, since we have $V \subset \{x \in \operatorname{dom} k \colon j(x) \preceq 0\}$, it is also clear from (6.13) that $\inf_V k \geq \mu$. This completes the proof of (a).

(b) As before, let $W := \{w \in C \colon k(w) < \mu\}$. As we have already observed, if $W = \emptyset$ then 0 is a Lagrange multiplier and there is nothing to prove, so we can and will assume that $W \neq \emptyset$. The result now follows from Theorem 6.4(c) and Lemma 6.5. $\qquad\square$

We now explore the connections between Lagrange multipliers as discussed above and generalized Karush–Kuhn–Tucker theory as discussed in Luenberger, [59, Theorem 9.1, pp. 249–250] and [59, Problem 9.7.9, p. 267]. We suppose that C is (for simplicity) a vector space, and $x_0 \in C$ is a "base point". Let $G \colon C \to E$ and $f \colon C \to \mathbb{R}$, and suppose that

$$\text{for all } x \in C, \quad \mathrm{d}^+ G(x) := \lim_{\alpha \to 0+} \frac{G(x_0 + \alpha x) - G(x_0)}{\alpha} \quad \text{exists in } E,$$

$$\text{for all } x \in C, \quad \mathrm{d}^+ f(x) := \lim_{\alpha \to 0+} \frac{f(x_0 + \alpha x) - f(x_0)}{\alpha} \quad \text{exists in } \mathbb{R},$$

$\mathrm{d}^+ G \colon C \to E$ is $\preceq$–convex and $\mathrm{d}^+ f \colon C \to \mathbb{R}$ is convex. Suppose, finally, that

$$G(x_0) \preceq 0 \quad \text{and} \quad \min \{f(x) \colon x \in C, \; G(x) \preceq 0\} = f(x_0). \quad (6.14)$$

Let $V := \{v \in C \colon G(x_0) + \mathrm{d}^+ G(v) \in \operatorname{int} N\}$. Our analysis depends on the following critical lemma:

Lemma 6.7. *Let $v \in V$. Then $\mathrm{d}^+ f(v) \geq 0$.*

Proof. It is clear from the definitions of V and $\mathrm{d}^+ G(v)$ that, for all sufficiently small $\alpha \in \,]0,1[$,

$$G(x_0) + \frac{G(x_0 + \alpha v) - G(x_0)}{\alpha} \in N,$$

consequently $G(x_0 + \alpha v) \in (1 - \alpha)G(x_0) + N \in (1 - \alpha)N + N = N$. (6.14) now implies that $f(x_0 + \alpha v) \geq f(x_0)$, and so

$$\frac{f(x_0 + \alpha v) - f(x_0)}{\alpha} \geq 0.$$

The result follows by letting $\alpha \to 0$. $\qquad\square$

Definition 6.8. A *KKT functional* for the minimization problem (6.14) is a $\preceq$–positive element z^* of E^* (i.e., (6.3) is satisfied) such that

$$\text{for all } x \in C, \quad \langle G(x_0) + \mathrm{d}^+G(x), z^* \rangle + \mathrm{d}^+f(x) \geq 0. \qquad (6.15)$$

From (6.14), $G(x_0) \in N$, and the $\preceq$–positivity of z^* gives $\langle G(x_0), z^* \rangle \leq 0$. Since $\mathrm{d}^+G(0) = 0$ and $\mathrm{d}^+f(0) = 0$, (6.15) implies that $\langle G(x_0), z^* \rangle = 0$ and

$$\inf_{x \in C} \left[\langle G(x_0) + \mathrm{d}^+G(x), z^* \rangle + \mathrm{d}^+f(x) \right] = 0. \qquad (6.16)$$

Of course, the converse of this is true: *if z^* is a $\preceq$–positive element of E^* and (6.16) is satisfied then z^* is a KKT functional for the minimization problem (6.14).* In the special case when G and f are Gâteaux differentiable, d^+G and d^+f are linear, and so (6.15) implies that $z^* \circ \mathrm{d}^+G + \mathrm{d}^+f = 0$, from which $z^* \circ G + f$ is stationary at x_0.

Theorem 6.9. *Suppose that*

$$V = \left\{ v \in C\colon G(x_0) + \mathrm{d}^+G(v) \in \mathrm{int}\, N \right\} \neq \emptyset,$$

$$W := \left\{ w \in C\colon \mathrm{d}^+f(w) < 0 \right\} \neq \emptyset,$$

$$G(x_0) + \mathrm{d}^+G(W) \subset E \setminus \overline{N},$$

and

$$\sup_{W} \frac{-\mathrm{d}^+f}{D_N \circ (G(x_0) + \mathrm{d}^+G)} < \infty.$$

Then there exists a KKT functional for the minimization problem (6.14), and

$$\min \left\{ \|z^*\|\colon z^* \text{ is a KKT functional} \right\} = \sup_{W} \frac{-\mathrm{d}^+f}{D_N \circ (G(x_0) + \mathrm{d}^+G)}$$

$$\leq \inf_{V} \frac{\mathrm{d}^+f}{D_{E \setminus N} \circ (G(x_0) + \mathrm{d}^+G)}.$$

Proof. Define $j\colon C \to E$ and $k\colon C \to \mathbb{R}$ by $j := G(x_0) + \mathrm{d}^+G$ and $k := \mathrm{d}^+f$. By assumption, j is $\preceq$–convex and k is convex. (6.14) gives

$$j(0) = G(x_0) \preceq 0,$$

and so $0 \in j^{-1}N$. Defining $\mu = \inf_{j^{-1}N} k$ as in (6.2), we derive that

$$\mu \leq k(0) = 0.$$

On the other hand, Theorem 6.6(a) and Lemma 6.7 imply that

$$\mu = \inf_{V} k \geq 0.$$

Thus $\mu = 0$, and so (6.16) is equivalent to (6.4). The result now follows from Theorem 6.4(b,c) and Theorem 6.6(b). $\square$

Remark 6.10. Now let X be a nonempty set, $j\colon X \to E$, $k\colon X \to {]-\infty, \infty]}$ be proper and, for all $x_1, x_2 \in \operatorname{dom} k$, there exists $u \in \operatorname{dom} k$ such that

$$j(u) \preceq \tfrac{1}{2}j(x_1) + \tfrac{1}{2}j(x_2) \quad \text{and} \quad k(u) \leq \tfrac{1}{2}k(x_1) + \tfrac{1}{2}k(x_2). \tag{6.17}$$

Let $\mu \in \mathbb{R}$ be the constrained infimum

$$\mu = \inf_{j^{-1}N} k = \inf \left\{ k(x)\colon\ x \in X,\ j(x) \preceq 0 \right\}. \tag{6.18}$$

A *Lagrange multiplier* for this infimization problem is a $\preceq$–positive element z^* of E^* such that

$$\inf_{x \in X} \left[\langle j(x), z^* \rangle + k(x) \right] = \mu.$$

Let $\inf_X k < \mu$,

$$W := \left\{ w \in X\colon\ k(w) < \mu \right\}$$

so that $\emptyset \neq W \subset X$, and

$$V := \left\{ v \in \operatorname{dom} k\colon\ j(v) \in \operatorname{int} N \right\} \subset C.$$

The argument of Lemma 6.2 and (6.17) give (1.6), and so the argument of Theorem 6.4 with Theorem 1.13 replacing Theorem 1.11 give (a) and (b) below, which generalize (b) and (c) of Theorem 6.4:

(a) *If z^* is a Lagrange multiplier then $j(W) \subset E \setminus \overline{N}$ and*

$$0 < \sup_W \frac{\mu - k}{D_N \circ j} \leq \|z^*\| < \infty.$$

(b) *If $j(W) \subset E \setminus \overline{N}$ and $0 < M := \sup_W \dfrac{\mu - k}{D_N \circ j} < \infty$ then*

$$\min \left\{ \|z^*\|\colon\ z^* \text{ is a Lagrange multiplier} \right\} = M.$$

We now consider the following generalization of Lemma 6.5 and Theorem 6.6(b):

(c) *Let $w \in W$ and $v \in V$. Then*

$$D_N \circ j(w)\big(k(v) - \mu\big) \geq D_{E \setminus N} \circ j(v)\big(\mu - k(w)\big) > 0. \tag{6.11}$$

(d) *There exists a Lagrange multiplier z^* such that*

$$\|z^*\| \leq \inf_V \frac{k - \mu}{D_{E \setminus N} \circ j}.$$

Now once (c) has been established, (d) follows exactly as in Theorem 6.6(b), so it remains to establish (c), which is somewhat more technical. Exactly as in the proof of Lemma 6.5, $k(w) < \mu \leq k(v) < \infty$, $D_{E \setminus N} \circ j(v) > 0$ and $j(w) \notin N$. Let $z \in N$, $0 < \eta < D_{E \setminus N} \circ j(v)$ and $\alpha := \|j(w) - z\| > 0$. Since the map $\theta \mapsto \theta / (\alpha + \theta)$ from $]0, \infty[$ into $]0, 1[$ is increasing,

$$\frac{\eta}{\alpha + \eta} < \frac{D_{E \setminus N} \circ j(v)}{\alpha + D_{E \setminus N} \circ j(v)},$$

and so there exists a dyadic rational δ such that

$$\frac{\eta}{\alpha + \eta} < \delta < \frac{D_{E \setminus N} \circ j(v)}{\alpha + D_{E \setminus N} \circ j(v)}.$$

Since the map $\theta \mapsto \theta/(1 - \theta)$ from $]0, 1[$ into $]0, \infty[$ is increasing,

$$\frac{\eta}{\alpha} < \frac{\delta}{1 - \delta} < \frac{D_{E \setminus N} \circ j(v)}{\alpha}. \tag{6.19}$$

In particular,

$$\left\| \frac{\delta}{1 - \delta}(j(w) - z) \right\| = \frac{\alpha \delta}{1 - \delta} < D_{E \setminus N} \circ j(v)$$

and so $\dfrac{\delta}{1 - \delta}(j(w) - z) + j(v) \in N$, from which $\delta j(w) + (1 - \delta)j(v) \preceq \delta z \preceq 0$.
Since δ is dyadic, (6.17) and recurrence give us $u \in \operatorname{dom} k$ such that

$$j(u) \preceq \delta j(w) + (1 - \delta)j(v) \quad \text{and} \quad k(u) \le \delta k(w) + (1 - \delta)k(v).$$

Thus $j(u) \preceq 0$ and so, from (6.18), $\delta k(w) + (1 - \delta)k(v) \ge k(u) \ge \mu$. If we now combine this with (6.19), we obtain

$$k(v) - \mu \ge \frac{\delta}{1 - \delta}(\mu - k(w)) \ge \frac{\eta}{\alpha}(\mu - k(w)),$$

i.e., $\alpha(k(v) - \mu) \ge \eta(\mu - k(w))$. The rest of the proof of (6.11) now follows exactly as in Lemma 6.5.

The results sketched above can obviously be parlayed into an analogous result on KKT functionals.

II Fenchel duality

7 A sharp version of the Fenchel Duality theorem

This section is similar in spirit to Section 6, but this time we apply the Hahn–Banach–Lagrange theorem to problems in convex analysis. So suppose that E is a nonzero normed space with dual E^*, $k \in \mathcal{PC}(E)$ and $x \in E$. Then the *subdifferential* of k at x is defined by

$$\partial k(x) := \{z^* \in E^*: \quad y \in E \Longrightarrow k(x) + \langle y - x, z^* \rangle \leq k(y)\}.$$

We first consider the question of when $\partial k(x) \neq \emptyset$. This question is answered in Example 7.1 below. The justification, using Theorem 1.11, follows similar lines to those used in Section 6(exercise!). In the sketch below, the "slope" of the subtangent at $\big(x, k(x)\big)$ is z^*.

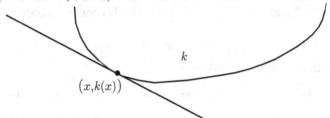

Example 7.1. $\partial k(x) \neq \emptyset$ if, and only if, $x \in \operatorname{dom} k$ and there exists $M \geq 0$ such that

$$y \in E \quad \Longrightarrow \quad k(x) - M\|y - x\| \leq k(y).$$

We now come to a more interesting example. Again, suppose that E is a nonzero normed space with dual E^*. When can a concave function and a convex function be separated by a continuous affine function? More precisely, given $f, g \in \mathcal{PC}(E)$, when do there exist $z^* \in E^*$ and $\beta \in \mathbb{R}$ such that

$$-f \leq z^* + \beta \leq g \quad \text{on} \quad E? \tag{7.1}$$

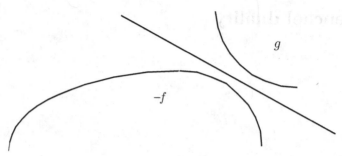

Now there exists $\beta \in \mathbb{R}$ such that $-f \le z^* + \beta \le g$ on E if, and only if, for all $x, y \in E$, $f(x) + g(y) + \langle x - y, z^* \rangle \ge 0$. The same technique as above (using Theorem 1.11, with $C = E \times E$, $k(x, y) = f(x) + g(y)$ and $j(x, y) = x - y$) then leads rapidly to the result below (exercise!):

Example 7.2. *Let E be a nonzero normed space with dual E^*, and $f, g \in \mathcal{PC}(E)$. Then there exist $z^* \in E^*$ and $\beta \in \mathbb{R}$ such that (7.1) is satisfied if, and only if,*

$$\left. \begin{array}{l} \text{there exists } M \ge 0 \text{ such that,} \\ \quad x, \ y \in E \quad \implies \quad f(x) + g(y) + M\|x - y\| \ge 0. \end{array} \right\} \tag{7.2}$$

In this case, there exist z^* and β satisfying (7.1) such that $\|z^*\| \le M$.

Remark 7.3. We note that (7.1) can also be split up into the two statements "$f^*(-z^*) \le \beta$" and "$g^*(z^*) \le -\beta$", where the *Fenchel conjugate* f^* is defined by

$$f^*(x^*) := \sup_E (x^* - f).$$

It follows that (7.2) is equivalent to:

$$\text{there exists } z^* \in E^* \text{ such that} \quad f^*(-z^*) + g^*(z^*) \le 0. \tag{7.3}$$

This is an old condition in convex analysis, due to Fenchel in the finite dimensional case. For this reason, we will say that $z^* \in E^*$ is a *Fenchel functional* for f and g if $f^*(-z^*) + g^*(z^*) \le 0$. Theorem 7.4 below, the *sharp Fenchel duality theorem*, contains a result for Fenchel functionals analogous to the result proved for Lagrange multipliers in Theorem 6.4.

The reader will note that we have defined Fenchel conjugate and Fenchel functional for proper convex functions on a normed space. This presents an impediment for some situations that will arise in our later discussions on multifunctions. Accordingly, we will redefine Fenchel conjugate and Fenchel functional *with respect to a bilinear form* in Section 8, where these issues will be discussed in greater detail. These later definitions will be entirely consistent with what we have introduced above.

Theorem 7.4. *Let E be a nonzero normed space with dual E^*, and $f, g \in \mathcal{PC}(E)$. Then:*
(a) *f and g have a Fenchel functional if, and only if, (7.2) is satisfied.*
(b) *If $z^* \in E^*$ is a Fenchel functional for f and g then*

$$\sup_{x, y \in E, \; x \neq y} \frac{-f(x) - g(y)}{\|x - y\|} \leq \|z^*\| < \infty.$$

(c) *If $f + g \geq 0$ on E and* $\displaystyle \sup_{x, y \in E, \; x \neq y} \frac{-f(x) - g(y)}{\|x - y\|} < \infty$ *then*

$$\left. \begin{aligned}
\min \left\{ \|z^*\| \colon z^* \text{ is a Fenchel functional for } f \text{ and } g \right\} \\
= \sup_{x, y \in E, \; x \neq y} \frac{-f(x) - g(y)}{\|x - y\|} \vee 0.
\end{aligned} \right\} \quad (7.4)$$

Proof. (a) is a restatement of Example 7.2.
(b) The analysis preceding Example 7.2 implies that

$$x, y \in E \quad \Longrightarrow \quad -f(x) - g(y) \leq \langle x - y, z^* \rangle \leq \|x - y\| \|z^*\|,$$

which gives the required result.

(c) The inequality "$\geq$" in (7.4) follows from (b) and the fact that $\|z^*\| \geq 0$. Now write M for the right hand side of (7.4). Then $M \geq 0$. Let $x, y \in E$. We have

$$x \neq y \quad \Longrightarrow \quad \frac{-f(x) - g(y)}{\|x - y\|} \leq M \quad \Longrightarrow \quad f(x) + g(y) + M\|x - y\| \geq 0,$$

and

$$x = y \quad \Longrightarrow \quad f(x) + g(y) + M\|x - y\| = (f + g)(x) \geq 0.$$

The result now follows from Example 7.2. $\qquad \square$

Example 7.5. The purpose of this example is to show that the "$\vee$" in (7.4) is necessary. Let $E = \mathbb{R}$ and $f = g = |\cdot|$. Then (exercise!) $f + g \geq 0$ on $\mathbb{R}$ and $\displaystyle \sup_{x, y \in \mathbb{R}, \; x \neq y} \frac{-|x| - |y|}{|x - y|} = -1$.   However, there cannot exist $z^* \in E^*$ such that $\|z^*\| = -1$.

Remark 7.6. In this remark, we discuss a geometric interpretation of Theorem 7.4(c). We write $\mathcal{CA}(E)$ for the set of all continuous affine real functions on E. If $a \in \mathcal{CA}(E)$ then a can be written uniquely in the form $a = z^* + \beta$, where $z^* \in E^*$ and $\beta \in \mathbb{R}$. Since z^* is the derivative of a in any reasonable sense, we shall write $z^* = a'$. Turning now to Theorem 7.4(c), write $h = -f$, so that h is proper and concave and $h \leq g$ on E. Let $a \in \mathcal{CA}(E)$. Remark 7.3 tells us that if $h \leq a \leq g$ on E then $f^*(-a') + g^*(a') \leq 0$ and, conversely, if $f^*(-z^*) + g^*(z^*) \leq 0$ then there exists $a \in \mathcal{CA}(E)$ such that $a' = z^*$ and $h \leq a \leq g$ on E. Furthermore,

$$\sup_{x,\,y\in E,\ x\neq y} \frac{-f(x)-g(y)}{\|x-y\|} = \sup_{x,\,y\in E,\ x\neq y} \frac{h(x)-g(y)}{\|x-y\|}$$

Now suppose, in addition, that $\sup_E h > \inf_E g$, to avoid the "$\vee$" in (7.4). Then the conclusion of Theorem 7.4(c) is that

$$\min\left\{\|a'\|\colon a\in\mathcal{CA}(E),\ h\leq a\leq g\text{ on }E\right\} = \sup_{x,\,y\in E,\ x\neq y} \frac{h(x)-g(y)}{\|x-y\|}.$$

The quotient on the right hand side of the equality above is, of course, the slope of the line–segment going from the point $\big(y,g(y)\big)$ on the graph of g to the point $\big(x,h(x)\big)$ on the graph of h.

8 Fenchel duality with respect to a bilinear form — locally convex spaces

Let E and E^* be nonzero real vector spaces, and $\langle\cdot,\cdot\rangle\colon E\times E^*\to\mathbb{R}$ be a bilinear form that separates the points of E and also separates the points of E^*. (This means that if $x\in E\setminus\{0\}$ then there exists $x^*\in E^*$ such that $\langle x,x^*\rangle\neq 0$ and that if $x^*\in E^*\setminus\{0\}$ then there exists $x\in E$ such that $\langle x,x^*\rangle\neq 0$.) If $f\in\mathcal{PC}(E)$, the *Fenchel conjugate* f^* with respect to $\langle\cdot,\cdot\rangle$ is defined by

$$f^*(x^*) := \sup_{x\in E}\big[\langle x,x^*\rangle - f(x)\big]. \tag{8.1}$$

We note that this definition implies the *Fenchel–Young inequality*

$$(x,x^*)\in E\times E^* \implies \langle x,x^*\rangle\leq f(x)+f^*(x^*). \tag{8.2}$$

If $k\colon E^*\to\,]-\infty,\infty]$ is convex in the sense of Definition 1.8, the function $^*k\colon E\to[-\infty,\infty]$ is defined by

$$^*k(x) := \sup_{x^*\in E^*}\big[\langle x,x^*\rangle - k(x^*)\big]. \tag{8.3}$$

If $f,g\in\mathcal{PC}(E)$, a *Fenchel functional* for f and g is an element z^* of E^* such that $f^*(-z^*)+g^*(z^*)\leq 0$. The definitions of f^* and "Fenchel functional" are compatible with those introduced in Remark 7.3 for normed spaces, if we take $\langle\cdot,\cdot\rangle$ to be the canonical bilinear form on $E\times E^*$. If $E^*=E$, we will write $f^{@}$ instead of f^*. (See Definition 19.1.)

This is an appropriate point to make some comments about our formulation of the definition of Fenchel conjugate and Fenchel functional. It will become clear in Lemma 22.1 that when we consider the theory of monotone multifunctions on nonreflexive Banach spaces, we need a version of the Fenchel duality theorem that falls outside the scope of Theorem 7.4(a). This need can be met by proving such a version for locally convex spaces. However, an inspection of (8.1) shows that it is the bilinear form $\langle\cdot,\cdot\rangle$ that is important for the definition of f^*, and not the topology on E that gives E^* as dual. This is why we have opted to make the definition as above. In some cases, the bilinear form is determined by a given topology on E and, in other cases, the topology is determined by a given bilinear form.

We say that a locally convex topology $\mathcal{T}$ on E is E^*–compatible if the $\mathcal{T}$–dual of E is exactly $\{\langle \cdot, x^* \rangle \colon x^* \in E^*\}$. In this case, we write $\mathcal{S}(E, \mathcal{T})$ for the family of all $\mathcal{T}$–continuous seminorms on E. The facts that we shall need about locally convex spaces are that (i) if $z^* \in E^*$ then $|z^*| \in \mathcal{S}(E, \mathcal{T})$, (ii) if $P \in \mathcal{S}(E, \mathcal{T})$ and L is a linear functional on E such that $L \leq P$ on E then $L \in \{\langle \cdot, x^* \rangle \colon x^* \in E^*\}$ and (iii) the sets $\{x \in E \colon Q(x) < 1\}$ $(Q \in \mathcal{S}(E, \mathcal{T}))$ form a $\mathcal{T}$–base for the neighborhoods of 0.

The main results of this section are Theorem 8.1 and Theorem 8.4. In Theorem 8.1, we show how Theorem 1.11 leads to a necessary and sufficient condition for there to exist a Fenchel functional for f and g in this context, while in Theorem 8.4 we give a sufficient condition (in which neither function satisfies a semicontinuity condition) implying the results that are used in practice. Corollary 8.5 is a special case (which will be bootstrapped in Theorem 10.1) that leads to Corollary 8.6, a classical result due to Rockafellar. We will also show in Theorem 15.1 how to deduce the Attouch-Brezis version of the Fenchel duality theorem from Theorem 8.4.

We emphasize that Theorem 8.1 and Theorem 7.4 give *necessary and sufficient* conditions for the existence of Fenchel functionals, and not merely sufficient conditions.

Theorem 8.1. *Let f, $g \in \mathcal{PC}(E)$ and $\mathcal{T}$ be an E^*–compatible topology on E. Then there exists a Fenchel functional for f and g if, and only if,*

$$\left. \begin{array}{l} \text{there exists } P \in \mathcal{S}(E, \mathcal{T}) \text{ such that} \\ x, y \in E \quad \Longrightarrow \quad f(x) + g(y) + P(x - y) \geq 0. \end{array} \right\} \tag{8.4}$$

Proof. Suppose first that z^* is a Fenchel functional for f and g. Then, for all $x, y \in E$, $\langle x, -z^* \rangle - f(x) + \langle y, z^* \rangle - g(y) \leq f^*(-z^*) + g^*(z^*) \leq 0$. Consequently, $f(x) + g(y) + \langle x - y, z^* \rangle \geq 0$, and (8.4) follows with $P := |z^*|$. See the remarks preceding Example 7.2 for an indication of how to prove the converse (using the Hahn–Banach–Lagrange theorem, Theorem 1.11). $\square$

Remark 8.2. In this remark we sketch an "intrinsic" version of Theorem 8.1, i.e., a version which depends only on the duality and does not involve an additional topology, $\mathcal{T}$. We have the following result: *Let f, $g \in \mathcal{PC}(E)$. Then there exists a Fenchel functional z^* for f and g if, and only if, there exists a nonempty convex $w(E^*, E)$–compact subset K of E^* such that*

$$x, y \in E \quad \Longrightarrow \quad f(x) + g(y) + \sup\langle x - y, K \rangle \geq 0.$$

"Only if" is obvious by taking $K := \{z^*\}$, and the converse follows by using the minimax theorem, Theorem 3.2, on the function defined on the set

$(\operatorname{dom} f \times \operatorname{dom} g) \times K$ by $((x, y), z^*) \mapsto f(x) + g(y) + \langle x - y, z^* \rangle$. Readers familiar with the theory of the Mackey topology will recognize the connection between this result and Theorem 8.1.

Notation 8.3. Let E be a nonzero vector space and $f, g \in \mathcal{PC}(E)$. If $w \in E$, we write $(f \ominus g)(w) := \inf_{z \in E} [f(z) + g(z - w)]$. Readers familiar with the definition of episum ($=$ inf–convolution) will recognize that the definition of $f \ominus g$ is the same as the definition of the episum of f and g except for a change of sign in the argument of g. We note that

$$\{x \in E \colon (f \ominus g)(x) < \infty\} = \operatorname{dom} f - \operatorname{dom} g. \tag{8.5}$$

Before embarking on Theorem 8.4, we should explain in broad terms what it achieves. Theorem 8.1 gives a *necessary and sufficient* condition for there to exist a Fenchel functional in terms of certain expressions in f and g being bounded *below*. Theorem 8.4 transforms this into a (more useful) *sufficient* condition for there to exist a Fenchel functional in terms of certain expressions in f and g being bounded *above*. Theorem 8.4 is sometimes known as a *decoupling* result. We will use Theorem 8.4 explicitly in Corollary 8.5, Theorem 10.1 and Theorem 15.1.

Theorem 8.4. *Let f, $g \in \mathcal{PC}(E)$, $f + g \geq 0$ on E and*

$$F := \bigcup_{\lambda > 0} \lambda(\operatorname{dom} g - \operatorname{dom} f) \ni 0.$$

Suppose that $\mathcal{T}$ is an E^*–compatible topology on E and $f \ominus g$ is (finitely) bounded above in some $\mathcal{T}$–neighborhood of 0 relative to F. Then:
(a) *(8.4) is satisfied.*
(b) *There exists a Fenchel functional for f and g.*

Proof. Choose $Q \in \mathcal{S}(E, \mathcal{T})$ and $M \geq 0$ such that

$$w \in F \text{ and } Q(w) < 1 \implies (f \ominus g)(w) < M. \tag{8.6}$$

We shall prove that (8.4) is satisfied with $P := MQ \in \mathcal{S}(E, \mathcal{T})$. Now the inequality in (8.4) is immediate if $x \notin \operatorname{dom} f$ or $y \notin \operatorname{dom} g$, so we can and will assume that $x \in \operatorname{dom} f$ and $y \in \operatorname{dom} g$. Let $\lambda > Q(x - y) \geq 0$ and $w := (y - x)/\lambda \in F$. Since $Q(w) < 1$, (8.6) gives $z \in E$ such that

$$f(z) + g(z - w) < M. \tag{8.7}$$

Now $x + \lambda z = y + \lambda(z - w)$, hence, since $f + g \geq 0$ on E,

$$f\left(\frac{x + \lambda z}{1 + \lambda}\right) + g\left(\frac{y + \lambda(z - w)}{1 + \lambda}\right) \geq 0.$$

Thus, using the convexity of f and g,

$$f(x) + \lambda f(z) + g(y) + \lambda g(z - w) \geq 0.$$

Combining this with (8.7), we derive that

$$f(x) + g(y) + M\lambda \geq 0.$$

Letting $\lambda \to Q(x - y)$ gives $f(x) + g(y) + MQ(x - y) \geq 0$,   that is to say

$$f(x) + g(y) + P(x - y) \geq 0.$$

This completes the proof of (a), and (b) then follows from Theorem 8.1.   □

Corollary 8.5. *Let* $f, g \in \mathcal{PC}(E)$, $f + g \geq 0$ *on* E, $\mathcal{T}$ *be an* E^**-compatible topology on* E, *and* g *be (finitely) bounded above in some* $\mathcal{T}$*-neighborhood of a point of* dom f. *Then there exists a Fenchel functional for* f *and* g.

Proof. In this case, $\bigcup_{\lambda > 0} \lambda(\text{dom } g - \text{dom } f) = E \ni 0$. Choose $z \in \text{dom } f$, $N \in \mathbb{R}$ and $Q \in \mathcal{S}(E, \mathcal{T})$ such that $w \in E$ and $Q(w) < 1 \implies g(z - w) < N$, and define

$$M := f(z) + N > f(z) + g(z) = (f + g)(z) \geq 0.$$

If $w \in E$ and $Q(w) < 1$ then $(f \ominus g)(w) \leq f(z) + g(z - w) < M$, and so (8.6) is satisfied. The result now follows from Theorem 8.4(b).   □

Corollary 8.6 is an immediate consequence of Corollary 8.5 (see Rockafellar, [77, Theorem 1, pp. 82–83], Zălinescu, [119, Theorem 2.8.3(iii), p. 123] or Borwein–Zhu, [24, Sections 4.3.1–2, pp. 127–129]). We will use this result explicitly in Theorem 9.3, the transversality theorem, Theorem 19.16, Lemma 22.1 and Lemma 35.5. We refer the reader to Remark 15.3 for a comparison of Corollary 8.6 with the Attouch–Brezis version of the Fenchel duality theorem, Theorem 15.1.

Corollary 8.6. *Let* $f, g \in \mathcal{PC}(E)$, $f + g \geq 0$ *on* E, $\mathcal{T}$ *be an* E^**-compatible topology on* E, *and* g *be finite and* $\mathcal{T}$*-continuous at a point of* dom f. *Then there exists a Fenchel functional for* f *and* g.

We have presented Corollary 8.6 as a consequence of Corollary 8.5. However, they are in fact equivalent, as is evident from the following result, which will be used in Lemma 13.3 and Definition 38.1. The argument is an adaptation to the seminorm case of that of Phelps, [68, Proposition 1.6, p. 4] or Borwein–Zhu, [24, Section 4.1.2, pp. 112–113].

Theorem 8.7. *Let* E *be a nonzero vector space,* $f \in \mathcal{PC}(E)$, $z_0 \in E$, $K \in \mathbb{R}$, *and* $P: E \to \mathbb{R}$ *be a seminorm such that*

$$z \in E \text{ and } P(z - z_0) \leq 1 \implies f(z) \leq K. \tag{8.8}$$

Then

$$\left.\begin{array}{l} x, y \in E, P(x - z_0) \leq \frac{1}{2} \text{ and } P(y - z_0) \leq \frac{1}{2} \implies \\ \left|f(x) - f(y)\right| \leq 4\big(K - f(z_0)\big)P(x - y). \end{array}\right\} \tag{8.9}$$

Proof. Let $x, y \in E$, $P(x - z_0) \le \frac{1}{2}$ and $P(y - z_0) \le \frac{1}{2}$. (8.8) implies that $f(x), f(y) \in \mathbb{R}$, and we can and will suppose that $f(x) - f(y) \ge 0$. Let $\lambda > 2P(x - y) \ge 0$, and $z := x + (x - y)/\lambda$. Then

$$P(z - z_0) \le P(x - z_0) + P(x - y)/\lambda \le \tfrac{1}{2} + \tfrac{1}{2} = 1,$$

and thus (8.8) gives

$$f(z) \le K. \tag{8.10}$$

On the other hand, since $P\big((2z_0 - y) - z_0\big) = P(y - z_0) \le \frac{1}{2}$, (8.8) also gives $f(2z_0 - y) \le K$, from which

$$f(z_0) \le \tfrac{1}{2}f(y) + \tfrac{1}{2}f(2z_0 - y) \le \tfrac{1}{2}f(y) + \tfrac{1}{2}K.$$

Thus $f(y) \ge 2f(z_0) - K$. If we combine this with (8.10), we obtain

$$f(z) - f(y) \le K - \big(2f(z_0) - K\big) = 2\big(K - f(z_0)\big). \tag{8.11}$$

Now $x = (y + \lambda z)/(1 + \lambda)$, from which $f(x) \le \big(f(y) + \lambda f(z)\big)/(1 + \lambda)$ and so, combining with (8.11),

$$(1 + \lambda)\big(f(x) - f(y)\big) \le \lambda\big(f(z) - f(y)\big) \le 2\lambda\big(K - f(z_0)\big).$$

Thus, since $1 \le 1 + \lambda$ and $f(x) - f(y) \ge 0$,

$$f(x) - f(y) \le 2\lambda\big(K - f(z_0)\big).$$

(8.9) now follows by letting $\lambda \to 2P(x - y)$. $\square$

We conclude this section with a result (the "bipolar theorem for locally convex spaces") which we will use in Theorem 45.12. The proof is an obvious adaptation of that of Theorem 4.4.

Theorem 8.8. *Let C be a nonempty convex subset of E, $\mathcal{T}$ be an E^*- compatible topology on E, $C^{\mathcal{T}}$ be the closure of C with respect to $\mathcal{T}$ and $x \in E$. Then $x \in C^{\mathcal{T}}$ if, and only if,*

$$x^* \in E^* \implies \langle x, x^* \rangle \le \sup\langle C, x^* \rangle. \tag{8.12}$$

Proof. "Only if" is immediate since all the functions $\{\langle \cdot, x^* \rangle \colon x^* \in E^*\}$ are $\mathcal{T}$–continuous on E. Suppose, conversely, that $x \in E \setminus C^{\mathcal{T}}$. Then there exists $P \in \mathcal{S}(E, \mathcal{T})$ such that $\inf_{c \in C} P(x - c) > 0$. From the Mazur–Orlicz theorem, Lemma 1.6, with $D := x - C$, there exists a linear function L on E such that $L \le P$ on E and $\inf_D L = \inf_D P > 0$. Thus there exists $x^* \in E^*$ such that $\inf_{c \in C} \langle x - c, x^* \rangle > 0$, from which $\langle x, x^* \rangle > \sup\langle C, x^* \rangle$. So (8.12) fails, completing the proof of the theorem. $\square$

9 Some properties of $\frac{1}{2}\|\cdot\|^2$

The main result of this section is Theorem 9.3, a sharp version of the Fenchel duality theorem for normed spaces that has proved to be very useful in the investigation of monotone multifunctions. This result was first established in Simons–Zălinescu, [109, Theorem 2.1, pp. 5–6] with a proof that is more direct but also much more computational. We start off with two preliminary lemmas.

Lemma 9.1. *Let E be a nonzero normed space with dual E^*, $x, y \in E$ and $-\infty < c \leq \frac{1}{2}\|x\|^2$. Then*

$$c - \tfrac{1}{2}\|y\|^2 \leq \|x - y\| \left[\|x\| - \sqrt{\|x\|^2 - 2c} \right].$$

Proof. The triangle inequality gives $\big| \|x\| - \|x - y\| \big| \leq \|y\|$. Squaring and dividing by 2, $\frac{1}{2}\|x\|^2 + \frac{1}{2}\|x - y\|^2 - \|x - y\|\|x\| \leq \frac{1}{2}\|y\|^2$, from which

$$\tfrac{1}{2}\|x\|^2 + \tfrac{1}{2}\|x - y\|^2 - \tfrac{1}{2}\|y\|^2 \leq \|x - y\|\|x\|.$$

Thus

$$
\begin{aligned}
c - \tfrac{1}{2}\|y\|^2 &\leq c - \tfrac{1}{2}\|y\|^2 + \tfrac{1}{2}\left[\sqrt{\|x\|^2 - 2c} - \|x - y\| \right]^2 \\
&= \tfrac{1}{2}\|x\|^2 + \tfrac{1}{2}\|x - y\|^2 - \tfrac{1}{2}\|y\|^2 - \|x - y\|\sqrt{\|x\|^2 - 2c} \\
&\leq \|x - y\|\|x\| - \|x - y\|\sqrt{\|x\|^2 - 2c}. \qquad \square
\end{aligned}
$$

Lemma 9.2. *Let E be a nonzero normed space with dual E^*, $x \in E$ and $-\infty \leq c \leq \frac{1}{2}\|x\|^2$. Then*

$$\sup_{y \in E \setminus \{x\}} \frac{c - \tfrac{1}{2}\|y\|^2}{\|x - y\|} \vee 0 = \left[\|x\| - \sqrt{\|x\|^2 - 2c} \right] \vee 0. \tag{9.1}$$

Proof. Since both sides of the above equation are zero when $c = -\infty$, we can and will suppose that $c \in \mathbb{R}$. Furthermore, the inequality "$\leq$" in (9.1) follows easily from Lemma 9.1. Define $g \colon E \to \mathbb{R}$ by $g(z) := \frac{1}{2}\|z\|^2$ and let

$$M := \sup_{y \in E \setminus \{x\}} \frac{c - \tfrac{1}{2}\|y\|^2}{\|x - y\|} \vee 0.$$

Then $0 \leq M < \infty$ and $y \neq x \implies M\|x - y\| + g(y) \geq c$. Since this inequality holds trivially if $y = x$, the Hahn–Banach–Lagrange theorem, Theorem 1.11, gives $z^* \in E^*$ such that $\|z^*\| \leq M$ and

$$y \in E \implies \langle x - y, z^* \rangle + g(y) \geq c \iff 2\langle x, z^* \rangle \geq 2\langle y, z^* \rangle - 2g(y) + 2c.$$

Taking the supremum of the latter inequality over $y \in E$ and using the (well known) fact that $g^*(z^*) = \frac{1}{2}\|z^*\|^2$ (exercise!), we have

$$2\langle x, z^* \rangle \geq 2g^*(z^*) + 2c = \|z^*\|^2 + 2c.$$

Thus

$$\|x\|^2 - 2\|x\|\|z^*\| + \|z^*\|^2 \leq \|x\|^2 - 2\langle x, z^* \rangle + \|z^*\|^2 \leq \|x\|^2 - 2c.$$

Taking the square root gives $\|x\| - \|z^*\| \leq \sqrt{\|x\|^2 - 2c}$, and so

$$M \geq \|z^*\| \geq \|x\| - \sqrt{\|x\|^2 - 2c}.$$

The inequality "$\geq$" in (9.1) now follows immediately.   □

Our next result will be used explicitly in Theorem 21.4.

Theorem 9.3. *Let E be a nonzero normed space with dual E^*, $f \in \mathcal{PC}(E)$ and*

$$x \in E \quad \Longrightarrow \quad f(x) + \tfrac{1}{2}\|x\|^2 \geq 0.$$

Then:
(a) *There exists $x^* \in E^*$ such that*

$$f^*(x^*) + \tfrac{1}{2}\|x^*\|^2 \leq 0.$$

(b)

$$\min\left\{\|x^*\|: \ x^* \in E^*, \ f^*(x^*) + \tfrac{1}{2}\|x^*\|^2 \leq 0\right\} = \sup_{x,y \in E, \ x \neq y} \frac{-f(x) - \tfrac{1}{2}\|y\|^2}{\|x - y\|} \vee 0.$$

(c) *There exists $x^* \in E^*$ such that $f^*(x^*) + \tfrac{1}{2}\|x^*\|^2 \leq 0$ and*

$$\|x^*\| = \sup_{x \in E}\left[\|x\| - \sqrt{2f(x) + \|x\|^2}\right] \vee 0.$$

(d) *Let $x^* \in E^*$ and $f^*(x^*) + \tfrac{1}{2}\|x^*\|^2 \leq 0$. Then*

$$\sup_{x \in E}\left[\|x\| - \sqrt{2f(x) + \|x\|^2}\right] \vee 0 \leq \|x^*\| \leq \inf_{x \in E}\left[\|x\| + \sqrt{2f(x) + \|x\|^2}\right].$$

Proof. (a) Define $g\colon E \to \mathbb{R}$ by $g(z) := \tfrac{1}{2}\|z\|^2$. Since g is continuous and convex, Rockafellar's version of the Fenchel duality theorem, Corollary 8.6, implies that there exists a Fenchel functional, z^*, for f and g, and (a) follows with $x^* := -z^*$.

(b) This follows from Theorem 7.4(b,c).

(c) It is clear that

$$\sup_{x,y \in E, \ x \neq y} \frac{-f(x) - \tfrac{1}{2}\|y\|^2}{\|x - y\|} = \sup_{x \in E} \ \sup_{y \in E \setminus \{x\}} \frac{-f(x) - \tfrac{1}{2}\|y\|^2}{\|x - y\|}.$$

For a given $x \in E$, we use Lemma 9.2 with $c = -f(x)$, then take the supremum over $x \in E$ and appeal to (b).

(d) The Fenchel–Young inequality, (8.2), implies that, for all $x \in E$,

$$\left(\|x^*\| - \|x\|\right)^2 = \|x^*\|^2 - 2\|x\|\|x^*\| + \|x\|^2 \le \|x^*\|^2 + 2\langle x, x^*\rangle + \|x\|^2$$
$$\le 2f(x) + 2f^*(x^*) + \|x^*\|^2 + \|x\|^2 \le 2f(x) + \|x\|^2,$$

thus $\|x\| - \sqrt{2f(x) + \|x\|^2} \le \|x^*\| \le \|x\| + \sqrt{2f(x) + \|x\|^2}$, and (d) is now immediate from the fact that $\|x^*\| \ge 0$. $\qquad\square$

10 The conjugate of a sum in the locally convex case

As in Section 8, E and E^* are nonzero real vector spaces, and $\langle \cdot, \cdot \rangle : E \times E^* \to \mathbb{R}$ is a bilinear form that separates the points of E and also separates the points of E^*. We now bootstrap Theorem 8.4(b) to obtain Theorem 10.1. The conclusion of Theorem 10.1 and its two corollaries is that $(f + g)^*$ is the *exact episum* or *exact inf–convolution* of f^* and g^*. Corollary 10.4 will be applied later to the existence of autoconjugates in SSDB spaces.

Theorem 10.1. Let $f, g \in \mathcal{PC}(E)$ and $F := \bigcup_{\lambda > 0} \lambda[\operatorname{dom} g - \operatorname{dom} f] \ni 0$. Let $\mathcal{T}$ be a E^*–compatible topology on E, $x^* \in E^*$ and $(f - x^*) \ominus g$ be (finitely) bounded above in some $\mathcal{T}$–neighborhood of 0 relative to F. Then

$$(f + g)^*(x^*) = \min_{z^* \in E^*} \left[f^*(x^* - z^*) + g^*(z^*) \right]. \tag{10.1}$$

Proof. If $z^* \in E^*$ and $x \in E$ then

$$\langle x, x^* \rangle - (f + g)(x) = \langle x, x^* - z^* \rangle - f(x) + \langle x, z^* \rangle - g(x) \le f^*(x^* - z^*) + g^*(z^*).$$

Taking the supremum over $x \in E$, we have $(f+g)^*(x^*) \le f^*(x^* - z^*) + g^*(z^*)$. Consequently,

$$(f + g)^*(x^*) \le \inf_{z^* \in E^*} \left[f^*(x^* - z^*) + g^*(z^*) \right].$$

In order to prove the opposite inequality, we can and will suppose that $(f + g)^*(x^*) \in \mathbb{R}$. The Fenchel–Young inequality, (8.2), implies that

$$\left(f - x^* + (f + g)^*(x^*)\right) + g = (f + g) + (f + g)^*(x^*) - x^* \ge 0 \text{ on } E.$$

Since $\operatorname{dom}\left(f - x^* + (f + g)^*(x^*)\right) = \operatorname{dom} f$, Theorem 8.4(b) implies that there exists a Fenchel functional, z^*, for $f - x^* + (f+g)^*(x^*)$ and g. However,

$$\left(f - x^* + (f + g)^*(x^*)\right)^*(-z^*) = f^*(x^* - z^*) - (f + g)^*(x^*),$$

and so $f^*(x^* - z^*) - (f + g)^*(x^*) + g^*(z^*) \le 0$, that is to say

$$f^*(x^* - z^*) + g^*(z^*) \le (f + g)^*(x^*).$$

This completes the proof of (10.1). $\qquad\square$

Corollary 10.2. *Let* $f, g \in \mathcal{PC}(E)$, $\mathcal{T}$ *be a* E^**-compatible topology on* E, *and* g *be (finitely) bounded above in some* $\mathcal{T}$*-neighborhood of a point of* dom f *and* $x^* \in E^*$. *Then*

$$(f + g)^*(x^*) = \min_{z^* \in E^*} \left[f^*(x^* - z^*) + g^*(z^*) \right]. \qquad (10.1)$$

Proof. In this case, $F = E \ni 0$. Choose $u \in$ dom f, $N \in \mathbb{R}$ and $Q \in \mathcal{S}(E, \mathcal{T})$ such that $w \in E$ and $Q(w) < 1 \Longrightarrow g(u - w) < N$, and define

$$M := f(u) - \langle u, x^* \rangle + N.$$

If $w \in E$ and $Q(w) < 1$ then $((f - x^*) \ominus g)(w) \le f(u) - \langle u, x^* \rangle + g(u - w) < M$, and so the result now follows from Theorem 10.1.   □

Corollary 10.3 is an immediate consequence of Corollary 10.2 (see Rockafellar, [77, Theorem 3(a), p. 85] or Zălinescu, [119, Theorem 2.8.7(iii), p. 127]):

Corollary 10.3. *Let* $f, g \in \mathcal{PC}(E)$, $\mathcal{T}$ *be a* E^**-compatible topology on* E, *and* g *be finite and* $\mathcal{T}$*-continuous at a point of* dom f. *Then, for all* $x^* \in E^*$,

$$(f + g)^*(x^*) = \min_{z^* \in E^*} \left[f^*(x^* - z^*) + g^*(z^*) \right]. \qquad (10.1)$$

The following "symmetric" result will be used in our discussion of the existence of autoconjugates in SSDB spaces in Theorem 21.10. It is based on some of the results established by Bauschke–Wang in [13] for "kernel averages" in spaces of the form $E \times E^*$ (where E is a reflexive Banach space).

Corollary 10.4. *Let* $f_1, f_2, g \in \mathcal{PC}(E)$, $\mathcal{T}$ *be a* E^**-compatible topology on* E, *and* g *be finite and* $\mathcal{T}$*-continuous at a point of* dom $f_1 -$ dom f_2. *Suppose that, for all* $x \in E$,

$$h(x) := \inf_{z \in E} \left[\tfrac{1}{2} f_1(x + z) + \tfrac{1}{2} f_2(x - z) + \tfrac{1}{4} g(2z) \right] > -\infty.$$

It is easily seen that $h \in \mathcal{PC}(E)$. *Then, for all* $x^* \in E^*$,

$$h^*(x^*) = \min_{z^* \in E^*} \left[\tfrac{1}{2} f_1^*(x^* + z^*) + \tfrac{1}{2} f_2^*(x^* - z^*) + \tfrac{1}{4} g^*(-2z^*) \right].$$

Proof. Define $f_3, g_1 \colon E \times E \to \,]-\infty, \infty]$ by

$$f_3(x_1, x_2) := \tfrac{1}{2} f_1(x_1) + \tfrac{1}{2} f_2(x_2) \quad \text{and} \quad g_1(x_1, x_2) := \tfrac{1}{4} g(x_1 - x_2).$$

Then

$$h^*(x^*) = \sup_{x, z \in E} \left[\langle x, x^* \rangle - f_3(x + z, x - z) - \tfrac{1}{4} g(2z) \right].$$

Now make the substitution $(x_1, x_2) = (x + z, x - z)$, so that $x = \tfrac{1}{2}(x_1 + x_2)$ and $2z = x_1 - x_2$. Thus

$$h^*(x^*) = \sup_{(x_1,x_2)\in E\times E} \left[\langle x_1 + x_2, \tfrac{1}{2}x^* \rangle - f_3(x_1,x_2) - \tfrac{1}{4}g(x_1 - x_2) \right]$$

$$= \sup_{(x_1,x_2)\in E\times E} \left[\langle (x_1,x_2), (\tfrac{1}{2}x^*, \tfrac{1}{2}x^*) \rangle - f_3(x_1,x_2) - g_1(x_1,x_2) \right]$$

$$= (f_3 + g_1)^*(\tfrac{1}{2}x^*, \tfrac{1}{2}x^*), \tag{10.2}$$

where the conjugate is computed using the bilinear form defined on $(E \times E) \times (E^* \times E^*)$ by $((x_1,x_2),(x_1^*,x_2^*)) \mapsto \langle x_1,x_1^* \rangle + \langle x_2,x_2^* \rangle$. We note then that the topology $\mathcal{T} \times \mathcal{T}$ on $E \times E$ is $E^* \times E^*$–compatible. We now compute $f_3{}^*$ and $g_1{}^*$. For all $(y_1^*,y_2^*) \in E^* \times E^*$, we have

$$f_3{}^*(y_1^*,y_2^*) = \sup_{(x_1,x_2)\in E\times E} \left[\langle x_1,y_1^* \rangle + \langle x_2,y_2^* \rangle - \tfrac{1}{2}f_1(x_1) - \tfrac{1}{2}f_2(x_2) \right]$$

$$= \tfrac{1}{2} \sup_{x_1,x_2\in E} \left[\langle x_1,2y_1^* \rangle + \langle x_2,2y_2^* \rangle - f_1(x_1) - f_2(x_2) \right]$$

$$= \tfrac{1}{2}f_1{}^*(2y_1^*) + \tfrac{1}{2}f_2{}^*(2y_2^*) \right] \tag{10.3}$$

and, for all $(z_1^*,z_2^*) \in E^* \times E^*$, we have

$$g_1{}^*(z_1^*,z_2^*) = \sup_{(x_1,x_2)\in E\times E} \left[\langle x_1,z_1^* \rangle + \langle x_2,z_2^* \rangle - \tfrac{1}{4}g(x_1 - x_2) \right]$$

$$= \sup_{x_1,x_3\in E} \left[\langle x_1,z_1^* \rangle + \langle x_1 - x_3,z_2^* \rangle - \tfrac{1}{4}g(x_3) \right]$$

$$= \sup_{x_1,x_3\in E} \left[\langle x_1,z_1^* + z_2^* \rangle + \langle -x_3,z_2^* \rangle - \tfrac{1}{4}g(x_3) \right]$$

$$= \sup_{x_1\in E} \left[\langle x_1,z_1^* + z_2^* \rangle + \sup_{x_3\in E} \left[\langle x_3,-z_2^* \rangle - \tfrac{1}{4}g(x_3) \right] \right]$$

$$= \sup_{x_1\in E} \langle x_1,z_1^* + z_2^* \rangle + \tfrac{1}{4}g^*(-4z_2^*)$$

$$= \begin{cases} \tfrac{1}{4}g^*(-4z_2^*), & \text{if } z_1^* + z_2^* = 0; \\ \infty, & \text{otherwise.} \end{cases} \tag{10.4}$$

Since $\operatorname{dom} f_3 = \operatorname{dom} f_1 \times \operatorname{dom} f_2$, g_1 is $\mathcal{T} \times \mathcal{T}$–continuous at a point of $\operatorname{dom} f_3$, and so we derive from Corollary 10.3 and (10.4) that

$$(f_3 + g_1)^*(\tfrac{1}{2}x^*, \tfrac{1}{2}x^*) = \min_{z_1^* + z_2^* = 0} \left[f_3{}^*(\tfrac{1}{2}x^* - z_1^*, \tfrac{1}{2}x^* - z_2^*) + \tfrac{1}{4}g^*(-4z_2^*) \right].$$

If we now put $z_1^* = -\tfrac{1}{2}z^*$ and $z_2^* = \tfrac{1}{2}z^*$, we obtain from (10.2) and (10.3) that

$$h^*(x^*) = \min_{z^*\in E^*} \left[f_3{}^*(\tfrac{1}{2}x^* + \tfrac{1}{2}z^*, \tfrac{1}{2}x^* - \tfrac{1}{2}z^*) + \tfrac{1}{4}g^*(-2z^*) \right]$$

$$= \min_{z^*\in E^*} \left[\tfrac{1}{2}f_1{}^*(x^* + z^*) + \tfrac{1}{2}f_2{}^*(x^* - z^*) + \tfrac{1}{4}g^*(-2z^*) \right]. \qquad \square$$

11 Fenchel duality vs the conjugate of a sum

The results of Section 10 indicate the (well known) fact that results on the conjugate of a sum are very close to the Fenchel duality theorem. The purpose of this section is to draw a distinction between these two kinds of result. Examples 11.1 and 11.2 were worked out in collaboration with Regina Burachik, and Example 11.3 is based on a suggestion of Jonathan Borwein. There is an example similar to Example 11.1 in Boţ–Wanka, [27, pp. 2798–2799]. Let E be a nonzero Banach space and $f, g \in \mathcal{PC}(E)$. We say that f and g satisfy *Fenchel duality* if there exists $z^* \in E^*$ such that

$$f^*(-z^*) + g^*(z^*) = (f+g)^*(0).$$

Example 11.1. We give an example of proper, convex lower semicontinuous functions f and g on $\mathbb{R}^2$ that satisfy Fenchel duality but, for most $r \in (\mathbb{R}^2)^* = \mathbb{R}^2$, it is not true that there exist $p, q \in \mathbb{R}^2$ such that $p + q = r$ and $f^*(p) + g^*(q) = (f+g)^*(r)$.

Let $C = \{x \in \mathbb{R}^2 : \|x\| \le 1\}$ and $x_0 = (1, 0) \in \mathbb{R}^2$. Write $A := x_0 - C$, $B := C - x_0$, $f := \mathbb{I}_A$ and $g := \mathbb{I}_B$, where $\mathbb{I}_X$ is the *indicator function of X*, that is to say

$$\mathbb{I}_X(x) := \begin{cases} 0, & \text{if } x \in X; \\ \infty, & \text{otherwise.} \end{cases}$$

We note then that $f + g = \mathbb{I}_{\{0\}}$. Since $f^*(0) = g^*(0) = (f+g)^*(0) = 0$, f and g satisfy Fenchel duality.

Now, for all $p, q \in \mathbb{R}^2$, $f^*(p) = \|p\| + p_1$ and $g^*(q) = \|q\| - q_1$. Consequently

$$f^*(p) \ge 0 \quad \text{and} \quad (f^*(p) = 0 \iff p_1 \le 0 \text{ and } p_2 = 0), \tag{11.1}$$

and

$$g^*(q) \ge 0 \quad \text{and} \quad (g^*(q) = 0 \iff q_1 \ge 0 \text{ and } q_2 = 0). \tag{11.2}$$

If $p, q \in \mathbb{R}^2$ are such that $p + q = r$ and $f^*(p) + g^*(q) = (f+g)^*(r)$ then, since $(f+g)^*(r) = 0$, (11.1) and (11.2) imply that $f^*(p) = 0$ and $g^*(q) = 0$, consequently $p_2 = 0$ and $q_2 = 0$, from which $r_2 = 0$. Thus if $r_2 \ne 0$ then there do not exist $p, q \in \mathbb{R}^2$ such that $p + q = r$ and $f^*(p) + g^*(q) = (f+g)^*(r)$.

We can look at this another way: if $r \in \mathbb{R}^2$, and f and $g - r$ satisfy Fenchel duality then there exists $p \in \mathbb{R}^2$ such that

$$(f + g - r)^*(0) = f^*(p) + (g - r)^*(-p),$$

that is to say $f^*(p) + g^*(r - p) = 0$, and the analysis above shows that $r_2 = 0$. This argument can easily be reversed: if $r \in \mathbb{R}^2$ and $r_2 = 0$ then there exist $p, q \in \mathbb{R}^2$ such that $p + q = r$ and $f^*(p) + g^*(q) = (f+g)^*(r)$, and f and $g - r$ satisfy Fenchel duality. At any rate, f and g fail "stable Fenchel–Rockafellar duality" in the sense of [34, Theorem 3.2(i)].

Example 11.2. [34, Theorem 3.2(ii)] tells us that epi $f^* + $ epi g^* is not closed in $\mathbb{R}^2 \times \mathbb{R}$ in Example 11.1. We now confirm this by giving an explicit description of this set. If $p_1 < 0 < q_1$ then

$$f^*(p) = \frac{\|p\|^2 - p_1^2}{\|p\| - p_1} \leq \frac{p_2^2}{2|p_1|} \quad \text{and} \quad g^*(q) = \frac{\|q\|^2 - q_1^2}{\|q\| + q_1} \leq \frac{q_2^2}{2|q_1|}. \tag{11.3}$$

Let r be an arbitrary element of $\mathbb{R}^2$ and $n > |r_1|$. Then, from (11.3),

$$f^*\left(\frac{r}{2} - ne_1\right) \leq \frac{r_2^2}{4(2n - r_1)} \quad \text{and} \quad g^*\left(\frac{r}{2} + ne_1\right) \leq \frac{r_2^2}{4(2n + r_1)}.$$

Thus

$$\left(\frac{r}{2} - ne_1, \frac{r_2^2}{4(2n - r_1)}\right) \in \text{epi } f^* \quad \text{and} \quad \left(\frac{r}{2} + ne_1, \frac{r_2^2}{4(2n + r_1)}\right) \in \text{epi } g^*,$$

and so

$$\left(r, \frac{r_2^2}{4(2n - r_1)} + \frac{r_2^2}{4(2n + r_1)}\right) \in \text{epi } f^* + \text{epi } g^*.$$

Since epi $f^* + $ epi g^* recedes vertically, it follows by letting $n \to \infty$ that

$$\{(r_1, r_2, \lambda) \colon r_2 = 0, \lambda \geq 0\} \cup \{(r_1, r_2, \lambda) \colon r_2 \neq 0, \lambda > 0\} \subset \text{epi } f^* + \text{epi } g^*.$$

It is also clear from (11.1) and (11.2) that epi $f^* + $ epi $g^* \subset \mathbb{R}^2 \times \mathbb{R}^+$. Suppose now that $(r, 0) \in$ epi $f^* + $ epi g^*. Then there exist $(p, \lambda) \in$ epi f^* and $(q, \mu) \in$ epi g^* such that $(p + q, \lambda + \mu) = (r, 0)$. Then $0 = \lambda + \mu \geq f^*(p) + g^*(q)$ so, from (11.1) and (11.2), $f^*(p) = 0$ and $g^*(q) = 0$. Arguing as in Example 11.1, $r_2 = 0$.

Combining all this together, we have

$$\text{epi } f^* + \text{epi } g^* = \{(r_1, r_2, \lambda) \colon r_2 = 0, \lambda \geq 0\} \cup \{(r_1, r_2, \lambda) \colon r_2 \neq 0, \lambda > 0\}$$

(which is obviously not closed).

We now investigate an even more unstable case of Fenchel duality. However, the analysis is a little more technical. Let E be a nonzero Banach space and $f, g \in \mathcal{PC}(E)$. We shall say that the pair f, g is *totally Fenchel unstable* if f and g satisfy Fenchel duality but

$$y^*, z^* \in E^* \text{ and } f^*(y^*) + g^*(z^*) = (f + g)^*(y^* + z^*) \implies y^* + z^* = 0.$$

Example 11.3. We recall that if C is a convex subset of a Banach space E and $x \in C$ then x is a *support point* of C if there exists $x^* \in E^* \setminus \{0\}$ such that $\langle x, x^* \rangle = \sup \langle C, x^* \rangle$. We will give an example below of a nonempty $w(E, E^*)$–compact convex subset C of a Banach space E (actually ℓ^2) such that there exists an extreme point x_0 of C which is not a support point of C. Again, write $A := x_0 - C$, $B := C - x_0$, $f := \mathbb{I}_A$ and $g := \mathbb{I}_B$. The fact that x_0 is an extreme point of C implies that $f + g = \mathbb{I}_{\{0\}}$. As in Example 11.1, f and g satisfy Fenchel duality.

Now, for all $y^*, z^* \in E^*$,

$$f^*(y^*) = \langle x_0, y^* \rangle - \inf \langle C, y^* \rangle \geq 0 \quad \text{and} \quad g^*(z^*) = \sup \langle C, z^* \rangle - \langle x_0, z^* \rangle \geq 0.$$

Let $y^*, z^* \in E^*$ be such that

$$y^* + z^* = x^* \text{ and } f^*(y^*) + g^*(z^*) = (f + g)^*(x^*).$$

Thus $f^*(y^*) + g^*(z^*) = 0$, from which $f^*(y^*) = 0$ and $g^*(z^*) = 0$. Consequently, $\langle x_0, y^* \rangle = \inf \langle C, y^* \rangle$ and $\langle x_0, z^* \rangle = \sup \langle C, z^* \rangle$. Since x_0 is not a support point of C, $y^* = 0$ and $z^* = 0$, thus $x^* = y^* + z^* = 0$. So we have established that f and g are totally Fenchel unstable.

By analogy with the result established in Example 11.2, one is tempted to ask whether

$$\operatorname{epi} f^* + \operatorname{epi} g^* = \{(0,0)\} \cup (E^* \times]0, \infty[). \tag{11.4}$$

The inclusion "$\subset$" is clear from the discussion above, and it is also clear that $(0,0) = (0,0) + (0,0) \in \operatorname{epi} f^* + \operatorname{epi} g^*$. Thus (11.4) is equivalent to:

$$\operatorname{epi} f^* + \operatorname{epi} g^* \supset E^* \times]0, \infty[. \tag{11.5}$$

We now prove that this is the case, using an adaptation of a very nice argument provided by Radu Ioan Boţ (personal communication). Let $y^* \in E^*$. Let $h \colon E^* \to \mathbb{R}$ and $k \colon E^* \to \mathbb{R}$ be defined by $h := f^*$ and $k(z^*) := g^*(y^* - z^*)$. Since h and k are continuous and convex on E^*, it follows from Rockafellar's formula for the conjugate of a sum, Corollary 10.3, that

$$-\inf_{E^*} [h + k] = (h + k)^*(0) = \min_{z^{**} \in E^{**}} [h^*(z^{**}) + k^*(-z^{**})].$$

Since $\widehat{A}$ and $\widehat{B}$ are $w(E^{**}, E^*)$–compact and $w(E^{**}, E^*)$ is an E^*–compatible topology on E^{**}, it follows from Theorem 8.8 that, for all $z^{**} \in E^{**}$, $h^*(z^{**}) = \mathbb{I}_{\widehat{A}}(z^{**})$ and $k^*(-z^{**}) = \mathbb{I}_{\widehat{B}}(z^{**}) - \langle y^*, z^{**} \rangle$. Consequently, if $h^*(z^{**}) + k^*(-z^{**}) < \infty$ then $z^{**} \in \widehat{A} \cap \widehat{B}$, from which $z^{**} = 0$. Thus

$$-\inf_{E^*} [h + k] = h^*(0) + k^*(-0) = 0,$$

and so, for all $\varepsilon > 0$, there exists $z^* \in E^*$ such that $h(z^*) + k(z^*) \leq \varepsilon$, that is to say $f^*(z^*) + g^*(y^* - z^*) \leq \varepsilon$. It is clear from this that $(y^*, \varepsilon) \in \operatorname{epi} f^* + \operatorname{epi} g^*$, which gives (11.5), as required.

Here is the promised example, which was suggested by Jonathan Borwein. Let $E = \ell^2$, $1 < p < 2$, and $C := \{x \in \ell^2 : \|x\|_p \leq 1\}$. Since the function $\|\cdot\|_p$ is lower semicontinuous on ℓ^2, C is closed, and obviously C is convex (and $C = -C$). Then x is an extreme point of C if, and only if, $\|x\|_p = 1$.

Let $x \in C$ and $\|x\|_p = 1$. We shall prove that x *is a support point of C if, and only if,* $x \in \ell^{2(p-1)}$. Suppose first that x is a support point of C. Then there exists $y \in \ell^2 = (\ell^2)^*$ such that $y \neq 0$ and (assuming that $\frac{1}{p} + \frac{1}{q} = 1$)

$$\langle x, y \rangle = \sup\langle C, y \rangle = \|y\|_q = \|x\|_p \|y\|_q.$$

Thus we have equality in Hölder's inequality, and so there exists $\lambda > 0$ such that, for all $n \geq 1$, $|y_n|^q = (\lambda |x_n|)^p$. Since $y \in \ell^2$, $\sum_{n \geq 1} (\lambda |x_n|)^{2p/q} < \infty$, that is to say, $x \in \ell^{2(p-1)}$, as required. Suppose, conversely, that $x \in \ell^{2(p-1)}$. For all $n \geq 1$, let $y_n = \operatorname{sgn} x_n |x_n|^{p-1}$. Then $y \in \ell^2 = (\ell^2)^*$. Further,

$$\langle x, y \rangle = \sum_{n \geq 1} x_n y_n = \sum_{n \geq 1} x_n \operatorname{sgn} x_n |x_n|^{p-1} = \sum_{n \geq 1} |x_n|^p = 1$$

and

$$\sup\langle C, y \rangle = \|y\|_q = \left(\sum_{n \geq 1} |x_n|^{q(p-1)} \right)^{1/q} = \left(\sum_{n \geq 1} |x_n|^p \right)^{1/q} = 1^{1/q} = 1,$$

so x is a support point of C. Since $2(p-1) < p$, there are plenty of extreme points of C that are not support points.

Remark 11.4. What we have actually shown above is that if C is a $w(E, E^*)$–compact convex subset of a Banach space E, x_0 is an extreme point of C, $f := \mathbb{I}_{x_0 - C}$, $g := \mathbb{I}_{C - x_0}$, $y^* \in E^*$ and $\varepsilon > 0$ then there exists $z^* \in E^*$ such that $f^*(z^*) + g^*(y^* - z^*) \leq \varepsilon$. This last inequality is equivalent to the statement that there exists $z^* \in E^*$ such that, for all $x, y \in E$, $f(x) + g(y) + \langle y - x, z^* \rangle \geq \langle y, y^* \rangle - \varepsilon$. From the Hahn–Banach–Lagrange theorem, Theorem 1.11, this is in turn equivalent to the statement that there exists $M \geq 0$ such that, for all $x, y \in E$, $f(x) + g(y) + M\|y - x\| \geq \langle y, y^* \rangle - \varepsilon$, that is to say there exists $M \geq 0$ such that, for all $u, v \in C$, $M\|u + v - 2x_0\| \geq \langle v - x_0, y^* \rangle - \varepsilon$. This observation leads to the following problem (which only makes sense if E is not reflexive):

Problem 11.5. Let C be a bounded closed convex subset of a Banach space E, x_0 be an extreme point of C, $y^* \in E^*$ and $\varepsilon > 0$. Then does there always exist $M \geq 0$ such that, for all $u, v \in C$, $M\|u + v - 2x_0\| \geq \langle v - x_0, y^* \rangle - \varepsilon$? If the answer to this question is in the affirmative then

$$\operatorname{epi}\left(\mathbb{I}_{x_0 - C}\right)^* + \operatorname{epi}\left(\mathbb{I}_{C - x_0}\right)^* \supset E^* \times]0, \infty[.$$

Problem 11.6. Do there exist a nonzero *finite dimensional* Banach space E and $f, g \in \mathcal{PC}(E)$ such that the pair f, g is totally Fenchel unstable?

12 The restricted biconjugate and Fenchel–Moreau points

We now return to the more general considerations of Section 8. Let E and E^* be nonzero real vector spaces, and $\langle \cdot, \cdot \rangle \colon E \times E^* \to \mathbb{R}$ be a bilinear form that separates the point of E and also separates the points of E^*. We define *the restricted biconjugate* of f to be $^*(f^*)\colon E \to [-\infty, \infty]$ (see (8.3)). To simplify notation, we shall abbreviate this to $^*f^*$. It follows easily from the definition of f^* in (8.1) that, for all $x \in E$,

$$f(x) \geq {}^*f^*(x). \tag{12.1}$$

One of the fundamental results in convex analysis is the *Fenchel–Moreau theorem* that if $f \in \mathcal{PC}(E)$ is lower semicontinuous with respect to a E^*–compatible topology on E then $f = {}^*f^*$ on E. We will revisit this result in Corollary 12.4.

Now suppose that $\mathcal{T}$ is a E^*–compatible topology on E and f is not necessarily $\mathcal{T}$–lower semicontinuous. Let us say that $x \in E$ is a *Fenchel–Moreau point* of f if equality holds in (12.1). It is very tempting to speculate that every point of $\mathcal{T}$–lower semicontinuity of f is a Fenchel–Moreau point of f. Example 12.1 below shows that this is false. However, we establish in Theorem 12.2 that every point of $\mathcal{T}$–lower semicontinuity of f is a Fenchel–Moreau point provided that f is bounded below in a $\mathcal{T}$–neighborhood of at least one point in its effective domain. Putting this another way, if there is a point of $\mathcal{T}$–lower semicontinuity of f that is not a Fenchel–Moreau point then f is unbounded below in every $\mathcal{T}$–neighborhood of every point of dom f.

Example 12.1. Let E be an infinite–dimensional normed space. Fix $x^* \in E^* \setminus \{0\}$ and a discontinuous linear functional L on E. Define

$$f(x) := \begin{cases} \infty, & \text{if } \langle x, x^* \rangle < 1; \\ L(x), & \text{if } \langle x, x^* \rangle \geq 1. \end{cases}$$

Clearly, $f \in \mathcal{PC}(E)$ and f is lower semicontinuous at 0. Let y^* be an arbitrary element of E^*. Since x^* and $y^* - L$ are linearly independent, there exist $y, z \in E$ such that

$$\langle y, x^* \rangle = 1, \ \langle z, x^* \rangle = 0, \ (y^* - L)(y) = 0, \ \text{and} \ (y^* - L)(z) = 1.$$

Let $\lambda \in \mathbb{R}$, and set $x := y + \lambda z$. Then $\langle x, x^* \rangle = \langle y, x^* \rangle = 1$, and so $f(x) = L(x)$. Thus

$$f^*(y^*) \geq \langle x, y^* \rangle - f(x) = (y^* - L)(x) = \lambda(y^* - L)(z) = \lambda.$$

Since this holds for all $\lambda \in \mathbb{R}$, $f^*(y^*) = \infty$. Thus we have

$$f(0) = \infty > -\infty = \sup\nolimits_{y^* \in E^*} \left[\langle 0, y^* \rangle - f^*(y^*) \right],$$

and so 0 is not a Fenchel–Moreau point of f. (This example can also be justified using Corollary 10.3.)

Theorem 12.2 contains a positive result on Fenchel–Moreau points. The subtlety in its proof is that we can do arithmetic with the expression $f(x) - f(z)$, but we cannot do arithmetic with the expression $f(x) - f(y)$, which may well have the value $-\infty$.

Theorem 12.2. *Let $f \in \mathcal{PC}(E)$ be (finitely) bounded below in a $\mathcal{T}$–neighborhood of an element z of $\mathrm{dom}\, f$, and f be $\mathcal{T}$–lower semicontinuous at an element y of E. Then y is a Fenchel–Moreau point of f, and $f^* \in \mathcal{PC}(E^*)$.*

Proof. Let $\lambda \in \mathbb{R}$ and $\lambda < f(y)$. Choose $\nu \in \mathbb{R}$ and $Q \in \mathcal{S}(E, \mathcal{T})$ such that

$$Q(x - z) \leq 1 \quad \Longrightarrow \quad f(x) > \nu \qquad (12.2)$$

and

$$Q(x - y) \leq 1 \quad \Longrightarrow \quad f(x) > \lambda. \qquad (12.3)$$

Write $\rho := f(z) - \nu > 0$. We first prove that

$$x \in E \quad \Longrightarrow \quad f(x) + \rho Q(x - y) \geq \nu - \rho Q(y - z). \qquad (12.4)$$

To this end, let x be an arbitrary element of E. If $Q(x - z) \leq 1$ then (12.2) implies that

$$f(x) + \rho Q(x - z) \geq f(x) > \nu \geq \nu - \rho Q(y - z).$$

If, on the other hand, $Q(x - z) > 1$, let $\gamma := 1/Q(x - z) \in \,]0, 1[$ and put $u := \gamma x + (1 - \gamma)z$. Then $Q(u - z) = \gamma Q(x - z) = 1$ and so, from the convexity of f, and (12.2) with x replaced by u,

$$\gamma f(x) + (1 - \gamma)f(z) \geq f\big(\gamma x + (1 - \gamma)z\big) = f(u) > \nu,$$

thus the definition of ρ implies that $\gamma\big(f(x) - f(z)\big) + \rho \geq 0$. Substituting in the formula for γ and clearing of fractions yields $f(x) + \rho Q(x - z) \geq f(z)$. Consequently, using (12.2) with $x = z$ and the fact that $Q(z - z) \leq 1$,

$$f(x) + \rho Q(x - y) \geq f(x) + \rho Q(x - z) - \rho Q(y - z)$$
$$\geq f(z) - \rho Q(y - z) > \nu - \rho Q(y - z).$$

This completes the proof of (12.4). Now let $\sigma := \big[\lambda - \nu + \rho Q(y - z)\big] \vee 0 \geq 0$. We will prove that

$$x \in E \quad \Longrightarrow \quad f(x) + (\rho + \sigma)Q(x - y) \geq \lambda. \qquad (12.5)$$

To this end, let x be an arbitrary element of E. If $Q(x - y) \leq 1$ then (12.3) implies that $f(x) + (\rho + \sigma)Q(x - y) \geq f(x) > \lambda$. If, on the other hand, $Q(x - y) > 1$ then, from (12.4),

$$f(x) + (\rho + \sigma)Q(x - y) = f(x) + \rho Q(x - y) + \sigma Q(x - y)$$
$$\geq \nu - \rho Q(y - z) + \sigma \geq \lambda.$$

This completes the proof of (12.5). It now follows from the Hahn–Banach–Lagrange theorem, Theorem 1.11 that there exists a linear functional L on E such that $L \leq (\rho + \sigma)Q$ on E and

$$x \in E \implies f(x) + L(x - y) \geq \lambda.$$

Let $z^* = -L \in E^*$. Then, for all $x \in E$, $\langle y, z^* \rangle - [\langle x, z^* \rangle - f(x)] \geq \lambda$. Taking the infimum over $x \in E$, $\langle y, z^* \rangle - f^*(z^*) \geq \lambda$. It follows by letting $\lambda \to f(y)$ that y is a Fenchel–Moreau point of f. Now $f^* \colon E^* \to]-\infty, \infty]$ is obviously convex. If z^* is a functional constructed as above for some $\lambda < f(y)$ then the inequality $\langle y, z^* \rangle - f^*(z^*) \geq \lambda$ implies that $f^*(z^*) \in \mathbb{R}$, and so $f^* \in \mathcal{PC}(E^*)$. $\qquad\square$

Definition 12.3. If E is a nonzero Hausdorff locally convex space, we write $\mathcal{PCLSC}(E)$ for the set

$$\{f \in \mathcal{PC}(E) \colon f \text{ is lower semicontinuous on } E\}.$$

Corollary 12.4 is the original Fenchel–Moreau result, which follows immediately from Theorem 12.2. See Moreau, [64, Section 5–6, pp. 26–39] or Zălinescu, [119, Theorem 2.3.3, pp. 77–78]. Corollary 12.4 will be used explicitly in Theorem 18.7, (19.9), Lemma 35.1, Lemma 45.9, Theorem 48.4 and Lemma 48.9.

Corollary 12.4. Let $f \in \mathcal{PCLSC}(E, \mathcal{T})$. Then $f^* \in \mathcal{PC}(E^*)$ and $^*f^* = f$ on E.

13 Surrounding sets and the dom lemma

In this and the next section, we collect together some results on convex lower semicontinuous functions that we shall need for our later work. In this section, we give the "dom lemma", Lemma 13.3, which is a "quantitative" result, and the "dom corollary", Corollary 13.5, which is a "qualitative" result. The dom lemma will be of use in Lemma 22.7. Both the dom lemma and the dom corollary are subsumed by the results of the next section — we have treated them independently for essentially pedagogical reasons.

Let E be a Banach space, $x \in E$ and $A \subset E$. A is said to be *absorbing* if $\bigcup_{\lambda>0} \lambda A = E$. Any neighborhood of 0 is absorbing (exercise!). We write "$x \in \operatorname{sur} A$" and say that "$A$ surrounds x" if, for each $w \in E \setminus \{0\}$, there exists $\delta > 0$ such that $x + \delta w \in A$. The statement "$x \in \operatorname{sur} A$" is related to x being an "absorbing point" of A (see Phelps, [68, Definition 2.27(b), p. 28]), but differs in that we do not require that $x \in A$. We also note that, if A is convex then $\operatorname{sur} A \subset A$, and so $\operatorname{sur} A$ is identical with the "core" or algebraic interior of A. In particular:

if A is convex then $(0 \in \operatorname{sur} A \iff A$ is absorbing$).$ (13.1)

In terms of these concepts, we have the following useful algebraic result about convex functions:

Lemma 13.1. *Let E be a nonzero vector space, $f \in \mathcal{PC}(E)$ and $\operatorname{dom} f$ surround 0. Then there exists $n \geq 1$ such that $\{z \in E \colon f(z) \leq n\}$ is absorbing.*

Proof. From (13.1),

$$\operatorname{dom} f \quad \text{is absorbing.} \tag{13.2}$$

In particular, $0 \in \operatorname{dom} f$. Let $n \geq f(0) \vee 0 + 1$. We will show that n has the required property. To this end, let y be an arbitrary element of E. (13.2) now provides $\lambda > 0$ and $x \in \operatorname{dom} f$ such that $\lambda y = x$. Choose $\mu \in \,]0, 1]$ so that $\mu\big(f(x) - n + 1\big) \leq 1$. Then

$$
\begin{aligned}
f(\mu\lambda y) = f(\mu x) &\leq \mu f(x) + (1 - \mu)f(0) \\
&\leq \mu f(x) + (1 - \mu)(n - 1) \\
&= \mu\big(f(x) - n + 1\big) + n - 1 \leq n.
\end{aligned}
$$

Consequently, $\{z \in E \colon f(z) \leq n\}$ is absorbing, as required.   □

Our next result depends ultimately on Baire's theorem:

Lemma 13.2. *Let E be a nonzero Banach space and C be a closed convex absorbing set in E. Then C is a neighborhood of 0.*

Proof. Let $D := C \cap -C$. Then D is closed, convex and absorbing (exercise!) and $D = -D$, i.e., D is a "barrel". The result follows by applying Kelley–Namioka, [52, p. 104] to D.   □

Lemma 13.3. *Let E be a nonzero Banach space, $f \in \mathcal{PCLSC}(E)$ and $\operatorname{dom} f$ surround 0. Then there exist $\eta > 0$ and $n \geq 1$ such that*

$$z \in E \text{ and } \|z\| \leq \eta \quad \Longrightarrow \quad f(z) \leq n. \tag{13.3}$$

Furthermore, f is continuous at 0.

Proof. Choose $n \geq 1$ as in Lemma 13.1. Lemma 13.2 now implies that $\{z \in E \colon f(z) \leq n\}$ is a neighborhood of 0, and it follows from Theorem 8.7 that f is continuous at 0.   □

Remark 13.4. The dom lemma, Lemma 13.3, can also be deduced from Rockafellar, [76, Corollary 7C, p. 61] (see also Moreau, [64, Proposition 5.f, p. 30] for a simpler proof of Rockafellar's result).

Corollary 13.5. *Let E be a nonzero Banach space and $f \in \mathcal{PCLSC}(E)$. Then*

$$\operatorname{sur}(\operatorname{dom} f) = \operatorname{int}(\operatorname{dom} f).$$

Proof. Exercise!

Remark 13.6. The classical "uniform boundedness theorem" can easily be deduced from the dom lemma. Here are the details: *Let E be a nonzero Banach space, F be a normed space and $\mathcal{B}$ be a nonempty pointwise bounded set of continuous linear operators from E into F. Then $\mathcal{B}$ is bounded in norm.*

Proof. Define $f\colon E \to \mathbb{R}$ by

$$f(x) := \sup_{T \in \mathcal{B}} \|Tx\|.$$

Since dom $f = E$, we can apply the dom lemma. It then follows from (13.3) that

$$T \in \mathcal{B} \quad \Longrightarrow \quad \|T\| \leq \frac{n}{\eta}. \qquad \qquad \square$$

The proof of the uniform boundedness theorem given above can be found in Holmes, [51, §17, p. 134]. Lemma 13.3 also implies the result that a convex lower semicontinuous function is locally bounded on the interior of its domain. (See, for instance, Phelps, [68, Proposition 3.3, p. 39].)

14 The ⊖–theorem

We now come to the "⊖–theorem", Theorem 14.2, which will be crucial for our analysis of the sums of maximally monotone operators in reflexive spaces. The ⊖–theorem is a "quantitative" result that also has a "qualitative" version, the "⊖–corollary", Corollary 14.3. Both of these results will have their uses, the ⊖–theorem in our proof of the Attouch–Brezis theorem, Theorem 15.1, and the ⊖–corollary in the local transversality theorem, Theorem 21.12, and also in Corollary 22.6. The ⊖–theorem, which generalizes the open mapping theorem (see Remark 14.4) can itself be generalized considerably. (In this connection, we refer the reader to Robinson, [75], Ursescu, [112], and Borwein, [16]). Here we confine our attention to what we will need in these notes. The idea for the proof of Lemma 14.1 is taken from Aubin–Ekeland, [3, Lemma 3.3.9, p. 136]. The dom lemma is an immediate consequence of the ⊖–theorem with $g := \mathbb{I}_{\{0\}}$.

Lemma 14.1 is the ⊖–theorem under more restrictive hypotheses. It will be bootstrapped in Theorem 14.2. We remind the reader that the function $f \ominus g\colon F \to [-\infty, \infty[$ was defined in Notation 8.3.

Lemma 14.1. *Let F be a nonzero Banach space, $f,\ g \in \mathcal{PCLSC}(F)$, $f \geq \|\cdot\|$ and $g \geq \|\cdot\|$ on F, and dom f − dom g surround 0. Then*

$$f \ominus g \text{ is (finitely) bounded above in a neighborhood of } 0 \text{ in } F. \qquad (14.1)$$

Proof. We first observe that, for all $w \in F$,

$$(f \ominus g)(w) = \inf_{z \in E}\left[f(z) + g(z - w)\right] \geq \inf_{z \in E}\left[\|z\| + \|w - z\|\right] \geq \|w\| > -\infty,$$

from which it follows easily that $f \ominus g \in \mathcal{PC}(F)$. (8.5) implies that $\mathrm{dom}\,(f \ominus g) = \mathrm{dom}\,f - \mathrm{dom}\,g$, and so $\mathrm{dom}\,(f \ominus g)$ surrounds 0. We now deduce from Lemma 13.1 that there exists $m > 1$ such that $\{w \in F \colon (f \ominus g)(w) < m\}$ is absorbing. Let $W := \{w \in F \colon (f \ominus g)(w) < m\}$. Since $\overline{W}$ is closed, convex and absorbing, Lemma 13.2 gives us that $\overline{W}$ is a neighborhood of 0 in F. Choose $\eta > 0$ so that

$$w \in F \text{ and } \|w\| \leq 2\eta \quad \Longrightarrow \quad w \in \overline{W}. \tag{14.2}$$

We shall prove that

$$w \in F \text{ and } \|w\| \leq \eta \quad \Longrightarrow \quad (f \ominus g)(w) \leq m, \tag{14.3}$$

which will give (14.1). So let $w \in F$ and $\|w\| \leq \eta$. Then, from (14.2), $2w \in \overline{W}$, consequently

$$\text{there exists } w_1 \in W \quad \text{such that} \quad \|2w - w_1\| \leq \eta.$$

From (14.2) again, $4w - 2w_1 = 2(2w - w_1) \in \overline{W}$, thus

$$\text{there exists } w_2 \in W \quad \text{such that} \quad \|4w - 2w_1 - w_2\| \leq \eta.$$

Continuing this argument, we find $w_1,\ w_2,\ w_3,\ \ldots \in W$ such that, for all $k \geq 1$,

$$\|2^k w - 2^{k-1}w_1 - \cdots - w_k\| \leq \eta,$$

from which

$$\|w - 2^{-1}w_1 - \cdots - 2^{-k}w_k\| \leq 2^{-k}\eta,$$

hence $\sum_{k=1}^{\infty} 2^{-k} w_k = w$. For all $n \geq 1$, since $w_n \in W$, we can choose $u_n \in F$ such that

$$f(u_n) + g(u_n - w_n) < m. \tag{14.4}$$

This implies that $\|u_n\| \leq \|u_n\| + \|u_n - w_n\| < m$. Since F is complete, there exists $u \in F$ such that $\sum_{k=1}^{\infty} 2^{-k} u_k = u$, from which

$$\sum_{k=1}^{\infty} 2^{-k}(u_k - w_k) = u - w.$$

(14.4) and the lower semicontinuity of f and g now imply that

$$f(u) + g(u - w) \leq m,$$

from which $(f \ominus g)(w) \leq m$. This completes the proof of (14.3), and hence also that of Lemma 14.1. $\qquad\square$

Theorem 14.2. *Let F be a nonzero Banach space, $h, k \in \mathcal{PCLSC}(F)$, and $\mathrm{dom}\,h - \mathrm{dom}\,k$ surround 0. Then $h \ominus k$ is (finitely) bounded above in a neighborhood of 0 in F.*

Proof. This is immediate from Lemma 14.1 with $f = h \vee \|\cdot\|$ and $g = k \vee \|\cdot\|$, since $\operatorname{dom} f = \operatorname{dom} h$, $\operatorname{dom} g = \operatorname{dom} k$ and $h \ominus k \leq f \ominus g$ on F. $\qquad\square$

Corollary 14.3. *Let F be a nonzero Banach space and $f, k \in \mathcal{PCLSC}(F)$. Then* $\operatorname{sur}(\operatorname{dom} f - \operatorname{dom} k) = \operatorname{int}(\operatorname{dom} f - \operatorname{dom} k)$, *and so* $\operatorname{sur}(\operatorname{dom} f - \operatorname{dom} k)$ *is open.*

Proof. We shall prove that

$$\operatorname{sur}(\operatorname{dom} f - \operatorname{dom} k) \subset \operatorname{int}(\operatorname{dom} f - \operatorname{dom} k). \tag{14.5}$$

This gives the desired result, since the reverse inclusion is trivial. So let x be an arbitrary element of $\operatorname{sur}(\operatorname{dom} f - \operatorname{dom} k)$. Define $h \in \mathcal{PCLSC}(F)$ by $h(y) := f(y + x)$ $(y \in F)$. Then $\operatorname{dom} h = \operatorname{dom} f - x$, which implies that $0 \in \operatorname{sur}(\operatorname{dom} h - \operatorname{dom} k)$. Theorem 14.2 now gives $\eta > 0$ and $m > 1$ such that if $w \in F$ and $\|w\| \leq \eta$ then $(h \ominus k)(w) \leq m$, from which $w \in \operatorname{dom} h - \operatorname{dom} k$. Thus we have proved that $0 \in \operatorname{int}(\operatorname{dom} h - \operatorname{dom} k)$. Since $\operatorname{dom} h - \operatorname{dom} k = \operatorname{dom} f - x - \operatorname{dom} k$, we have $x \in \operatorname{int}(\operatorname{dom} f - \operatorname{dom} k)$, which completes the proof of (14.5). $\qquad\square$

Remark 14.4. The classical "open mapping theorem" can easily be deduced from the $\ominus$-theorem. Here are the details. We first observe that *if C and D are closed convex subsets of a Banach space F and $C - D$ surrounds 0 then there exist $\eta > 0$ and $m > 1$ such that if $w \in F$ and $\|w\| \leq \eta$ then*

there exist $c \in C$ and $d \in D$ such that $w = c - d$ and $\|c\| \leq m$.

We obtain this by applying Theorem 14.2 with $h := \mathbb{I}_C \vee \|\cdot\|$ and $k := \mathbb{I}_D$. If now E and H are Banach spaces and $T \in B(E, H)$ is surjective then, for all $(x, y) \in E \times H$, there exists $z \in E$ such that that $y = Tz$, and consequently

$$(x, y) = (x, Tz) = (z, Tz) - (z - x, 0) \in G(T) - \big(E \times \{0\}\big).$$

We now define $F := E \times H$ with norm $\|(x, y)\| := \sqrt{\|x\|^2 + \|y\|^2}$, $C := G(T)$ and $D := E \times \{0\}$. From the result above, there exist $\eta > 0$ and $m > 1$ such that if $y \in H$ and $\|y\| \leq \eta$ then there exist $x, z \in E$ such that $(0, y) = (x, Tx) - (z, 0)$ and $\|(x, Tx)\| \leq m$. This implies that $Tx = y$ and $\|x\| \leq m$, and it follows that T is an open mapping. $\qquad\square$

Thus the $\ominus$-theorem is both a generalization of the open mapping theorem and, in some sense, a "second order" generalization of the uniform boundedness theorem.

Remark 14.5. As we have observed, Lemma 14.1 is a generalization of Lemma 13.3. In this remark, we shall sketch a generalization of Lemma 13.3 in a totally different direction. Let E be a nonzero Banach space.
(a) *Let B be a nonmeager Borel set in E (that is, a Borel set of the second category). Then $B - B$ is a neighborhood of 0.*

(b) *Let D be a convex absorbing Borel set in E and D be symmetric, i.e., $D = -D$. Then D is a neighborhood of 0.*

(c) *Let C be a convex absorbing Borel set in E. Then C is a neighborhood of 0.*

(d) *Let C be a convex Borel set in E. Then $\operatorname{sur} C = \operatorname{int} C$.*

(e) *Let $f \in \mathcal{PC}(E)$ be a Borel function and $\operatorname{dom} f$ surround 0. Then there exist $\eta > 0$ and $n \geq 1$ such that*

$$w \in E \text{ and } \|w\| \leq \eta \implies f(w) \leq n.$$

Proof. (a) Any Borel set satisfies the "condition of Baire", that is to say, there exists an open set U such that $U \setminus B$ and $B \setminus U$ are meager, and so (a) follows from the "difference theorem". See Kelley–Namioka, [52, 10.4, p. 92] and the discussion preceding.

(b) It follows from Baire's theorem that E, being a complete metric space, is nonmeager. Since $\bigcup_{n \geq 1} nD = E$ there exists $n \geq 1$ such that nD is non-meager, from which $\frac{1}{2}D$ is nonmeager. Since D is convex and symmetric,

$$D = \tfrac{1}{2}D + \tfrac{1}{2}D = \tfrac{1}{2}D - \tfrac{1}{2}D,$$

thus it follows from (a) that D is a neighborhood of 0.

(c) Let $D := C \cap -C$. Then D is a convex absorbing Borel set and $D = -D$. From (b), D is a neighborhood of 0, from which C is a neighborhood of 0 also.

(d) is immediate from (c), a translation argument and (13.1).

(e) From Lemma 13.1, there exists $n \geq 1$ such that $\{x \in E : f(x) \leq n\}$ is absorbing. The result now follows from (c).

Remark 14.6. Theorem 14.2 and Remark 14.5 suggest the following question:

Problem 14.7. Let F be a Banach space, h, $k \in \mathcal{PC}(F)$ be Borel functions and $\operatorname{dom} h - \operatorname{dom} k$ surround 0. Is $h \ominus k$ necessarily (finitely) bounded above in some neighborhood of 0 in F? In particular: Let C and D be convex Borel sets in F and $C - D$ be absorbing. Is $C - D$ necessarily a neighborhood of 0 in F?

15 The Attouch–Brezis theorem

This section is devoted to a single result, the Attouch–Brezis version of the Fenchel duality theorem, which we will use explicitly in Lemma 16.2 and the local transversality theorem, Theorem 21.12. As stated below, this result also follows from [1, Corollary 2.3, pp. 131–132] (a much more general result was established in [119, Theorem 2.8.6, pp. 125–126]):

Theorem 15.1. *Let E be a nonzero Banach space, f, $g \in \mathcal{PCLSC}(E)$,*

$$F := \bigcup_{\lambda>0} \lambda[\operatorname{dom} f - \operatorname{dom} g] \quad \text{be a closed subspace of } E$$

and

$$f + g \geq 0 \text{ on } E.$$

Then there exists a Fenchel functional for f and g.

Proof. Since $0 \in F$, there exists $z \in \operatorname{dom} f \cap \operatorname{dom} g$. Define h, k: $E \to$ $]-\infty, \infty]$ by $h(x) := f(x+z)$ and $k(x) := g(x+z)$ $(x \in E)$. Then $\operatorname{dom} h \subset F$, $\operatorname{dom} k \subset F$ and $\operatorname{dom} h - \operatorname{dom} k$ surrounds 0 in F. From the $\ominus$–theorem, Theorem 14.2, there exist $\eta > 0$ and $m > 1$ such that if $w \in F$ and $\|w\| \leq \eta$ then

there exist $u, v \in F$ such that $w = u - v$ and $h(u) + k(v) \leq m$.

But then $w = (u + z) - (v + z)$ and $f(u + z) + g(v + z) \leq m$, and so $(f \ominus g)(w) \leq m$. The result now follows from Theorem 8.4(b). $\qquad\square$

Remark 15.2. Theorem 15.1 can easily be bootstrapped into the following result (which is [1, Theorem 1.1, pp. 126–130]): *Let E be a nonzero Banach space, f, $g \in \mathcal{PCLSC}(E)$ and $\bigcup_{\lambda>0} \lambda(\operatorname{dom} f - \operatorname{dom} g)$ be a closed subspace of E. Then, for all $x^* \in E^*$,*

$$(f + g)^*(x^*) = \min_{z^* \in E^*} \left[f^*(x^* - z^*) + g^*(z^*) \right]. \tag{10.1}$$

Remark 15.3. It is often said that, in the normed case, Theorem 15.1 is a "generalization" of Rockafellar's version of the Fenchel duality theorem, Corollary 8.6. This is inaccurate, since Theorem 15.1 requires both f and g to be lower semicontinuous.

In the two cases in these notes in which Corollary 8.6 is used explicitly in a normed space (Theorem 9.3 and Lemma 35.5), we cannot substitute Theorem 15.1 because of the lack of this semicontinuity.

Corollary 8.6 is also used explicitly in a non–normed situation in the transversality theorem, Theorem 19.16, and also in Lemma 22.1.

The Attouch–Brezis theorem is, however, a very powerful result, which enables us to consider Fenchel duality in which $\operatorname{int} \operatorname{dom} f = \operatorname{int} \operatorname{dom} g = \emptyset$. We will investigate a bivariate version of the Attouch–Brezis theorem in the next section.

16 A bivariate Attouch–Brezis theorem

The main result of this section is the bivariate version of the Attouch–Brezis theorem that will appear in Theorem 16.4. Apart from some minor changes of notation, this result was first proved in Simons–Zălinescu [109, Theorem 4.2, pp. 9–10]. The proof given here using Lemma 16.2 is somewhat simpler, and first appeared in [106].

Notation 16.1. If E and F are nonzero Banach spaces, we norm $E \times F$ by

$$\|b\| := \sqrt{\|b_1\|^2 + \|b_2\|^2} \quad (b = (b_1, b_2) \in E \times F).$$

The dual of $E \times F$ is $F^* \times E^*$ under the pairing

$$\langle b, v \rangle := \langle b_1, v_2 \rangle + \langle b_2, v_1 \rangle \quad (b = (b_1, b_2) \in E \times F, \ v = (v_1, v_2) \in F^* \times E^*),$$

and the dual norm of $F^* \times E^*$ is given by $\|(v_1, v_2)\| = \sqrt{\|v_1\|^2 + \|v_2\|^2}$. We define the *projection maps* π_1, π_2 by $\pi_1(x, y) := x$ and $\pi_2(x, y) := y$.

Lemma 16.2 is a stepping–stone to Theorem 16.4. It will also be used explicitly in Theorem 46.3, in our proof of the maximal monotonicity of the sum of maximally monotone multifunctions with convex graph.

Lemma 16.2. *Let E and F be nonzero Banach spaces, $p, q \in \mathcal{PCLSC}(E \times F)$,*

$$L := \bigcup_{\lambda > 0} \lambda [\pi_1 \mathrm{dom}\, p - \pi_1 \mathrm{dom}\, q] \text{ be a closed subspace of } E$$

and

$$(x, y, z) \in E \times F \times F \implies p(x, y) + q(x, z) \geq 0.$$

Then

there exists $x^* \in E^*$ such that $p^*(0, -x^*) + q^*(0, x^*) \leq 0$.

Proof. For all $(x, y, z) \in E \times F \times F$, let $f(x, y, z) := p(x, y)$ and $g(x, y, z) := q(x, z)$. We first prove that

$$\bigcup_{\lambda > 0} \lambda [\mathrm{dom}\, f - \mathrm{dom}\, g] = L \times F \times F. \tag{16.1}$$

To this end, let $(x, y, z) \in L \times F \times F$. Then there exist $\lambda > 0$, $(a_1, a_2) \in \mathrm{dom}\, p$ and $(b_1, b_2) \in \mathrm{dom}\, q$ such that $x = \lambda(a_1 - b_1)$. Thus

$$(x, y, z) = \lambda [(a_1, a_2, b_2 + z/\lambda) - (b_1, a_2 - y/\lambda, b_2)] \in \lambda [\mathrm{dom}\, f - \mathrm{dom}\, g].$$

This establishes "$\supset$" in (16.1), and (16.1) now follows since the inclusion "$\subset$" is obvious. Also,

$$(x, y, z) \in E \times F \times F \implies (f + g)(x, y, z) = p(x, y) + q(x, z) \geq 0.$$

Now represent the dual of $E \times F \times F$ by $E^* \times F^* \times F^*$ under the pairing $\langle (x, y, z), (x^*, y^*, z^*) \rangle := \langle x, x^* \rangle + \langle y, y^* \rangle + \langle z, z^* \rangle$. Since $L \times F \times F$ is a closed subspace of $E \times F \times F$, Theorem 15.1 gives $(x^*, y^*, z^*) \in E^* \times F^* \times F^*$ such that

$$f^*(-x^*, -y^*, -z^*) + g^*(x^*, y^*, z^*) \le 0. \tag{16.2}$$

So $f^*(-x^*, -y^*, -z^*) < \infty$, from which $f^*(-x^*, -y^*, -z^*) = p^*(-y^*, -x^*)$ and $z^* = 0$. Similarly, $g^*(x^*, y^*, z^*) = q^*(z^*, x^*)$ and $y^* = 0$. Thus (16.2) reduces to

$$p^*(0, -x^*) + q^*(0, x^*) \le 0. \qquad \square$$

Before discussing the promised bivariate version of the Attouch–Brezis theorem, we make some preliminary definitions:

Definition 16.3. Let E and F be nonzero Banach spaces, $B := E \times F$ and $f, g \in \mathcal{PC}(B)$. For all $b \in B$, let

$$(f \oplus_2 g)(b) := \inf \{ f(a) + g(c) \colon a, c \in B, \; a_1 = c_1 = b_1, \; a_2 + c_2 = b_2 \}.$$

So $(f \oplus_2 g)(x, \cdot)$ is the *inf–convolution* of $f(x, \cdot)$ and $g(x, \cdot)$. Similarly, for all $b \in B$, let

$$(f \oplus_1 g)(b) := \inf \{ f(a) + g(c) \colon a, c \in B, \; a_1 + c_1 = b_1, \; a_2 = c_2 = b_2 \}.$$

The bivariate version of the Attouch–Brezis theorem that appears in Theorem 16.4 below will be used explicitly in Lemma 22.9 and Theorem 35.8. This latter result on $\widetilde{BC}$–functions will be pivotal for our investigation of the different classes of maximally monotone multifunctions on a nonreflexive Banach space. The conclusion of Theorem 16.4(a) is that $(f \oplus_2 g)^*(y^*, \cdot)$ is the *exact inf–convolution* of $f^*(y^*, \cdot)$ and $g^*(y^*, \cdot)$. A similar comment can be made about Theorem 16.4(b).

Theorem 16.4. Let E and F be nonzero Banach spaces, $B := E \times F$ and $f, g \in \mathcal{PCLSC}(B)$. Write $B^* = F^* \times E^*$
(a) Let

$$\bigcup_{\lambda > 0} \lambda [\pi_1 \mathrm{dom}\, f - \pi_1 \mathrm{dom}\, g] \text{ be a closed subspace of } E$$

and, for all $b \in B$, $(f \oplus_2 g)(b) > -\infty$. Then, for all $v \in B^* = (E \times F)^*$,

$$(f \oplus_2 g)^*(v) = \min \{ f^*(u) + g^*(w) \colon u, w \in B^*, \; u_1 = w_1 = v_1, \; u_2 + w_2 = v_2 \}.$$

In particular, $(f \oplus_2 g)^* = f^* \oplus_2 g^*$ on B^*.
(b) Let

$$\bigcup_{\lambda > 0} \lambda [\pi_2 \mathrm{dom}\, f - \pi_2 \mathrm{dom}\, g] \text{ be a closed subspace of } F$$

and, for all $b \in B$, $(f \oplus_1 g)(b) > -\infty$. Then, for all $v \in B^* = (E \times F)^*$,

$$(f \oplus_1 g)^*(v) = \min \{ f^*(u) + g^*(w) \colon u, w \in B^*, \; u_1 + w_1 = v_1, \; u_2 = w_2 = v_2 \}.$$

In particular, $(f \oplus_1 g)^* = f^* \oplus_1 g^*$ on B^*.

Proof. Let $h := f \oplus_2 g$, Then h is convex and, since $\pi_1 \mathrm{dom}\, f \cap \pi_1 \mathrm{dom}\, g \neq \emptyset$, h is proper. Let $v \in B^*$. It is easy to see that

$$h^*(v) \leq \inf \left\{ f^*(u) + g^*(w) \colon\ u, w \in B^*,\ u_1 = w_1 = v_1,\ u_2 + w_2 = v_2 \right\}.$$

So what we have to prove for (a) is that there exists $x^* \in E^*$ such that

$$f^*(v_1, v_2 - x^*) + g^*(v_1, x^*) \leq h^*(v). \tag{16.3}$$

Since h is proper, $h^*(v) > -\infty$, so we can and will suppose that $h^*(v) \in \mathbb{R}$. Define $p, q \in \mathcal{PCLSC}(B)$ by $p(x, y) := h^*(v) + f(x, y) - \langle x, v_2 \rangle - \langle y, v_1 \rangle$ and $q(x, z) := g(x, z) - \langle z, v_1 \rangle$. Then, for all $(x, y, z) \in E \times F \times F$, the Fenchel–Young inequality, (8.2), implies that

$$\begin{aligned}
p(x, y) + q(x, z) &= h^*(v) + f(x, y) - \langle x, v_2 \rangle - \langle y, v_1 \rangle + g(x, z) - \langle z, v_1 \rangle \\
&\geq h^*(v) + h(x, y + z) - \langle x, v_2 \rangle - \langle y + z, v_1 \rangle \geq 0.
\end{aligned}$$

Lemma 16.2 now gives $x^* \in E^*$ such that $p^*(0, -x^*) + q^*(0, x^*) \leq 0$. By direct computation,

$$p^*(0, -x^*) = f^*(v_1, v_2 - x^*) - h^*(v) \quad \text{and} \quad q^*(0, x^*) = g^*(v_1, x^*),$$

which implies (16.3), and completes the proof of (a). The proof of (b) is similar. $\qquad \square$

III Multifunctions, SSD spaces, monotonicity and Fitzpatrick functions

17 Multifunctions, monotonicity and maximality

We now introduce some general notation for "multifunctions" or "set–valued maps". If X and Y are nonempty sets, we write $S\colon X \rightrightarrows Y$ if, for all $x \in X$, Sx is a (possibly empty) subset of Y. We define

$$G(S) := \{(x,y)\colon x \in X,\ y \in Sx\} \quad \text{and} \quad G^{-1}(S) := \{(y,x)\colon x \in X,\ y \in Sx\}.$$

$G(S)$ is the *graph* of S and $G^{-1}(S)$ is the *inverse graph* of S. We shall always suppose that $G(S) \neq \emptyset$ — we shall emphasize this by saying that S is *nontrivial*. We write

$$D(S) := \{x \in X\colon Sx \neq \emptyset\} = \pi_1 G(S).$$

$D(S)$ is the *domain* of S. We write

$$R(S) := \bigcup_{x \in X} Sx = \pi_2 G(S).$$

$R(S)$ is the *range* of S. (The projection maps π_1 and π_2 were defined in Notation 16.1.) Finally, if $S\colon X \rightrightarrows Y$, we define $S^{-1}\colon Y \rightrightarrows X$ by

$$S^{-1}y := \{x \in X\colon Sx \ni y\}.$$

S^{-1} is the *inverse* of S. Obviously $D(S^{-1}) = R(S)$, $R(S^{-1}) = D(S)$ and $G(S^{-1}) = G^{-1}(S)$. We point to the books [3] by Aubin, [4] by Aubin–Frankowska and [38] by Deimling as general references on multifunctions. We shall be concerned here with multifunctions from one Banach space into another, in which case additional operations can be defined. If $S\colon E \rightrightarrows F$ and $T\colon E \rightrightarrows F$ are nontrivial, we define $S + T\colon E \rightrightarrows F$ $\big($with $D(S+T) = D(S) \cap D(T)\big)$ by

$$(S+T)x := Sx + Tx \quad (x \in E), \tag{17.1}$$

where $Sx + Tx$ is the "Minkowski sum" $\{y + z\colon y \in Sx,\ z \in Tx\}$.

Let E be a nonzero Banach space and A be a nonempty subset of $E \times E^*$. We say that A is *monotone* if

$$(x,x^*) \text{ and } (y,y^*) \in A \quad \Longrightarrow \quad \langle x - y, x^* - y^* \rangle \geq 0.$$

A is said to be *maximally monotone* if S is monotone, and S has no proper monotone superset. Now let $S\colon E \rightrightarrows E^*$. S is said to be *monotone* if

$$(x, x^*) \text{ and } (y, y^*) \in G(S) \quad \Longrightarrow \quad \langle x - y, x^* - y^* \rangle \geq 0.$$

S is said to be *maximally monotone* if S is monotone, and S has no proper monotone extension. We point to the notes [69] and the book [68] by Phelps, and the book [122] by Zeidler as general references on monotone multifunctions.

We now give some examples of maximally monotone multifunctions. The first one we consider is that of *positive linear operators*. Let $S\colon E \to E^*$ be linear and

$$x \in E \quad \Longrightarrow \quad \langle x, Sx \rangle \geq 0.$$

Then S is a *(single valued) maximally monotone operator*. More precisely, the multifunction T defined by $Tx := \{Sx\}$ is maximally monotone. The monotonicity is easy to see. To prove the maximality, suppose that $(z, z^*) \in E \times E^*$ and $G(S) \cup \{(z, z^*)\}$ is monotone. Then

$$\inf_{y \in E} \langle y - z, Sy - z^* \rangle \geq 0.$$

Let $x \in E$, $\lambda \in \mathbb{R}$, and put $y := z + \lambda x$ and deduce from this that $(z, z^*) \in G(S)$ (exercise!). As a special case of the above, we mention *skew linear operators*. These are linear operators $S\colon E \to E^*$ such that

$$x \in E \quad \Longrightarrow \quad \langle x, Sx \rangle = 0,$$

or equivalently,

$$x, y \in E \quad \Longrightarrow \quad \langle x, Sy \rangle = -\langle y, Sx \rangle.$$

See the papers [8] and [9] by Bauschke–Borwein and the paper [70] by Phelps–Simons for recent work on positive linear operators.

The second example that we consider is that of *subdifferentials*. (See Section 7.) If $f \in \mathcal{PCLSC}(E)$ then $\partial f\colon E \rightrightarrows E^*$ is maximally monotone. The monotonicity is easy to see. The maximality is *Rockafeller's maximal monotonicity theorem* (see Theorem 18.7). It is easy to see in this situation that $D(\partial f) \subset \operatorname{dom} f$ (exercise!), however this inclusion may be proper: let $E := \mathbb{R}$ and $f\colon \mathbb{R} \to]-\infty, \infty]$ be defined by

$$f(x) := \begin{cases} -\sqrt{1 - x^2}, & \text{if } x \in [-1, 1]; \\ \infty, & \text{otherwise}; \end{cases}$$

then $D(\partial f) =]-1, 1[$ but $\operatorname{dom} f = [-1, 1]$. The Brøndsted–Rockafellar theorem (see Corollary 18.5) establishes a close connection between $D(\partial f)$ and $\operatorname{dom} f$: Let $f \in \mathcal{PCLSC}(E)$, $\alpha, \beta > 0$, $y \in \operatorname{dom} f$ and $f(y) \leq \inf_E f + \alpha\beta$. Then there exist $x \in E$ and $z^* \in \partial f(x)$ such that $\|x - y\| \leq \alpha$, $f(x) \leq f(y)$ and $\|z^*\| \leq \beta$. In particular, $D(\partial f)$ is dense in $\operatorname{dom} f$. Incidentally, if S is linear, nonzero and skew then S is not a subdifferential so, provided that E has dimension > 1, there always exist maximally monotone multifunctions that are not subdifferentials. (If $E = \mathbb{R}$ then every maximally monotone multifunction on E is a subdifferential (exercise!).)

The final example that we mention here is that of the *normality multi-function*. Let C be a nonempty closed convex subset of E and $N_C \colon E \rightrightarrows E^*$ be defined by

$$(x, x^*) \in G(N_C) \iff x \in C \text{ and } \langle x, x^* \rangle = \max\langle C, x^* \rangle. \qquad (17.2)$$

Then N_C is maximally monotone. Again, the monotonicity is easy to see. The maximality can be seen in two ways. First, if we define $\mathbb{I}_C \colon E \to \left]-\infty, \infty\right]$ to be the *indicator function of C*, that is to say

$$\mathbb{I}_C(x) := \begin{cases} 0, & \text{if } x \in C; \\ \infty, & \text{otherwise}; \end{cases}$$

then $N_C = \partial \mathbb{I}_C$. Since $\mathbb{I}_C$ is convex and lower semicontinuous, it follows from the result of Rockafellar mentioned above that N_C is maximally monotone. Alternatively, one can proceed directly from the definition of N_C and use the consequence of the Bishop–Phelps theorem that *C is the intersection of the closed half–spaces defined by its supporting hyperplanes*. See the remarks preceding Theorem 18.10 for more details of this.

It is worth noting that the Bishop–Phelps theorem and the Brøndsted–Rockafellar theorem mentioned above were both precursors (and are consequences of) Ekeland's variational principle (see Theorem 18.4).

Remark 17.1. Let $B := E \times E^*$. If $S \colon E \rightrightarrows E^*$ then $G(S) \subset B$. It is sometimes very convenient to formulate problems on the monotonicity of S in terms of subsets of B by using $G(S)$ as the link. For all $b = (b_1, b_2)$ and $c = (c_1, c_2) \in B$, we set $\lfloor b, c \rfloor := \langle b_1, c_2 \rangle + \langle c_1, b_2 \rangle$. Then $\lfloor \cdot, \cdot \rfloor \colon B \times B \to \mathbb{R}$ is a symmetric bilinear form that separates the points of B. We define the quadratic form q on B by $q(b) := \frac{1}{2}\lfloor b, b \rfloor$. Then

$$q(b_1, b_2) = \tfrac{1}{2}\big[\langle b_1, b_2 \rangle + \langle b_1, b_2 \rangle\big] = \langle b_1, b_2 \rangle.$$

Consequently, if $b = (b_1, b_2)$ and $c = (c_1, c_2) \in B$ then

$$\langle b_1 - c_1, b_2 - c_2 \rangle = q(b_1 - c_1, b_2 - c_2) = q\big((b_1, b_2) - (c_1, c_2)\big) = q(b - c).$$

So the monotonicity of $S \colon E \rightrightarrows E^*$ is equivalent to the statement:

$$b, c \in G(S) \implies q(b - c) \geq 0.$$

Furthermore, if S is monotone then S is maximally monotone exactly when:

$$\text{if } b \in B \quad \text{and} \quad \big(a \in G(S) \implies q(b - a) \geq 0\big) \quad \text{then} \quad b \in G(S).$$

Let $A \subset B$ and $b \in B$. We say that b is *monotonically related to* A if

$$a \in A \quad \Longrightarrow \quad q(b - a) \geq 0.$$

So, if S is monotone then S is maximally monotone exactly when:

if $b \in B$ and b is monotonically related to $G(S)$ then $b \in G(S)$.

These considerations lead us to the more abstract situation that will be considered starting in Section 19. We will return to this specific example in Section 22.

18 Subdifferentials are maximally monotone

In Theorem 18.7, we establish Rockafellar's maximal monotonicity theorem (first proved in [81]) that the subdifferential of a proper, convex lower semicontinuous function on a nonzero Banach space is maximally monotone. Our proof is based on a very elegant argument found recently by M. Marques Alves and B. F. Svaiter in [60]. We refer the reader to Remark 18.8 for comparisons between the argument given here and those of [60] and [99], and to Remark 18.9 for an explanation of how the argument can be shortened when E is reflexive.

We start off by developing in the formula for the subdifferential of the sum of two convex functions under two different hypotheses, and the Brøndsted–Rockafellar theorem. We recall from Section 7 that if E is a nonzero normed space with dual E^*, $k \in \mathcal{PC}(E)$, $x \in E$ and $x^* \in E^*$ then

$$x^* \in \partial k(x) \iff \sup_{y \in E} \left[\langle y, x^* \rangle - k(y) \right] \leq \langle x, x^* \rangle - k(x)$$
$$\iff k^*(x^*) \leq \langle x, x^* \rangle - k(x) \iff k(x) + k^*(x^*) \leq \langle x, x^* \rangle.$$

Thus, by virtue of the Fenchel–Young inequality, (8.2),

$$x^* \in \partial k(x) \iff k(x) + k^*(x^*) = \langle x, x^* \rangle.$$

Theorem 18.1. *Let E be a nonzero normed space, f, $g \in \mathcal{PC}(E)$, and g be (finitely) bounded above in some neighborhood of a point of $\mathrm{dom}\, f$. Then $\partial(f + g) = \partial f + \partial g$.*

Proof. We leave as an exercise the proof of the implication

$$x^* \in (\partial f + \partial g)(x) \quad \Longrightarrow \quad x^* \in \partial(f + g)(x).$$

To prove the opposite implication, suppose that $x^* \in \partial(f + g)(x)$. So $(f + g)(x) + (f + g)^*(x^*) = \langle x, x^* \rangle$. Corollary 10.2 now gives $z^* \in E^*$ such that $(f + g)(x) + f^*(x^* - z^*) + g^*(z^*) = \langle x, x^* \rangle$, that is to say

$$\left[f(x) + f^*(x^* - z^*)\right] + \left[g(x) + g^*(z^*)\right] = \langle x, x^* - z^* \rangle + \langle x, z^* \rangle.$$

It now follows from the Fenchel–Young inequality, (8.2), that

$$f(x) + f^*(x^* - z^*) = \langle x, x^* - z^* \rangle \quad \text{and} \quad g(x) + g^*(z^*) = \langle x, z^* \rangle,$$

and so $x^* - z^* \in \partial f(x)$ and $z^* \in \partial g(x)$, from which $x^* = (x^* - z^*) + z^* \in \partial f(x) + \partial g(x) = (\partial f + \partial g)(x)$. $\square$

Our next result appears in [1, Corollary 2.1, pp. 130–131]. In fact, more general results using a so–called *closedness–type regularity condition* have been established recently by Boţ–Wanka in [27], even for Fréchet spaces. See the references in [27] for other work in this direction.

Theorem 18.2. *Let E be a nonzero Banach space, f, $g \in \mathcal{PCLSC}(E)$ and $\bigcup_{\lambda>0} \lambda[\operatorname{dom} f - \operatorname{dom} g]$ be a closed subspace of E. Then $\partial(f+g) = \partial f + \partial g$.*

Proof. This follows from Remark 15.2 and an argument similar to that employed above. $\square$

Either Theorem 18.1 or Theorem 18.2 can be used for the results in this section. One important consequence of Theorem 18.2 which will be used in our proof of Voisei's theorem, Theorem 51.1, is the following:

Corollary 18.3. *Let E be a nonzero Banach space, C, D be nonempty closed convex subset of E and $\bigcup_{\lambda>0} \lambda[C - D]$ be a closed subspace of E. Then $N_{C \cap D} = N_C + N_D$.*

Proof. Let $f := \mathbb{I}_C$ and $g := \mathbb{I}_D$. Then $f + g = \mathbb{I}_{C \cap D}$. The result follows from Theorem 18.2 since $\partial(f+g) = N_{C \cap D}$, $\partial f = N_C$ and $\partial g = N_D$. $\square$

Ekeland's variational principle is a result on complete metric spaces which has had a large number of applications to nonlinear analysis. We refer the reader to Ekeland's survey article, [39], and also to Borwein–Zhu, [24, Chapter 2, pp. 5–36], for a description of the Flower–petal theorem, the Drop theorem, the Caristi–Kirk fixed–point theorem, the Borwein–Preiss smooth variational principle and the Deville–Godefroy–Zizler variational principle. The Banach space case, as in the statement of Theorem 18.4 below, is proved in Phelps, [68, Lemma 3.13, p. 45]. Note that it does *not* require f to be convex.

Theorem 18.4. *Let E be a nonzero Banach space, $f \colon E \to]-\infty, \infty]$ be a lower semicontinuous function $\alpha, \beta > 0$, $y \in \operatorname{dom} f$ and $f(y) \leq \inf_E f + \alpha\beta$. Then there exists $x \in E$ such that $f(x) \leq f(y)$ (hence $x \in \operatorname{dom} f$), $\|x-y\| \leq \alpha$ and*

$$z \in \operatorname{dom} f \implies f(z) + \beta\|z - x\| - f(x) \geq 0.$$

We now use the Hahn–Banach–Lagrange theorem to deduce the "vanilla" version of the Brøndsted–Rockafellar theorem (see [31, p. 608] and Phelps, [68, Theorem 3.17, p. 48]).

Corollary 18.5. *Let E be a nonzero Banach space, $f \in \mathcal{PCLSC}(E)$, $\alpha, \beta > 0$, $y \in \operatorname{dom} f$ and $f(y) \leq \inf_E f + \alpha\beta$. Then there exists $(x, z^*) \in G(\partial f)$ such that $\|x - y\| \leq \alpha$, $f(x) \leq f(y)$ and $\|z^*\| \leq \beta$.*

Proof. Let x be as in Theorem 18.4, so that

$$z \in E \quad \Longrightarrow \quad f(z) + \beta\|z - x\| \geq f(x).$$

From the Hahn–Banach–Lagrange theorem, Theorem 1.11, there exists a linear functional L on E such that $L \leq \beta\|\cdot\|$ on E and

$$z \in E \quad \Longrightarrow \quad f(z) + L(z - x) \geq f(x).$$

It follows easily from this that $L \in E^*$ and, writing $z^* = -L$, $\|z^*\| \leq \beta$ and $(x, z^*) \in G(\partial f)$.   □

Here is the full version of the Brøndsted–Rockafellar theorem. We have slightly changed the notation to make it more consistent with that of the density result, Theorem 48.1, in which we will use Theorem 18.6. In particular, in the statement of Theorem 18.6, $q: E \times E^* \to \mathbb{R}$ is as in Remark 17.1. Theorem 18.6 will almost be generalized by Corollary 48.8.

Theorem 18.6. *Let E be a nonzero Banach space, $f \in \mathcal{PCLSC}(E)$, $\alpha, \beta > 0$, $b \in E \times E^*$ and $f(b_1) + f^*(b_2) \leq q(b) + \alpha\beta$. Then there exists $s \in G(\partial f)$ such that $\|s_1 - b_1\| \leq \alpha$ and $\|s_2 - b_2\| \leq \beta$.*

Proof. This is immediate from Corollary 18.5, with f replaced by $f - b_2$, $y := b_1$ and $s := (x, z^* + b_2)$.   □

We now come to the main result of this section.

Theorem 18.7. *Let E be a nonzero Banach space and $f \in \mathcal{PCLSC}(E)$. Then $\partial f: E \rightrightarrows E^*$ is maximally monotone.*

Proof. Let $b \in E \times E^*$ and

$$a \in G(\partial f) \quad \Longrightarrow \quad q(a - b) \geq 0. \tag{18.1}$$

We want to prove that
$$b \in G(\partial f). \tag{18.2}$$

Let $k := f(\cdot + b_1)$, $g := \frac{1}{2}\|\cdot\|^2$, and $h := k + g$. Since g is continuous, the formula for the subdifferential of a sum, Theorem 18.1, implies that, for all $x \in E$,
$$\partial h(x) = \partial k(x) + \partial g(x) = \partial f(x + b_1) + Jx, \tag{18.3}$$

where $J: E \rightrightarrows E^*$ is the *duality map* (see Definition 23.5). The properties of J that we will need (which are easy to check) are that

$$J0 = \{0\} \quad \text{and} \quad d \in G(J) \Longrightarrow q(d) = \|d_1\|^2. \tag{18.4}$$

From the Fenchel–Moreau Theorem, Corollary 12.4, there exists $y^* \in E^*$ such that $k^*(y^*) \in \mathbb{R}$. It is easily seen that, for all $x \in E$,

$$\left.\begin{aligned}
\langle x, b_2 \rangle - h(x) &= \langle x, b_2 \rangle - k(x) - g(x) \\
&\leq \langle x, b_2 \rangle - \langle x, y^* \rangle + k^*(y^*) - g(x) \\
&= k^*(y^*) + \langle x, b_2 - y^* \rangle - g(x) \\
&\leq k^*(y^*) + g^*(b_2 - y^*) \\
&= k^*(y^*) + \tfrac{1}{2}\|b_2 - y^*\|^2 < \infty,
\end{aligned}\right\} \tag{18.5}$$

and so $h^*(b_2) < \infty$. Consequently, for all $n \geq 1$, there exists $x_n \in E$ such that

$$\langle x_n, b_2 \rangle - h(x_n) \geq h^*(b_2) - 1/n^2. \tag{18.6}$$

The Brøndsted–Rockafellar theorem, Theorem 18.6, now gives $s_n \in G(\partial h)$ such that

$$\|s_{n1} - x_n\| \leq 1/n \quad \text{and} \quad \|s_{n2} - b_2\| \leq 1/n, \tag{18.7}$$

and (18.3) gives $d_n \in G(J)$ such that

$$d_{n1} = s_{n1} \quad \text{and} \quad (s_{n1} + b_1, s_{n2} - d_{n2}) \in G(\partial f).$$

From (18.1),

$$\langle s_{n1}, s_{n2} - d_{n2} - b_2 \rangle \geq 0,$$

and so $\langle s_{n1}, d_{n2} \rangle \leq \langle s_{n1}, s_{n2} - b_2 \rangle$. From (18.4), $\langle s_{n1}, d_{n2} \rangle = \|s_{n1}\|^2$, thus (18.7) implies that $\|s_{n1}\|^2 \leq \|s_{n1}\|/n$, from which $\|s_{n1}\| \leq 1/n$ and so, using (18.7) again, $\|x_n\| \leq 2/n$, thus $x_n \to 0$ as $n \to \infty$. Passing to the limit in (18.6) and using the lower semicontinuity of h, $h(0) + h^*(b_2) \leq 0$, from which $b_2 \in \partial h(0)$. Using (18.3) and (18.4) again, $b_2 \in \partial f(b_1) + J0 = \partial f(b_1)$. This completes the proof of (18.2) and, consequently, also that of Theorem 18.7.

$\square$

Remark 18.8. We now compare the argument that we have given above in Theorem 18.7 and those of [99, Chapter VII, pp 111–139] and [60]. The argument of Theorem 18.7 is completely analytic, while that of [99, Theorem 29.4, pp. 116–118] is much harder and has a much more geometric feel. On the other hand, this latter argument leads to the more general results in the later parts of [99, Chapter VII]. All these more general results will appear in Theorem 48.4, but they will be established using the results on maximally monotone multifunctions of type (ED) that we will prove in Sections 37–42. This being the case, there is obviously an incentive to give the simplest possible proof of Theorem 18.7. The proof given here is based on the very elegant one found recently by M. Marques Alves and B. F. Svaiter in [60], but is structurally simpler, and exploits the properties of subdifferentials and the duality multifunction to avoid some of the computations in [60].

We refer the reader to Borwein, [17, Theorem 3.1, p. 568] for a proof of the maximal monotonicity of subdifferentials in Fréchet smooth Banach spaces using Zagrodny's approximate mean value theorem (see Borwein–Zhu, [24, Section 3.4.2, pp. 81–82]).

Remark 18.9. In this remark, we indicate how the proof of Theorem 18.7 can be simplified if E is reflexive. (18.5) implies that, for all $x \in E$,

$$\langle x, b_2 \rangle - h(x) \geq h^*(b_2) - 1 \quad \Longrightarrow \quad \tfrac{1}{2} \|x\|^2 - \|x\| \|b_2 - y^*\| \leq 1 + k^*(y^*) - h^*(b_2).$$

Thus

$$\{ x \in E \colon \langle x, b_2 \rangle - h(x) \geq h^*(b_2) - 1 \} \text{ is a bounded subset of } E.$$

So if E is reflexive, it follows from a standard weak compactness argument that $b_2 - h$ attains its maximum on E, that is to say there exists $x \in E$ such that $\langle x, b_2 \rangle - h(x) = h^*(b_2)$, from which $b_2 \in \partial h(x)$, and so (18.3) gives $d \in G(J)$ such that $d_1 = x$ and $(x + b_1, b_2 - d_2) \in G(\partial f)$. Then (18.1) and (18.4) imply that $\langle x, -d_2 \rangle \geq 0$ and $\langle x, d_2 \rangle = \|x\|^2$, from which $x = 0$. Thus $b_2 \in \partial h(0)$, and the rest of the proof of Theorem 18.7 proceeds as before. Thus the Brøndsted–Rockafellar theorem is not needed in Theorem 18.7 if E is reflexive. See also Remark 21.6.

The equivalence of (18.8) and (18.10) in our next result is actually part of the Bishop–Phelps theorem (Phelps, [68, Proposition 3.20, p. 49]): *C is the intersection of the closed half–spaces defined by its supporting hyperplanes.* Theorem 18.10 will be used explicitly in Lemma 28.4.

Theorem 18.10. *Let C be a nonempty closed convex subset of a Banach space E and $x \in E$. Then the conditions (18.8)–(18.10) are equivalent.*

$$x \in C. \tag{18.8}$$

$$a \in G(N_C) \quad \Longrightarrow \quad \langle x, a_2 \rangle \leq \langle a_1, a_2 \rangle. \tag{18.9}$$

$$a \in G(N_C) \quad \Longrightarrow \quad \langle x, a_2 \rangle \leq \sup \langle C, a_2 \rangle. \tag{18.10}$$

Proof. (18.9) is equivalent to the statement

$$a \in G(N_C) \quad \Longrightarrow \quad \langle a_1 - x, a_2 - 0 \rangle \geq 0.$$

From Rockafellar's maximal monotonicity theorem, Theorem 18.7, this is equivalent to the assertion $(x, 0) \in G(N_C)$ which is, in turn, equivalent to (18.8). This establishes that (18.8) $\Longleftrightarrow$ (18.9), and it is immediate from the definition of N_C that (18.9) $\Longleftrightarrow$ (18.10). □

19 SSD spaces, q–positive sets and BC–functions

In this section, we introduce the concepts of a SSD space, q–positive set and BC–function, which we will use later in our investigation of monotonicity. The reader interested in a more detailed discussion of q–positive sets (with particular reference to the finite dimensional case) can find such a discussion in [108]. We also give a cursory discussion of the "inverted" concepts of q–negative set and TBC–function — we will need these in the important transversality theorem, Theorem 19.16.

Definition 19.1. We will say that B $\big($more precisely, $(B, \lfloor \cdot, \cdot \rfloor)\big)$ is a *symmetrically self–dual space (SSD space)* if B is a nonzero real vector space and $\lfloor \cdot, \cdot \rfloor \colon B \times B \to \mathbb{R}$ is a symmetric bilinear form that separates the points of B. We define the quadratic form q on B by $q(b) := \frac{1}{2} \lfloor b, b \rfloor$. If $f \in \mathcal{PC}(B)$, we write $f^{@}$ for the Fenchel conjugate of f with respect to the pairing $\lfloor \cdot, \cdot \rfloor$. We will say that a locally convex topology $\mathcal{T}$ on B is B–*compatible* if the $\mathcal{T}$–dual of B is exactly $\big\{ \lfloor \cdot, c \rfloor \colon c \in B \big\}$. This conforms with the usage in Section 8.

As a general notation, if E is a normed space then $\mathcal{T}_{\| \ \|}(E)$ stands for the norm topology of E.

We now give some examples of SSD spaces.

Examples 19.2. (a) Let E be a nonzero Banach space and B and $\lfloor \cdot, \cdot \rfloor$ be defined as in Remark 17.1. Then B is a SSD space. Let $\mathcal{T}_{\mathcal{NW}}(B)$ be the topology $\mathcal{T}_{\| \ \|}(E) \times w(E^*, E)$ on B. Then $\mathcal{T}_{\mathcal{NW}}(B)$ is B–compatible. As in Notation 16.1, we can introduce a norm on B by $\| b \| := \sqrt{\| b_1 \|^2 + \| b_2 \|^2}$. If E is not reflexive then the topology $\mathcal{T}_{\| \ \|}(B)$ is not B–compatible. We point out that any finite dimensional SSD space of the form described here must have *even* dimension. Thus odd dimensional cases of Examples (b,c), and Example (d) below cannot be of this form. Since this example will be discussed in much greater detail in Section 22, we will not discuss it any further in this section.

(b) If B is a Hilbert space with inner product $(b, c) \mapsto \langle b, c \rangle$ then B is a SSD space with $\lfloor b, c \rfloor := \langle b, c \rangle$, and $q(b) = \frac{1}{2} \| b \|^2$.

(c) If B is a Hilbert space with inner product $(b, c) \mapsto \langle b, c \rangle$ then B is a SSD space with $\lfloor b, c \rfloor := -\langle b, c \rangle$, and $q(b) = -\frac{1}{2} \| b \|^2$.

(d) $\mathbb{R}^3$ is a SSD space with

$$\lfloor (b_1, b_2, b_3), (c_1, c_2, c_3) \rfloor := b_1 c_2 + b_2 c_1 + b_3 c_3,$$

and $\quad q(b_1, b_2, b_3) = b_1 b_2 + \frac{1}{2} b_3^2$.

Examples 19.3. $\mathbb{R}^3$ is *not* a SSD space with

$$\lfloor (b_1, b_2, b_3), (c_1, c_2, c_3) \rfloor := b_1 c_2 + b_2 c_3 + b_3 c_1.$$

(The bilinear form $\lfloor \cdot, \cdot \rfloor$ is not symmetric.)

Before we turn to the main topic of the section, q–positive sets, we will prove a simple lemma which we will use explicitly in our investigation of maximally monotone multifunctions with convex graph in Theorem 46.1(c).

Lemma 19.4. *Let B be a SSD space and A be a nonempty subset of B. Suppose that*

$$b \in B \quad \Longrightarrow \quad \inf q(A - b) \le \inf q(A) \in \mathbb{R}. \tag{19.1}$$

Then

$$b \in B \quad \Longrightarrow \quad \sup\lfloor A, b\rfloor \ge 0.$$

Proof. Let $\lambda > 0$. Since $q(a - \lambda b) = q(a) - \lambda\lfloor a, b\rfloor + \lambda^2 q(b)$, we have

$$\inf q(A - \lambda b) \ge \inf q(A) - \lambda \sup\lfloor A, b\rfloor + \lambda^2 q(b).$$

Combining this with (19.1) with b replaced by λb gives $\lambda \sup\lfloor A, b\rfloor \ge \lambda^2 q(b)$, and the result follows by dividing by λ and then letting $\lambda \to 0$.

Definition 19.5. Let B be a SSD space and $\emptyset \ne A \subset B$. We say that A is q–positive if

$$b, c \in A \Longrightarrow q(b - c) \ge 0.$$

Examples 19.6. We now give some examples of q–positive sets. We first make the elementary observation that if $b \in B$ and $q(b) \ge 0$ then the linear span $\mathbb{R}b$ of $\{b\}$ is q–positive.

In Example 19.2(a), the q–positive sets are exactly the sets $G(S)$, where $S: E \rightrightarrows E^*$ is nontrivial and monotone. (See the discussion in Remark 17.1.)

In Example 19.2(b), every subset of B is q–positive, and in Example 19.2(c), the q–positive sets are the singletons.

In Example 19.2(d), If M is any nonempty monotone subset of $\mathbb{R} \times \mathbb{R}$ (in the obvious sense) then $M \times \mathbb{R}$ is a q–positive subset of B. The set $\mathbb{R}(1, -1, 2)$ is a q–positive subset of B which is not contained in a set $M \times \mathbb{R}$ for any monotone subset of $\mathbb{R} \times \mathbb{R}$. The helix $\{(\cos\theta, \sin\theta, \theta): \theta \in \mathbb{R}\}$ is a q–positive subset of B, but if $0 < \lambda < 1$ then the helix $\{(\cos\theta, \sin\theta, \lambda\theta): \theta \in \mathbb{R}\}$ is not.

We first discuss a simple convexity property of q–positive sets which we will use explicitly in our investigation of maximally monotone operators with convex graph in Theorem 46.1 and Theorem 46.3. This result is based on an observation of Heinz Bauschke (personal communication), and the statement has been simplified as a result of a comment of Radu Ioan Boţ.

Lemma 19.7. *Let B be a SSD space and A be a nonempty q–positive subset of B. Suppose that $a, c \in A$ and $\lambda \in]0, 1[$. Let $b \in B$. Then*

$$q\big(\lambda a + (1 - \lambda)c - b\big) \le \lambda q(a - b) + (1 - \lambda)q(c - b).$$

Proof. For simplicity in writing, let $\mu := 1 - \lambda \in]0,1[$. Then

$$q(\lambda a + \mu c - b) = q\big(\lambda(a-b)+\mu(c-b)\big) = \lambda^2 q(a-b)+\lambda\mu\lfloor a-b, c-b\rfloor + \mu^2 q(c-b).$$

Thus

$$\lambda q(a-b) + \mu q(c-b) - q(\lambda a + \mu c - b)$$
$$= (\lambda - \lambda^2)q(a-b) + (\mu - \mu^2)q(c-b) - \lambda\mu\lfloor a - b, c - b\rfloor$$
$$= \lambda\mu\big[q(a-b) - \lfloor a-b, c-b\rfloor + q(c-b)\big]$$
$$= \lambda\mu\big[q\big(a-b-(c-b)\big)\big] = \lambda\mu q(a-c) \geq 0. \qquad \square$$

The following simple lemma shows how q–positive sets can be obtained from convex functions. In this result, we use the "parallelogram law", which is a purely algebraic result. Lemma 19.8 will be used explicitly in Theorem 19.21, Theorem 21.4, and Lemma 23.1.

Lemma 19.8. *Let B be a SSD space, $f \in \mathcal{PC}(B)$ and $f \geq q$ on B. Let* pos $f := \{b \in B\colon f(b) = q(b)\}$. *If* pos $f \neq \emptyset$ *then* pos f *is a q–positive subset of B.*

Proof. Let $b, c \in$ pos f. Then, from the parallelogram law, the quadraticity of q, and the convexity of f,

$$\tfrac{1}{2}q(b-c) = q(b) + q(c) - \tfrac{1}{2}q(b+c)$$
$$= q(b) + q(c) - 2q\big(\tfrac{1}{2}(b+c)\big)$$
$$\geq f(b) + f(c) - 2f\big(\tfrac{1}{2}(b+c)\big) \geq 0. \qquad \square$$

Remark 19.9. q–positive sets of the special form described in Lemma 19.8 have been investigated in [108].

Remark 19.10. Let B be a SSD space. Let $\emptyset \neq A \subset B$. We say that A is *q–negative* if

$$b, c \in A \Longrightarrow q(b - c) \leq 0.$$

The results (a) and (b) below follow by observing that B is also a SSD space under the bilinear form $(b, c) \mapsto -\lfloor b, c\rfloor$, $\tfrac{1}{2}(-\lfloor b, b\rfloor) = (-q)(b)$ and "q–negative" means the same thing as "$(-q)$–positive".

(a) We have the following analog of Lemma 19.7: *Let A be a nonempty q–negative subset of B. Suppose that $a, c \in A$ and $\lambda \in]0,1[$. Then, for all $b \in B$, $q\big(\lambda a + (1-\lambda)c - b\big) \geq \lambda q(a-b) + (1-\lambda)q(c-b)$.*

(b) We have the following analog of Lemma 19.8: *Let $g \in \mathcal{PC}(B)$ and $g \geq -q$ on B. Let* neg $g := \{b \in B\colon g(b) = -q(b)\}$. *If* neg $g \neq \emptyset$ *then* neg g *is a q-negative subset of B.*

Definition 19.11. Let B be a SSD space. We say that $f \in \mathcal{PC}(B)$ is a *BC–function* if

$$b \in B \quad \Longrightarrow \quad f^{@}(b) \geq f(b) \geq q(b). \tag{19.2}$$

"BC" stands for "bigger conjugate". (We recall that $f^{@}$ was defined in Definition 19.1.) This concept was introduced in [106] in a special case under the name of "LC–function" – we have changed the notation in order to avoid confusion with the initials of "locally convex".

The following result is somewhat unexpected, and will be used explicitly in the numerical estimates in Theorem 21.4(c), and in our results on the maximal monotonicity of the sum of maximally monotone multifunctions in Theorem 24.1.

Lemma 19.12. *Let B be a SSD space and $f \in \mathcal{PC}(B)$ be a BC–function. Then*

$$\operatorname{pos} f^{@} = \operatorname{pos} f \subset \operatorname{dom} f.$$

Proof. Let $c \in \operatorname{pos} f$. Let b be an arbitrary element of B. Let $\lambda \in \,]0,1[$. For simplicity in writing, let $\mu := 1 - \lambda \in \,]0,1[$. Then

$$\lambda^2 q(b) + \lambda\mu\lfloor b,c \rfloor + \mu^2 q(c) = q(\lambda b + \mu c) \leq f(\lambda b + \mu c)$$
$$\leq \lambda f(b) + \mu f(c) = \lambda f(b) + \mu q(c).$$

Thus $\lambda\mu\lfloor b,c \rfloor - \lambda f(b) \leq \lambda\mu q(c) - \lambda^2 q(b)$. Dividing by λ and letting $\lambda \to 0$, we obtain $\lfloor b,c \rfloor - f(b) \leq q(c)$. It now follows by taking the supremum over $b \in B$ that $f^{@}(c) \leq q(c)$, and (19.2) implies that $c \in \operatorname{pos} f^{@}$. Thus we have proved that $\operatorname{pos} f \subset \operatorname{pos} f^{@}$. Of course, it is immediate from (19.2) that $\operatorname{pos} f^{@} \subset \operatorname{pos} f \subset \operatorname{dom} f$. $\qquad\square$

Lemma 19.13 on translating a BC–function by an element of B will be used explicitly in the transversality theorem, Theorem 19.16, Theorem 21.4, the local transversality theorem, Theorem 21.12, Lemma 22.9, Lemma 34.2 and Corollary 35.9.

Lemma 19.13. *Let B be a SSD space, $f \in \mathcal{PC}(B)$ be a BC–function and $c \in B$. We define $f_c \in \mathcal{PC}(B)$ by $f_c := f(\cdot + c) - \lfloor \cdot, c \rfloor - q(c)$. Then f_c is a BC–function, $\operatorname{dom} f_c = \operatorname{dom} f - c$ and $\operatorname{pos} f_c = \operatorname{pos} f - c$.*

Proof. For all $b \in B$,

$$f_c{}^{@}(b) = \sup_{d \in B} \big[\lfloor d,b \rfloor + \lfloor d,c \rfloor + q(c) - f(d+c)\big]$$
$$= \sup_{e \in B} \big[\lfloor e - c, b + c \rfloor + q(c) - f(e)\big]$$
$$= \sup_{e \in B} \big[\lfloor e, b + c \rfloor - \lfloor c, b \rfloor - f(e)\big] - q(c)$$
$$= f^{@}(b + c) - \lfloor c, b \rfloor - q(c).$$

It follows from (19.2) that $f_c{}^{@}(b) \geq f(b+c) - \lfloor b,c \rfloor - q(c) = f_c(b)$ and

$$f_c(b) = f(b+c) - \lfloor b,c \rfloor - q(c) \geq q(b+c) - \lfloor b,c \rfloor - q(c) = q(b).$$

Consequently, f_c is a BC–function. It is obvious that $\operatorname{dom} f_c = \operatorname{dom} f - c$. Further, since

$$b \in \operatorname{pos} f_c \iff f(b+c) - \lfloor b,c \rfloor - q(c) = q(b)$$
$$\iff f(b+c) = q(b+c) \iff b+c \in \operatorname{pos} f,$$

we have $\operatorname{pos} f_c = \operatorname{pos} f - c$, as required. $\qquad\square$

Here is the "analog" of Definition 19.11 in the sense of Remark 19.10.

Definition 19.14. Let B be a SSD space. We say that $g \in \mathcal{PC}(B)$ is a *TBC–function* if

$$b \in B \implies g^{@}(-b) \geq g(b) \geq -q(b). \tag{19.3}$$

"TBC" stands for "twisted bigger conjugate". A significant example of a TBC–function will be provided by (21.2).

Remark 19.15. Arguing as in Remark 19.10, we have the following analog of Lemma 19.12: *Let B be a SSD space and $g \in \mathcal{PC}(B)$ be a TBC–function. Then*

$$-\operatorname{neg} g^{@} = \operatorname{neg} g \subset \operatorname{dom} g.$$

We end this discussion of BC–functions and TBC–functions in the abstract by proving an important "transversality theorem", which will be used in our characterization of the maximality of a q–positive set in a SSDB space in Theorem 21.4, and also in Theorem 30.6 and Lemma 35.5.

Theorem 19.16. *Let B be a SSD space, $f \in \mathcal{PC}(B)$ be a BC–function and $g: B \to \mathbb{R}$ be a TBC–function that is continuous with respect to a B–compatible topology. Then* $\operatorname{pos} f - \operatorname{neg} g = B$.

Proof. Let c be an arbitrary element of B. From Lemma 19.13, f_c is a BC–function and so, using (19.2) and (19.3),

$$b \in B \implies f_c(b) + g(b) \geq q(b) - q(b) = 0.$$

Thus Rockafellar's version of the Fenchel duality theorem, Corollary 8.6, gives $a \in B$ such that $f_c^{@}(a) + g^{@}(-a) \leq 0$ so, from (19.2) and (19.3) again, $f_c(a) + g(a) \leq 0 = q(a) - q(a)$. From (19.2) and (19.3) for a third time, $f_c(a) \geq q(a)$ and $g(a) \geq -q(a)$. Consequently, $f_c(a) = q(a)$ and $g(a) = -q(a)$, that is to say, using Lemma 19.13, $a \in \operatorname{pos} f_c = \operatorname{pos} f - c$ and also $a \in \operatorname{neg} g$. But then $c = (c+a) - a \in \operatorname{pos} f - \operatorname{neg} g$. $\qquad\square$

We now consider the situation opposite to that in Lemma 19.8: how to obtain a convex function from a q–positive set.

Definition 19.17. Let B be a SSD space and A be a nonempty q–positive subset of B. We define the function $\Phi_A\colon B \to {]-\infty, \infty]}$ associated with A by

$$\Phi_A(b) := \sup_{a \in A} \left[\lfloor b, a \rfloor - q(a)\right] = q(b) - \inf_{a \in A} q(b-a). \qquad (19.4)$$

These definitions imply that $\Phi_A \in \mathcal{PC}(B)$,

$$\Phi_A = q \text{ on } A, \quad \text{and} \quad b \in B \text{ and } a \in A \Longrightarrow \lfloor b, a \rfloor \leq \Phi_A(b) + q(a). \qquad (19.5)$$

Now let $a \in A$. (19.5) implies that, for all $b \in B$, $\lfloor b, a \rfloor - \Phi_A(b) \leq q(a)$. Taking the supremum over $b \in B$, $\Phi_A{}^{@}(a) \leq q(a)$. Thus we have proved that

$$\Phi_A{}^{@} \leq q \text{ on } A. \qquad (19.6)$$

Let $c \in B$. Then we see from the definition of $\Phi_A{}^{@}$, the symmetry of $\lfloor \cdot, \cdot \rfloor$ and (19.5) that

$$\begin{aligned}
\Phi_A{}^{@}(c) &= \sup_{b \in B} \left[\lfloor b, c \rfloor - \Phi_A(b)\right] \\
&\geq \left[\lfloor c, c \rfloor - \Phi_A(c)\right] \vee \sup_{a \in A} \left[\lfloor c, a \rfloor - \Phi_A(a)\right] \\
&= \left[2q(c) - \Phi_A(c)\right] \vee \sup_{a \in A} \left[\lfloor c, a \rfloor - q(a)\right] \\
&= \left[2q(c) - \Phi_A(c)\right] \vee \Phi_A(c) \geq \min_{\lambda \in]-\infty, \infty]} \left[2q(c) - \lambda\right] \vee \lambda = q(c).
\end{aligned}$$

Thus it follows that

$$\Phi_A{}^{@} \geq \Phi_A \vee q \text{ on } B. \qquad (19.7)$$

Combining this with (19.6), we see that

$$A \subset \operatorname{pos} \Phi_A{}^{@}. \qquad (19.8)$$

In fact, $\operatorname{pos} \Phi_A{}^{@}$ is the largest q–positive subset C of B such that $\Phi_C = \Phi_A$ on B. See [108, Theorem 6.5(a), p. 309]. Clearly, Φ_A is $w(B, B)$–lower semicontinuous, and so the Fenchel–Moreau theorem, Corollary 12.4, gives us that

$$\Phi_A{}^{@} \in \mathcal{PC}(B) \quad \text{and} \quad \Phi_A{}^{@@} = \Phi_A. \qquad (19.9)$$

(The first of these assertions also follows from (19.8).) From (19.6), (19.7) and the fact that $\operatorname{dom} \Phi_A{}^{@}$ is convex, writing "co" for "convex hull",

$$A \subset \operatorname{co} A \subset \operatorname{dom} \Phi_A{}^{@} \subset \operatorname{dom} \Phi_A. \qquad (19.10)$$

Now let $c \in B$. Then $A - c$ is q–positive and, for all $b \in B$,

$$\begin{aligned}
\Phi_{A-c}(b) &= \sup_{a \in A} \left[\lfloor b, a - c \rfloor - q(a - c)\right] \\
&= \sup_{a \in A} \left[\lfloor b, a \rfloor - \lfloor b, c \rfloor - q(a) + \lfloor c, a \rfloor - q(c)\right] \\
&= \sup_{a \in A} \left[\lfloor b + c, a \rfloor - q(a)\right] - \lfloor b, c \rfloor - q(c) \\
&= \Phi_A(b + c) - \lfloor b, c \rfloor - q(c) = (\Phi_A)_c(b)
\end{aligned}$$

(see Lemma 19.13). It follows easily from this that

$$\operatorname{dom} \Phi_{A-c} = \operatorname{dom} \Phi_A - c. \tag{19.11}$$

Finally, suppose that $\Phi_A \geq q$ on B. Then (19.7) implies that Φ_A is a BC–function and so, from Lemma 19.12, $\operatorname{pos} \Phi_A{}^@ = \operatorname{pos} \Phi_A$. If we combine this with (19.5), we obtain:

$$\Phi_A \geq q \text{ on } B \implies \Phi_A \text{ is a BC–function and } \operatorname{pos} \Phi_A{}^@ = \operatorname{pos} \Phi_A \supset A. \tag{19.12}$$

Our next result will be used explicitly in Theorems 27.5 and 27.6.

Lemma 19.18. *Let B be a SSD space, A be a nonempty q–positive subset of B and $b \in \operatorname{dom} \Phi_A$. Let $f \colon A \to \mathbb{R}$ and $\inf_A f > 0$. Then*

$$\inf_{a \in A} \ q(b-a)/f(a) > -\infty.$$

Proof. Let

$$M := \frac{\left[\Phi_A(b) - q(b)\right] \vee 0}{\inf_A f} \in \mathbb{R}.$$

From (19.4), for all $a \in A$, $q(b-a) + Mf(a) \geq q(b) - \Phi_A(b) + M \inf_A f \geq 0$, which implies that $q(b-a)/f(a) > -M$. $\square$

Remark 19.19. Let B be a SSD space and A be a nonempty q–negative subset of B. We can define the function $\Psi_A \colon B \to \,]-\infty, \infty]$ associated with A by $\Psi_A(b) := \sup_{a \in A} \left[-\lfloor b, a \rfloor + q(a)\right] = -q(b) + \sup_{a \in A} q(b-a)$. Ψ_A has properties analogous to those of Φ_A. We leave the details to the reader.

The following simple example shows how much more general BC–functions are than functions of the form Φ_A for some nonempty q–positive subset A of B.

Example 19.20. Let $B = \mathbb{R}^2$ with $\lfloor(b_1, b_2), (c_1, c_2)\rfloor := b_1 c_2 + b_2 c_1$. (This SSD space is a special case of Example 19.2(a).) Define $h \colon B \to \mathbb{R}$ by $h(b_1, b_2) := \frac{1}{2}(b_1^2 + b_2^2)$. Then, by direct computation, $h^@ = h$ on B (compare Lemma 9.2). Since $h(b_1, b_2) - q(b_1, b_2) = \frac{1}{2}(b_1^2 + b_2^2) - b_1 b_2 = \frac{1}{2}(b_1 - b_2)^2$, h is a BC–function. Furthermore,

$$(a_1, a_2) \in \operatorname{pos} h \quad \Longleftrightarrow \quad a_1 = a_2. \tag{19.13}$$

Now suppose that A were a nonempty q–positive subset of B such that $\Phi_A = h$ on B. Then, from (19.5), $h = q$ on A, that is to say, $A \subset \operatorname{pos} h$, and (19.13) would imply that

$$(a_1, a_2) \in A \quad \Longrightarrow \quad a_1 = a_2.$$

Now let $b = (b_1, b_2)$ be an arbitrary element of B. Then we would have

$$\tfrac{1}{2}(b_1^2 + b_2^2) = h(b) = \Phi_A(b) = \sup_{a \in A} \left[\lfloor b, a \rfloor - q(a)\right]$$
$$\leq \sup_{\lambda \in \mathbb{R}} \left[\lfloor b, (\lambda, \lambda)\rfloor - q(\lambda, \lambda)\right] = \sup_{\lambda \in \mathbb{R}} \left[(b_1 + b_2)\lambda - \lambda^2\right]$$
$$= \tfrac{1}{4}(b_1 + b_2)^2.$$

It would follow from this that $2(b_1^2 + b_2^2) \leq (b_1 + b_2)^2$, that is to say, $(b_1 - b_2)^2 \leq 0$. Since this is manifestly false whenever $b_1 \neq b_2$, we have reached a contradiction.

We conclude this section with a precise characterization of functions of the form Φ_A (for some nonempty q–positive subset A of B).

Theorem 19.21. *Let B be a SSD space and $f \in \mathcal{PC}(B)$. Then there exists a nonempty q–positive subset A of B such that $f = \Phi_A$ on B if, and only if,*

$$f^@ \geq q \text{ on } B \quad \text{and} \quad b \in B \Longrightarrow f(b) \leq \sup_{c \in \text{pos } f^@} \left[\lfloor b, c \rfloor - q(c)\right]. \quad (19.14)$$

Proof. ($\Longrightarrow$) If A is a nonempty q–positive subset of B then, from (19.7) and (19.8), $\Phi_A{}^@ \geq q$ on B and $A \subset \text{pos } \Phi_A{}^@$. Thus, for all $b \in B$,

$$\Phi_A(b) = \sup_{a \in A} \left[\lfloor b, a \rfloor - q(a)\right] \leq \sup_{c \in \text{pos } \Phi_A{}^@} \left[\lfloor b, c \rfloor - q(c)\right].$$

This gives the desired result.

($\Longleftarrow$) If f satisfies (19.14) then clearly $\text{pos } f^@ \neq \emptyset$ thus, from Lemma 19.8 (with f replaced by $f^@$), $\text{pos } f^@$ is q–positive. If $b \in B$ and $c \in \text{pos } f^@$ then the Fenchel–Young inequality, (8.2), implies that $\lfloor b, c \rfloor \leq f(b) + f^@(c) = f(b) + q(c)$, so $\lfloor b, c \rfloor - q(c) \leq f(b)$. Combining this with (19.14), we obtain that

$$b \in B \quad \Longrightarrow \quad f(b) = \sup_{c \in \text{pos } f^@} \left[\lfloor b, c \rfloor - q(c)\right] = \Phi_{\text{pos } f^@}(b).$$

The desired result follows by taking $A = \text{pos } f^@$.   □

20 Maximally q–positive sets in SSD spaces

Definition 20.1. Let B be a SSD space and A be a nonempty q–positive subset of B. We say that $b \in B$ is *q–positively related* to A if

$$a \in A \quad \Longrightarrow \quad q(b - a) \geq 0.$$

It is clear from (19.4) that this is equivalent to the assertion that $\Phi_A(b) \leq q(b)$. We say that A is *maximally q–positive* if A is not properly contained in any other q–positive set. This is equivalent to the assertion that

$$b \in B \setminus A \quad \Longrightarrow \quad b \text{ is not } q\text{–positively related to } A.$$

Consequently, A is maximally q–positive exactly when

$$b \in B \setminus A \quad \Longrightarrow \quad \Phi_A(b) > q(b). \quad (20.1)$$

If $\Phi_A \geq q$ on B then, from (19.12), $\text{pos } \Phi_A{}^@ = \text{pos } \Phi_A$ is the unique maximally q–positive superset of A. For more on this topic, see [108, Theorem 5.4, p. 307 and Theorem 6.5(c), p. 309].

Now let A be a maximally q–positive subset of B. Combining (20.1) with (19.5) implies that $\Phi_A \geq q$ on B and $\text{pos } \Phi_A = A$. Thus, from (19.12) again,

$$\Phi_A \text{ is a BC–function and } \text{pos } \Phi_A{}^@ = \text{pos } \Phi_A = A. \quad (20.2)$$

Example 19.20 shows that there are BC–functions that are not of the form Φ_A for some *maximally* q–positive subset A of B. We now identify which BC–functions are of this form.

Definition 20.2. Let B be a SSD space. We say that $h \in \mathcal{PC}(B)$ is a *SBC–function* if $h \geq q$ on B and

$$b \in B \quad \Longrightarrow \quad h(b) = \sup_{a \in \text{pos}\,h} \left[\lfloor b, a \rfloor - q(a) \right]. \qquad (20.3)$$

"SBC" stands for "strongly bigger conjugate". Since

$$\sup_{a \in \text{pos}\,h} \left[\lfloor b, a \rfloor - q(a) \right] = \sup_{a \in \text{pos}\,h} \left[\lfloor b, a \rfloor - h(a) \right]$$
$$\leq \sup_{c \in B} \left[\lfloor b, c \rfloor - h(c) \right] = h^{@}(b),$$

a SBC–function is automatically a BC–function. (20.3) also implies that a SBC–function is automatically lower semicontinuous with respect to any B–compatible topology. Finally, it follows from (19.4) and (20.2) that if A is a maximally q–positive subset of B then

$$\Phi_A \text{ is a SBC–function.} \qquad (20.4)$$

Theorem 20.3. *Let B be a SSD space. Then the mapping $A \mapsto \Phi_A$ is a bijection from the maximally q–positive subsets of B onto the SBC–functions on B. The inverse mapping is* pos $(\cdot)$.

Proof. It is clear from (20.2) and (20.4) that the mapping $A \mapsto \Phi_A$ is an injection from the maximally q–positive subsets of B into the SBC–functions on B. Suppose, conversely, that $h \in \mathcal{PC}(B)$ is a SBC–function. (20.3) implies that $\text{pos}\,h \neq \emptyset$, and so, from Lemma 19.8, $\text{pos}\,h$ is a nonempty q–positive subset of B. From (20.3) again, $\Phi_{\text{pos}\,h} = h$. So if $b \in B$ and $\Phi_{\text{pos}\,h}(b) \leq q(b)$ then $h(b) \leq q(b)$ and thus, since $h \geq q$ on B, $b \in \text{pos}\,h$. From (20.1), $\text{pos}\,h$ is a maximally q–positive subset of B, so pos $(\cdot)$ is an injection from the SBC–functions on B into the maximally q–positive subsets of B. The result is now immediate. $\qquad \square$

The final result in this section is about q–positive sets that are "flattened" by a certain element of B. It will be used in our results on the linear and affine hulls of certain sets associated with maximally monotone multifunctions in Lemma 28.1 and Lemma 28.8.

Lemma 20.4. *Let B be a SSD space and $c \in B$ with $q(c) = 0$.*
(a) *Let A be a nonempty q–positive subset of B and $A + \mathbb{R}c \subset A$. Then there exists $\mu \in \mathbb{R}$ such that $\lfloor \text{dom}\,\Phi_A, c \rfloor = \lfloor A, c \rfloor = \{\mu\}$.*
(b) *Let A be a maximally q–positive subset of B. Suppose that $\mu \in \mathbb{R}$ and $\lfloor A, c \rfloor = \{\mu\}$. Then $A + \mathbb{R}c \subset A$ and $\lfloor \text{dom}\,\Phi_A, c \rfloor = \{\mu\}$.*

Proof. (a) Let $a \in A$ and $b \in \operatorname{dom} \Phi_A$. Then, for all $\lambda \in \mathbb{R}$,

$$\lfloor b, a \rfloor + \lambda \lfloor b - a, c \rfloor - q(a) = \lfloor b, a + \lambda c \rfloor - q(a + \lambda c).$$

Since $a + \lambda c \in A$, (19.4) implies that

$$\lfloor b, a \rfloor + \lambda \lfloor b - a, c \rfloor - q(a) \le \Phi_A(b),$$

from which

$$\lambda \lfloor b - a, c \rfloor \le \Phi_A(b) - \lfloor b, a \rfloor + q(a) < \infty.$$

Since this holds for all $\lambda \in \mathbb{R}$, $\lfloor b - a, c \rfloor = 0$, that is to say $\lfloor b, c \rfloor = \lfloor a, c \rfloor$. This gives (a).

(b) Let $a \in A$ be given. Let $\lambda \in \mathbb{R}$ and $b \in A$ be arbitrary. Then, since $\lfloor c, b \rfloor = \lfloor b, c \rfloor = \mu = \lfloor a, c \rfloor$, we have

$$\lfloor a + \lambda c, b \rfloor - q(b) = \lfloor a, b \rfloor - q(b) + \lambda \lfloor a, c \rfloor.$$

Taking the supremum over $b \in A$ and using (19.4) and (20.2),

$$\Phi_A(a + \lambda c) = \Phi_A(a) + \lambda \lfloor a, c \rfloor = q(a) + \lambda \lfloor a, c \rfloor + \lambda^2 q(c) = q(a + \lambda c).$$

From (20.2) again, $a + \lambda c \in A$. Thus $A + \mathbb{R}c \subset A$, and (b) follows from (a). $\qquad \square$

21 SSDB spaces

We now introduce the SSDB spaces, a subclass of the class of SSD spaces. Our treatment of the SSD spaces in the previous two sections has been essentially nontopological. The additional norm structure of the SSDB spaces is essentially what makes maximally monotone multifunctions on a reflexive Banach space much more tractable than those on a general Banach space.

In Theorem 21.4, we will show how BC–functions automatically give rise to maximally q–positive sets in the SSDB case; in Theorem 21.7 we will give a "transversal" criterion for a nonempty q–positive subset of a SSDB space to be maximally q–positive; in Theorem 21.10, we give a result on the existence of autoconjugates in a SSDB space; in Theorem 21.11, we apply this to obtain a description of a maximally q–positive superset of a given nonempty q–positive subset of a SSDB space; finally, in Theorem 21.12 we give a local transversality result for SSDB spaces that will eventually be applied to abstract Hammerstein equations.

Definition 21.1. We will say that B $\big($more precisely, $(B, \lfloor \cdot, \cdot \rfloor)\big)$ is a *symmetrically self–dual Banach space (SSDB space)* if B is a SSD space and a Banach space, $\mathcal{T}_{\| \cdot \|}(B)$ is B–compatible, and the norm of $\lfloor \cdot, c \rfloor$ as a functional on B is identical with $\|c\|$. In this case, the quadratic form q (defined as in Definition 19.1) is continuous and, for all $b \in B$,

$$|q(b)| = \tfrac{1}{2}\big|\lfloor b, b \rfloor\big| \leq \tfrac{1}{2}\|b\|^2. \tag{21.1}$$

Let $g_0 := \tfrac{1}{2}\| \cdot \|^2$ on B. From Lemma 9.2,

$$g_0 \text{ is a BC–function and a TBC–function.} \tag{21.2}$$

We have already used the first of these facts in Example 19.20. Since the continuity of q implies that the closure of a q–positive set is q–positive, any maximally q–positive set in a SSDB space is closed.

Examples 21.2. (a) If E is a nonzero reflexive Banach space, we can define B and its associated norm as in Remark 17.1 and Example 19.2(a). Then B is a SSDB space. Since this example will be considered in much greater detail in Section 29, we will not discuss it any further in this section, apart from a brief digression in Remark 21.6.

(b) In Example 19.2(b), B is a SSDB space under the Hilbert space norm, $q = g_0$ and neg $g_0 = \{0\}$. $\big($We recall that the set neg g_0 was defined in Remark 19.10(b).$\big)$

(c) In Example 19.2(c), B is a SSDB space under the Hilbert space norm, $q = -g_0$ and neg $g_0 = B$.

(d) In Example 19.2(d), B is a SSDB space under the Euclidean norm and

$$\text{neg } g_0 = \{(b_1, b_2, b_3) \in B \colon b_1 + b_2 = 0, \ b_3 = 0\}.$$

We now give a simple sufficient condition for maximality. In fact, we shall see in Theorem 21.7 that this condition is also necessary.

Lemma 21.3. *Let B be a SSDB space and and A be a nonempty q–positive subset of B such that $A - \text{neg } g_0 = B$. Then A is maximally q–positive.*

Proof. We will establish this result by using the criterion for maximal q–positivity in (20.1). To this end, suppose that $b \in B \setminus A$. By hypothesis, there exists $a \in A$ such that $a - b \in \text{neg } g_0$, that is to say $q(a - b) = -\tfrac{1}{2}\|a - b\|^2$. Since $a \neq b$, $q(a-b) < 0$, thus (19.4) implies that $\Phi_A(b) > q(b)$. Consequently, (20.1) is satisfied. $\qquad \square$

Theorem 21.4 below contain subtler property of BC–functions on a SSDB space, which will be used explicitly in Theorem 21.7, in our results on the maximal monotonicity of the sum of maximally monotone multifunctions in Theorem 24.1, in the explicit formula for the minimum of the norm of the resolvent of a maximally monotone multifunction on a reflexive space in Theorem 29.6, and in our preparation for the Brezis–Crandall–Pazy theorem in Lemma 34.2.

Theorem 21.4. *Let B be a SSDB space and $f \in \mathcal{PC}(B)$ be a BC–function. Then:*

(a) $\operatorname{pos} f - \operatorname{neg} g_0 = B$.

(b) $\operatorname{pos} f$ *is maximally q–positive.*

(c) *Let $c \in B$. Then*

$$\min\left\{\|a\|\colon a \in \operatorname{neg} g_0 \cap (\operatorname{pos} f - c)\right\} = \sup_{b \in B}\left[\|b\| - \sqrt{2f_c(b) + \|b\|^2}\right] \vee 0$$

and

$$\sup\left\{\|a\|\colon a \in \operatorname{neg} g_0 \cap (\operatorname{pos} f - c)\right\} \leq \inf_{b \in B}\left[\|b\| + \sqrt{2f_c(b) + \|b\|^2}\right].$$

Proof. (a) This follows from the transversality theorem, Theorem 19.16, and (21.2).

(b) This is immediate from Lemma 19.8, (a) and Lemma 21.3.

(c) Suppose first that $a \in \operatorname{neg} g_0 \cap (\operatorname{pos} f - c)$. Then, from Lemma 19.13 and Lemma 19.12, $a \in \operatorname{neg} g_0 \cap \operatorname{pos} f_c = \operatorname{neg} g_0 \cap \operatorname{pos} f_c^{@}$. Thus $f_c^{@}(a) = q(a) = -\frac{1}{2}\|a\|^2$. But then, for all $b \in B$, the Fenchel–Young inequality, (8.2), implies that

$$f_c(b) \geq \lfloor b, a \rfloor - f_c^{@}(a) \geq \tfrac{1}{2}\|a\|^2 - \|b\|\|a\|$$

and so $2f_c(b) + \|b\|^2 \geq \|a\|^2 - 2\|b\|\|a\| + \|b\|^2 = \left(\|a\| - \|b\|\right)^2$. Consequently, $\|b\| - \sqrt{2f_c(b) + \|b\|^2} \leq \|a\| \leq \|b\| + \sqrt{2f_c(b) + \|b\|^2}$. Taking the supremum and the infimum over $b \in B$, and using the fact that $\|a\| \geq 0$, we obtain

$$\sup_{b \in B}\left[\|b\| - \sqrt{2f_c(b) + \|b\|^2}\right] \vee 0 \leq \|a\| \leq \inf_{b \in B}\left[\|b\| + \sqrt{2f_c(b) + \|b\|^2}\right].$$

On the other hand, it is immediate from the proof of (a) and Theorem 9.3(c) that there exists $a \in \operatorname{neg} g_0$ such that $a \in \operatorname{pos} f_c = \operatorname{pos} f - c$ and

$$\|a\| = \sup_{b \in B}\left[\|b\| - \sqrt{2f_c(b) + \|b\|^2}\right] \vee 0. \qquad \square$$

Remark 21.5. Let B be a SSDB space, $f \in \mathcal{PC}(B)$ be a BC–function and $c \in B$. Then, from Lemma 19.13, $f_c(t - c) = f(t - c + c) - \lfloor t - c, c \rfloor - q(c) = f(t) - \lfloor t, c \rfloor + q(c)$. Consequently,

$$\sup_{b \in B}\left[\|b\| - \sqrt{2f_c(b) + \|b\|^2}\right] = \sup_{t \in B}\left[\|t - c\| - \sqrt{2f_c(t - c) + \|t - c\|^2}\right]$$

$$= \sup_{t \in B}\left[\|t - c\| - \sqrt{2f(t) - 2\lfloor t, c \rfloor + 2q(c) + \|t - c\|^2}\right]$$

and

$$\inf_{b \in B}\left[\|b\| + \sqrt{2f_c(b) + \|b\|^2}\right]$$

$$= \inf_{t \in B}\left[\|t - c\| + \sqrt{2f(t) - 2\lfloor t, c \rfloor + 2q(c) + \|t - c\|^2}\right].$$

Though these alternative expressions for the supremum and infimum appearing in Theorem 21.4(c) are more coumbersome, they are sometimes useful since they are expressed in terms of f instead of f_c.

Remark 21.6. If E is a nonzero reflexive Banach space, we define the SSDB space B and its associated norm as in Remark 17.1 and Example 19.2(a). If now $f \in PCLSC(E)$ and, for all $(x, x^*) \in B$, $h(x, x^*) := f(x) + f^*(x^*)$ then the Fenchel–Moreau theorem, Corollary 12.4, implies that h is a BC–function. Since $\text{pos}\, h^@ = G(\partial f)$, Theorem 21.4(b) gives us that ∂f is maximally monotone. Compare this with Remark 18.9.

We will use Theorem 21.7 to obtain the following results for reflexive spaces: the characterization of the maximality of a monotone multifunction in Theorem 29.2, Rockafellar's surjectivity theorem in Theorem 29.5, Torralba's theorem in Theorem 29.9, and the characterization of the closures of the domain and range of a maximally monotone multifunction in Theorem 31.2.

Theorem 21.7. *Suppose that B is a SSDB space and A is a nonempty q–positive subset of B. Then*

$$A \text{ is maximally } q\text{–positive} \iff A - \text{neg}\, g_0 = B.$$

Proof. ($\Longleftarrow$) was proved in Lemma 21.3, while ($\Longrightarrow$) follows from (20.2) amd Theorem 21.4(a) with $f := \Phi_A$. $\qquad\square$

The following definition is natural.

Definition 21.8. Let B be a SSDB space and $h \in PC(B)$. We say that h is autoconjugate if $h^@ = h$ on B.

Lemma 21.9. *Let B be a SSDB space and $h \in PC(B)$ be autoconjugate. Then h is a BC–function and $\text{pos}\, h^@$ is maximally q–positive.*

Proof. From the the Fenchel–Young inequality, (8.2), for all $b \in B$,

$$h(b) = \tfrac{1}{2}h(b) + \tfrac{1}{2}h^@(b) \geq \tfrac{1}{2}\lfloor b, b \rfloor = q(b),$$

and the result follows from Definition 19.11 and Theorem 21.4(b). $\qquad\square$

Theorem 21.10 shows that some of the results established by Bauschke–Wang in [13] for "kernel averages" in spaces of the form $E \times E^*$ (where E is a reflexive Banach space) can be generalized naturally to SSDB spaces. It will be used in Theorem 21.11.

Theorem 21.10. *Let B be a SSDB space, $f \in PCLSC(B)$ and $f^@ \geq f$ on B. For all $b \in B$, let*

$$h(b) := \inf_{c \in B} \left[\tfrac{1}{2}f(b + c) + \tfrac{1}{2}f^@(b - c) + g_0(c) \right].$$

Then:
(a) h is autoconjugate.
(b) $f \vee q \leq h \leq f^@$ on B and $\text{pos}\, h$ is a maximally q–positive superset of $\text{pos}\, f^@$.
(c) $b \in \text{pos}\, h$ if, and only if,

$$\text{there exists } d \in \text{pos}\, g_0 \text{ such that } (b - d, b + d) \in G(\partial f).$$

Proof. (a) From the the Fenchel–Young inequality, (8.2), and (21.1), for all $b \in B$,

$$h(b) \geq \inf_{c \in B} \left[\tfrac{1}{2} \lfloor b + c, b - c \rfloor + g_0(c) \right]$$
$$= \inf_{c \in B} \left[q(b) - q(c) + g_0(c) \right] \geq q(b) > -\infty.$$

Thus Corollary 10.4 and the Fenchel–Moreau Theorem, Corollary 12.4, imply that, for all $b \in B$,

$$h^@(b) = \min_{d \in B} \left[\tfrac{1}{2} f^@(b + d) + \tfrac{1}{2} f^{@@}(b - d) + g_0(d) \right]$$
$$= \min_{d \in B} \left[\tfrac{1}{2} f^@(b + d) + \tfrac{1}{2} f(b - d) + g_0(d) \right] = h(b), \qquad (21.3)$$

and so h is autoconjugate.

(b) Since $f^@ \geq f$ and $g_0 \geq 0$ on B, for all $b \in B$, the convexity of f gives

$$h(b) \geq \inf_{c \in B} \left[\tfrac{1}{2} f(b + c) + \tfrac{1}{2} f(b - c) + g_0(c) \right] \geq \inf_{c \in B} [f(b) + 0] = f(b).$$

Thus $h \geq f$ on B, and it follows by taking conjugates and using Lemma 21.9 that $f^@ \geq h^@ = h \geq q$ on B, from which, pos $f^@ \subset$ pos h. Another application of Lemma 21.9 gives us that pos h is maximally q–positive.

(c) It is clear from (21.3) that $b \in$ pos $h =$ pos $h^@$ if, and only if,

there exists $d \in B$ such that $\tfrac{1}{2} f(b - d) + \tfrac{1}{2} f^@(b + d) = q(b) - g_0(d)$.

From the Fenchel–Young inequality and (20.1) again,

$$\tfrac{1}{2} f(b - d) + \tfrac{1}{2} f^@(b + d) \geq \tfrac{1}{2} \lfloor b - d, b + d \rfloor = q(b) - q(d) \geq q(b) - g_0(d).$$

Thus $b \in$ pos $h^@$ if, and only if, there exists $d \in B$ such that

$$\tfrac{1}{2} f(b - d) + \tfrac{1}{2} f^@(b + d) = \tfrac{1}{2} \lfloor b - d, b + d \rfloor \quad \text{and} \quad g_0(d) = q(d),$$

that is to say, $(b - d, b + d) \in G(\partial f)$ and $d \in$ pos g_0. $\qquad \square$

Now it is easily seen from Zorn's lemma that every nonempty q–positive subset of a SSD space has a maximally q–positive superset. In the next result, we give a somewhat more explicit description of such a superset in a SSDB space, which will be applied to monotone multifunctions on a reflexive space in Theorem 29.8.

Theorem 21.11. *Let B be a SSDB space, A be a nonempty q–positive subset of B and, for all $b \in B$*

$$h(b) := \inf_{c \in B} \left[\tfrac{1}{2} \Phi_A(b + c) + \tfrac{1}{2} \Phi_A{}^@(b - c) + g_0(c) \right].$$

Then:
(a) h is autoconjugate.
(b) $\Phi_A \vee q \leq h \leq \Phi_A{}^@$ on B and pos h is a maximally q–positive superset of A. If A is maximally q–positive then pos $h = A$.
(c) $b \in$ pos h if, and only if,

there exists $d \in$ pos g_0 such that $(b - d, b + d) \in G(\partial \Phi_A)$.

Proof. It is clear from (19.7) that Theorem 21.10 can be applied with $f := \Phi_A$. The result now follows from (19.8). □

The final result in this section is a "local transversality theorem". This is an adaptation of Theorem 19.16 and will be used explicitly in Theorem 30.1, which leads to a number of surjectivity results, including an abstract Hammerstein theorem.

Theorem 21.12. Let B be a SSDB space, $f \in \mathcal{PCLSC}(B)$ be a BC–function and $g \in \mathcal{PCLSC}(B)$ be a TBC–function (see Definition 19.14). Then

$$\text{int}\big(\text{pos}\, f - \text{neg}\, g\big) = \text{sur}\big(\text{dom}\, f - \text{dom}\, g\big). \tag{21.4}$$

Consequently, $\text{int}\big(\text{pos}\, f - \text{neg}\, g\big)$ is convex and $\text{sur}\big(\text{dom}\, f - \text{dom}\, g\big)$ is open.

Proof. Let c be an arbitrary element of $\text{sur}\big(\text{dom}\, f - \text{dom}\, g\big)$. From Lemma 19.13, $f_c\big(\in \mathcal{PCLSC}(B)\big)$ is a BC–function and $0 \in \text{sur}\big(\text{dom}\, f_c - \text{dom}\, g\big)$. Using (19.2) and (19.3),

$$b \in B \implies f_c(b) + g(b) \geq q(b) - q(b) = 0.$$

Thus the Attouch-Brezis theorem, Theorem 15.1, gives $a \in B$ such that $f_c{}^@(a) + g^@(-a) \leq 0$ so, from (19.2) and (19.3) again, $f_c(a) + g(a) \leq 0 = q(a) - q(a)$. From (19.2) and (19.3) for a third time, $f_c(a) \geq q(a)$ and $g(a) \geq -q(a)$. Consequently, $f_c(a) = q(a)$ and $g(a) = -q(a)$, that is to say, using Lemma 19.13, $a \in \text{pos}\, f_c = \text{pos}\, f - c$ and also $a \in \text{neg}\, g$. But then $c = (c+a) - a \in \text{pos}\, f - \text{neg}\, g$. Since this holds for all $c \in \text{sur}\big(\text{dom}\, f - \text{dom}\, g\big)$, we have established that $\text{pos}\, f - \text{neg}\, g \supset \text{sur}\big(\text{dom}\, f - \text{dom}\, g\big)$ and the inclusion "$\supset$" in (21.4) now follows from the $\ominus$–corollary, Corollary 14.3. The opposite inclusion is obvious. □

22 The SSD space $E \times E^*$

We now turn our attention to the space of Remark 17.1 and Example 19.2(a). So let E be a nonzero Banach space and E^* be its topological dual space. Then $B := E \times E^*$ is a SSD space with

$$\lfloor b, c \rfloor := \langle b_1, c_2 \rangle + \langle c_1, b_2 \rangle \quad \big(b = (b_1, b_2), c = (c_1, c_2) \in B\big).$$

As we have already observed in Remark 17.1,

$$q(b) = \tfrac{1}{2}\lfloor (b_1, b_2), (b_1, b_2) \rfloor = \tfrac{1}{2}\big[\langle b_1, b_2 \rangle + \langle b_1, b_2 \rangle\big] = \langle b_1, b_2 \rangle. \tag{22.1}$$

The main thrust of this section is to relate (by means of the projection map π_1) certain convex functions defined on B with certain convex functions defined on E. In this direction, we mention Theorem 22.5 and Theorem 22.8.

These results and the other lemmas in this section will have wide applications to the theory of monotone multifunctions.

We start off by giving an application of Rockafellar's version of the Fenchel duality theorem, Corollary 8.6. All the results of this section up to and including Theorem 22.8 flow from Lemma 22.1 by making various substitutions for its parameters.

Lemma 22.1. *Let E be a nonzero Banach space, $B := E \times E^*$, $f \in \mathcal{PC}(B)$, $x, y \in E$, $M \in \mathbb{R}$ and $K \geq 0$. Then*

$$\text{there exists } a \in (\pi_1)^{-1}x \text{ such that } \|a_2\| \leq K \text{ and } f^@(a) \leq M + \langle y, a_2 \rangle$$

if, and only if,

$$b \in B \quad \Longrightarrow \quad f(b) + M + K\|b_1 - y\| - \langle x, b_2 \rangle \geq 0.$$

Proof. ($\Longrightarrow$) Using the Fenchel–Young inequality, (8.2), for all $b \in B$,

$$
\begin{aligned}
f(b) + M + K\|b_1 - y\| - \langle x, b_2 \rangle &= f(b) + M + K\|b_1 - y\| - \langle a_1, b_2 \rangle \\
&\geq f(b) + M + \langle y - b_1, a_2 \rangle - \langle a_1, b_2 \rangle \\
&= f(b) + M + \langle y, a_2 \rangle - \lfloor b, a \rfloor \\
&\geq f(b) + f^@(a) - \lfloor b, a \rfloor \geq 0.
\end{aligned}
$$

($\Longleftarrow$) Let $g(b) := M + K\|b_1 - y\| - \langle x, b_2 \rangle$, so that $f + g \geq 0$ on B. Since g is $\mathcal{T}_{\mathcal{NW}}(B)$–continuous on B and $\mathcal{T}_{\mathcal{NW}}(B)$ is a compatible topology for the SSD space B, Corollary 8.6 gives $a \in B$ such that $f^@(a) + g^@(-a) \leq 0$. By direct computation,

$$
g^@(-a) = \begin{cases} -M - \langle y, a_2 \rangle & (a_1 = x \text{ and } \|a_2\| \leq K); \\ \infty & (\text{otherwise}). \end{cases}
$$

Thus $\pi_1 a = a_1 = x$ and $\|a_2\| \leq K$, and $f^@(a) \leq -g^@(-a) = M + \langle y, a_2 \rangle$. □

Lemma 22.2. *Let E be a nonzero Banach space, $B := E \times E^*$, $f \in \mathcal{PC}(B)$, $x \in E$ and $N \geq 0$. Then*

$$\text{there exists } a \in (\pi_1)^{-1}x \text{ such that } f^@(a) \vee \|a_2\| \leq N \tag{22.2}$$

if, and only if,

$$b \in B \quad \Longrightarrow \quad N(1 + \|b_1\|) \geq \langle x, b_2 \rangle - f(b). \tag{22.3}$$

Proof. Immediate from Lemma 22.1, with $M = K = N$ and $y = 0$. □

Definition 22.3. Let E be a nonzero Banach space, $B := E \times E^*$ and $f \in \mathcal{PC}(B)$. We define the function $f_{\div} \colon E \to \,]-\infty, \infty]$ by

$$f_{\div}(x) := \sup_{b=(b_1,b_2)\in B} \frac{\langle x, b_2 \rangle - f(b)}{1 + \|b_1\|}.$$

Since $f_{\div}$ is the supremum of a family of continuous affine functions, $f_{\div}$ is convex and lower semicontinuous. We will give in Theorem 22.5 a necessary and sufficient condition that $f_{\div} \in \mathcal{PCLSC}(E)$.

The following relationship between $f^@$ and $f_÷$ will be critical.

Lemma 22.4. *Let E be a nonzero Banach space, $B := E \times E^*$, $f \in \mathcal{PC}(B)$, and $x \in E$. Then*

$$f_÷(x) \vee 0 = \min_{a \in (\pi_1)^{-1}x} \left(f^@(a) \vee \|a_2\| \right). \tag{22.4}$$

Proof. Suppose first that $\inf_{a \in (\pi_1)^{-1}x} \left(f^@(a) \vee \|a_2\| \right) < \infty$. Let $N \in \mathbb{R}$ and $N > \inf_{a \in (\pi_1)^{-1}x} \left(f^@(a) \vee \|a_2\| \right)$. Then (22.2) is satisfied. From Lemma 22.2, (22.3) is satisfied. It follows by dividing by $1 + \|b_1\|$ and taking the supremum over $b \in B$ that $f_÷(x) \leq N$. Letting $N \to \inf_{a \in (\pi_1)^{-1}x} \left(f^@(a) \vee \|a_2\| \right)$, we obtain the inequality "$\leq$" in (22.4). We now write $N := f_÷(x) \vee 0$, and we will prove the inequality "$\geq$" in (22.4). For this, we can and will suppose that $N \in \mathbb{R}$. Then the definition of $f_÷$ implies that (22.3) is satisfied. Using Lemma 22.2 in the opposite direction, we obtain (22.2). This gives the inequality "$\geq$" in (22.4), and shows that the infimum is, in fact, a minimum. If, on the other hand, $\inf_{a \in (\pi_1)^{-1}x} \left(f^@(a) \vee \|a_2\| \right) = \infty$ then, for all $a \in (\pi_1)^{-1}x$, $f^@(a) = \infty$, and so Lemma 22.2 implies that, $f_÷(x) = \infty$. This implies (22.4), and that the infimum is attained in this case also. $\quad\square$

Theorem 22.5 and Lemma 22.7 will both be used explicitly in Theorem 26.1 in our result on the local boundedness of monotone multifunctions.

Theorem 22.5. *Let E be a nonzero Banach space, $B := E \times E^*$ and $f \in \mathcal{PC}(B)$. Then $\pi_1 \mathrm{dom}\, f^@ = \mathrm{dom}\, f_÷$. Consequently, $f_÷$ is proper if, and only if, $f^@$ is proper.*

Proof. $x \in \mathrm{dom}\, f_÷ \iff f_÷(x) \vee 0 < \infty$. Thus, from Lemma 22.4, $x \in \mathrm{dom}\, f_÷ \iff$ there exists $a \in (\pi_1)^{-1}x$ such that $f^@(a) \vee \|a_2\| < \infty$ $\iff x \in \pi_1 \mathrm{dom}\, f^@$. $\quad\square$

Corollary 22.6 will be used explicitly in the "> 6 set theorem" for pairs of maximally monotone multifunctions on a reflexive Banach space in Theorem 33.1.

Corollary 22.6. *Let E be a nonzero Banach space, $B := E \times E^*$ and $f, g \in \mathcal{PC}(B)$. Then*

$$\mathrm{sur}\left[\pi_1 \mathrm{dom}\, f^@ - \pi_1 \mathrm{dom}\, g^@ \right] = \mathrm{int}\left[\pi_1 \mathrm{dom}\, f^@ - \pi_1 \mathrm{dom}\, g^@ \right].$$

Consequently, $\mathrm{sur}\left[\pi_1 \mathrm{dom}\, f^@ - \pi_1 \mathrm{dom}\, g^@ \right]$ is open.

Proof. We can and will assume that $f^@$ and $g^@$ are proper. Theorem 22.5 gives us that $f_÷, g_÷ \in \mathcal{PCLSC}(B)$ and then that

$$\pi_1 \mathrm{dom}\, f^@ - \pi_1 \mathrm{dom}\, g^@ = \mathrm{dom}\, f_÷ - \mathrm{dom}\, g_÷.$$

The result now follows from the "$\ominus$–corollary", Corollary 14.3. $\quad\square$

Lemma 22.7. *Let E be a nonzero real Banach space, $B := E \times E^*$, $f \in \mathcal{PC}(B)$ and $z \in \operatorname{sur} \operatorname{dom} f_{\div}$. Then there exist $K > 0$ and $\eta \in \,]0, 1[$ such that*

$$\left.\begin{array}{l} \|x - z\| \leq \eta \text{ and } b \in B \\ \qquad \Longrightarrow f(b) + K\|b_1 - x\| - \langle x, b_2 \rangle \geq \eta\big(\|b_2\| - K\big) \end{array}\right\} \quad (22.5)$$

and

$$b \in B, \ \|b_1 - z\| \leq \eta \text{ and } f(b) \leq q(b) \quad \Longrightarrow \quad \|b_2\| \leq K. \quad (22.6)$$

Proof. The dom lemma, Lemma 13.3 gives $\eta \in \,]0, 1[$ and $N > 0$ such that

$$y \in E \text{ and } \|y\| \leq 2\eta \quad \Longrightarrow \quad f_{\div}(z + y) \leq N,$$

from which $b \in B \Longrightarrow f(b) + N(1 + \|b_1\|) - \langle z + y, b_2 \rangle \geq 0$. Taking the infimum over y, we have

$$b \in B \Longrightarrow f(b) + N(1 + \|b_1\|) - \langle z, b_2 \rangle - 2\eta\|b_2\| \geq 0$$
$$\Longrightarrow f(b) + N(1 + \|z\| + \|b_1 - z\|) - \langle z, b_2 \rangle - 2\eta\|b_2\| \geq 0.$$

Now suppose that $\|x - z\| \leq \eta$. Since $\eta < 1$, we derive that

$$b \in B \Longrightarrow f(b) + N(2 + \|z\| + \|b_1 - x\|) - \langle x, b_2 \rangle - \eta\|b_2\| \geq 0,$$

and (22.5) now follows by setting $K := (2 + \|z\|)N/\eta \geq N$. If we substitute $x = b_1$ in (22.5), we obtain

$$b \in B \text{ and } \|b - z\| \leq \eta \quad \Longrightarrow \quad f(b) - \langle b_1, b_2 \rangle \geq \eta\big(\|b_2\| - K\big),$$

and (22.6) now follows since, from (22.1), $\langle b_1, b_2 \rangle = q(b)$. □

Theorem 22.8(a) will be used explicitly in the "six set theorem" for maximally monotone multifunctions, Theorem 27.1, and Theorem 22.8(b) will be used explicitly in the proof in Theorem 25.1 that maximally monotone multifunctions with bounded range have full domain.

Theorem 22.8. *Let E be a nonzero Banach space, $B := E \times E^*$, $f \in \mathcal{PC}(B)$, $f \geq q$ and $f^@ \geq q$ on B. Then:*

(a)
$$\operatorname{int} \pi_1 \operatorname{pos} f^@ = \operatorname{int} \pi_1 \operatorname{dom} f^@ = \operatorname{int} \operatorname{dom} f_{\div}$$
$$= \operatorname{sur} \pi_1 \operatorname{pos} f^@ = \operatorname{sur} \pi_1 \operatorname{dom} f^@ = \operatorname{sur} \operatorname{dom} f_{\div}.$$

Consequently, $\operatorname{int} \pi_1 \operatorname{pos} f^@$ is convex and $\operatorname{sur} \pi_1 \operatorname{pos} f^@$ is open.
(b) *Suppose now that, for all $x \in E$, there exists $K \geq 0$ such that*

$$b \in B \Longrightarrow f(b) + K(\|x\| + \|b_1\|) - \langle x, b_2 \rangle \geq 0.$$

Then $\pi_1 \operatorname{pos} f^@ = E$.

Proof. (a) From Theorem 22.5, $\pi_1 \text{pos}\, f^@ \subset \pi_1 \text{dom}\, f^@ = \text{dom}\, f_\div$, hence $\text{int}\, \pi_1 \text{pos}\, f^@ \subset \text{int}\, \pi_1 \text{dom}\, f^@ = \text{int}\, \text{dom}\, f_\div$ and $\text{sur}\, \pi_1 \text{pos}\, f^@ \subset \text{sur}\, \pi_1 \text{dom}\, f^@ = \text{sur}\, \text{dom}\, f_\div$. Since $\text{int}\, (\ldots) \subset \text{sur}\, (\ldots)$, it remains to prove that $\text{sur}\, \text{dom}\, f_\div \subset \text{int}\, \pi_1 \text{pos}\, f^@$. So suppose that $z \in \text{sur}\, \text{dom}\, f_\div$, and choose K and η as in Lemma 22.7. Let $\|x - z\| < \eta$. We shall prove that

$$b \in B \quad \Longrightarrow \quad f(b) + K\|b_1 - x\| - \langle x, b_2 \rangle \geq 0. \qquad (22.7)$$

This is immediate from (22.5) if $\|b_2\| \geq K$. If, on the other hand, $\|b_2\| < K$ then, since $f(b) \geq q(b) = \langle b_1, b_2 \rangle$,

$$f(b) + K\|b_1 - x\| - \langle x, b_2 \rangle \geq K\|b_1 - x\| + \langle b_1 - x, b_2 \rangle \geq 0,$$

which completes the proof of (22.7). From Lemma 22.1 with $y = x$ and $M = 0$, there exists $a \in (\pi_1)^{-1}x$ such that $f^@(a) \leq \langle x, a_2 \rangle$, that is to say $f^@(a) - q(a) \leq 0$. But then $a \in \text{pos}\, f^@$, and so $x \in \pi_1 \text{pos}\, f^@$. Since this holds whenever $\|x - z\| < \eta$, this establishes that $z \in \text{int}\, \text{pos}\, f^@$, and completes the proof of (a).

(b) Let x be an arbitrary element of E and $M := K(\|x\| \vee 1)$. Then

$$b \in B \Longrightarrow M(1 + \|b_1\|) \geq K(\|x\| + \|b_1\|) \geq \langle x, b_2 \rangle - f(b).$$

Consequently, $f_\div(x) \leq M$, from which $x \in \text{dom}\, f_\div$. Since this holds for all $x \in E$, $\text{dom}\, f_\div = E$ and so, from (a), $\pi_1 \text{pos}\, f^@ = E$. This completes the proof of (b). $\qquad \square$

The final result in this section will be used in our treatment of the sum theorem for maximally monotone operators in Theorem 24.1. We follow the notation of Notation 16.1. That is to say, we norm B by $\|(b_1, b_2)\| = \sqrt{\|b_1\|^2 + \|b_2\|^2}$. Then the dual of B is $B^* = E^{**} \times E^*$ under the pairing

$$\langle b, v \rangle := \langle b_1, v_2 \rangle + \langle b_2, v_1 \rangle \quad (b = (b_1, b_2) \in B, \; v = (v_1, v_2) \in B^*).$$

Though B is a SSD space and $q(x, x^*) = \langle x, x^* \rangle$, $\mathcal{T}_{\|\ \|}(B)$ is not a compatible topology unless E is reflexive. We note then that if $f \in \mathcal{PC}(B)$ then, for all $b = (b_1, b_2) \in B$,

$$f^@(b) = \sup_{d \in B} \left[\lfloor d, b \rfloor - f(d) \right] = \sup_{d \in B} \left[\langle d, (\widehat{b_1}, b_2) \rangle - f(d) \right] = f^*(\widehat{b_1}, b_2). \qquad (22.8)$$

We recall that *BC–functions* were defined in Definition 19.11, and the binary operator $\oplus_2$ was defined in Definition 16.3. Part of Lemma 22.9 can also be deduced from [67, Corollary 3.7].

Lemma 22.9. *Let E be a nonzero Banach space, $B := E \times E^*$, $f, g \in \mathcal{PCLSC}(B)$ be BC–functions and*

$$\bigcup_{\lambda > 0} \lambda [\pi_1 \text{dom}\, f - \pi_1 \text{dom}\, g] \; \text{be a closed subspace of } E.$$

Then $f \oplus_2 g$ is a BC–function. Furthermore, $b \in \text{pos}\, (f \oplus_2 g)^@ = \text{pos}\, (f \oplus_2 g)$ if, and only if, there exist $a \in \text{pos}\, f^@ = \text{pos}\, f$ and $c \in \text{pos}\, g^@ = \text{pos}\, g$ such that $a_1 = c_1 = b_1$ and $a_2 + c_2 = b_2$.

Proof. Let $h := f \oplus_2 g$. For all $b = (b_1, b_2) \in B$, we have

$$h(b) = \inf \{f(a) + g(c) \colon a, c \in B, \ a_1 = c_1 = b_1, \ a_2 + c_2 = b_2\}. \qquad (22.9)$$

Since $f \geq q$ and $g \geq q$ on B, (22.9) implies that, for all $b \in B$,

$$h(b) \geq \inf \{q(a) + q(c) \colon a, c \in B, \ a_1 = c_1 = b_1, \ a_2 + c_2 = b_2\} = q(b).$$

From Theorem 16.4(a), (22.8) and the fact that $f^@ \geq f$ and $g^@ \geq g$ on B,

$$\begin{aligned}
h^@(b) &= h^*(\widehat{b_1}, b_2) \\
&= \min \{f^*(u) + g^*(w) \colon u, w \in B^*, \ u_1 = w_1 = \widehat{b_1}, \ u_2 + w_2 = v_2\} \\
&= \min \{f^@(a) + g^@(c) \colon a, c \in B, \ a_1 = c_1 = b_1, \ a_2 + c_2 = b_2\} \qquad (22.10) \\
&\geq \inf \{f(a) + g(c) \colon a, c \in B, \ a_1 = c_1 = b_1, \ a_2 + c_2 = b_2\} = h(b).
\end{aligned}$$

Thus h is a BC–function, and the result follows from (22.10) and Lemma 19.12. $\qquad \square$

Remark 22.10. Theorem 16.4(a) actually implies that (22.10) can be improved to: for all $v = (v_1, v_2) \in B^* = E^{**} \times E^*$,

$$(f \oplus_2 g)^*(v) = \min \{f^*(u) + g^*(w) \colon u, w \in B^*, \ u_1 = w_1 = v_1, \ u_2 + w_2 = v_2\}.$$

Remark 22.11. Let E be a nonzero Banach space, $B := E \times E^*$, $f, g \in \mathcal{PCLSC}(B)$ be BC–functions and

$$\bigcup_{\lambda > 0} \lambda [\pi_2 \mathrm{dom}\, f - \pi_2 \mathrm{dom}\, g] \text{ be a closed subspace of } E^*.$$

Then, arguing as in Lemma 22.9, $f \oplus_1 g \geq q$ on B and, using Theorem 16.4(b), for all $v = (v_1, v_2) \in B^* = E^{**} \times E^*$,

$$(f \oplus_1 g)^*(v) = \min \{f^*(u) + g^*(w) \colon u, w \in B^*, \ u_1 + w_1 = v_1, \ u_2 = w_2 = v_2\}.$$

Hovever, we have the following problem:

Problem 22.12. Let E be a nonzero Banach space, $B := E \times E^*$, $f, g \in \mathcal{PCLSC}(B)$ be BC–functions and

$$\bigcup_{\lambda > 0} \lambda [\pi_2 \mathrm{dom}\, f - \pi_2 \mathrm{dom}\, g] \text{ be a closed subspace of } E^*.$$

Then is $f \oplus_1 g$ a BC–function? (From Remark 22.11, for all $b \in B$,

$$(f \oplus_1 g)^@(b) = \min \{f^*(u) + g^*(w) \colon u, w \in B^*, \ u_1 + w_1 = \widehat{b_1}, \ u_2 = w_2 = v_2\}.$$

The obstruction is that it is possible that $u_1, w_1 \in E^{**} \setminus \widehat{E}$.)

23 Fitzpatrick functions and fitzpatrifications

We continue with the situation of Section 22 and Example 19.2(a). We first discuss how some results from Sections 19 and 20 translate into our present situation.

Lemma 23.1 appears in Burachik–Svaiter, [33, Theorem 3.1, pp. 2381–2382] and Penot, [66, Proposition 4(h)$\Longrightarrow$(a), pp. 860–861]. See also Fitzpatrick, [42, Section 2]. Monotone multifunctions S of the special form described in Lemma 23.1 have been investigated thoroughly in Martínez-Legaz–Svaiter, [61], and subsequently in [108]. We also refer the reader to [61] and [108] for more results in the finite–dimensional case.

Lemma 23.1. *Let E be a nonzero Banach space, $f: E \times E^* \to]-\infty, \infty]$ be proper and convex and, for all $(x, x^*) \in E \times E^*$, $f(x, x^*) \geq \langle x, x^* \rangle$. Let*

$$\operatorname{pos} f = \big\{ (x, x^*) \in E \times E^*\colon f(x, x^*) = \langle x, x^* \rangle \big\} \neq \emptyset.$$

Then the multifunction defined by $G(S) := \operatorname{pos} f$ is monotone.

Proof. This is immediate from Lemma 19.8.   □

As in Definition 19.11, and using (22.1), we say that $f: E \times E^* \to]-\infty, \infty]$ is a *BC–function* if f is proper and convex and

$$(x, x^*) \in E \times E^* \quad \Longrightarrow \quad f^@(x, x^*) \geq f(x, x^*) \geq \langle x, x^* \rangle. \tag{23.1}$$

There are extensive discussions in [33] and Penot–Zălinescu, [67], of lower semicontinuous convex functions $f: E \times E^* \to]-\infty, \infty]$ such that

$$(x, x^*) \in E \times E^* \quad \Longrightarrow \quad f(x, x^*) \geq \langle x, x^* \rangle \text{ and } f^@(x, x^*) \geq \langle x, x^* \rangle.$$

Clearly, any lower semicontinuous BC–function has this property. As examples of the additional benefits of this BC–function assumption, it follows from Lemma 19.12 that *if E is a nonzero Banach space and $f: E \times E^* \to]-\infty, \infty]$ is a BC–function then $\operatorname{pos} f = \operatorname{pos} f^@$* and, from Theorem 21.4, that *if E is a nonzero reflexive Banach space then the multifunction defined by $G(S) := \operatorname{pos} f$ is maximally monotone.* Other examples have appeared in Theorem 21.12 and Lemma 22.9. We will see more examples in Theorem 30.1, Theorem 30.6, Lemma 34.2, and Theorem 34.3. See also Simons–Zălinescu, [109, Theorem 1.4(a), p. 4] or Penot–Zălinescu, [67, Proposition 2.1].

As in Definition 20.2, we say that $f: E \times E^* \to]-\infty, \infty]$ is a *SBC–function* if f is proper and convex, for all $(x, x^*) \in E \times E^*$, $f(x, x^*) \geq \langle x, x^* \rangle$ and

$$f(x, x^*) = \sup\nolimits_{(y, y^*) \in \operatorname{pos} f} \big[\langle x, y^* \rangle + \langle y, x^* \rangle - \langle y, y^* \rangle \big].$$

A SBC–function is automatically a lower semicontinuous BC–function. Example 19.20 (with $E = \mathbb{R}$) provides an example of a BC–function that is not a SBC–function.

Let $S\colon E \rightrightarrows E^*$ be nontrivial. As observed in Remark 17.1, S is monotone exactly when $G(S)$ is q–positive. It follows that

$$S \text{ is maximally monotone} \iff G(S) \text{ is maximally } q\text{–positive}. \qquad (23.2)$$

Definition 23.2. Let E be a nonzero Banach space and $S\colon E \rightrightarrows E^*$ be nontrivial and monotone. We define the *Fitzpatrick function* $\varphi_S\colon E \times E^* \to \,]-\infty, \infty]$ associated with S by

$$\varphi_S(x, x^*) := \sup_{(s,s^*)\in G(S)} \big[\langle s, x^*\rangle + \langle x, s^*\rangle - \langle s, s^*\rangle\big]. \qquad (23.3)$$

(The function φ_S was introduced by Fitzpatrick in [42, Definition 3.1, p. 61] under the notation L_S. φ_S has been computed in a number of important cases by Bauschke, McLaren and Sendov in [12].) It is clear from (22.1) that

$$\varphi_S = \Phi_{G(S)},$$

a fact which we will use without justification from now on. (19.5), (19.7), (19.8) and (19.10) translate to:

$$s \in G(S) \quad \Longrightarrow \quad \varphi_S(s) = \langle s_1, s_2\rangle, \qquad (23.4)$$

$$b \in E \times E^* \quad \Longrightarrow \quad \varphi_S{}^@(b) \geq \varphi_S(b) \vee \langle b_1, b_2\rangle, \qquad (23.5)$$

$$G(S) \subset \operatorname{pos}\varphi_S{}^@ \quad \text{and} \quad G(S) \subset \operatorname{co} G(S) \subset \operatorname{dom}\varphi_S{}^@ \subset \operatorname{dom}\varphi_S. \qquad (23.6)$$

$\big($See [42, Theorem 3.4, p. 61 and Proposition 4.2, p. 63]$\big)$. (19.9) gives us that

$$\varphi_S{}^@ \in \mathcal{PC}(E \times E^*) \quad \text{and} \quad \varphi_S{}^{@@} = \varphi_S. \qquad (23.7)$$

In fact, [108, Theorem 6.5(a), p. 309] implies that $\operatorname{pos}\varphi_S{}^@$ is the graph of the largest monotone multifunction $T\colon E \rightrightarrows E^*$ such that $\varphi_T = \varphi_S$ on $E \times E^*$.

Definition 23.3. Let E be a nonzero Banach space and $S\colon E \rightrightarrows E^*$ be nontrivial and monotone. We define the multifunction $S_\varphi\colon E \rightrightarrows E^*$ by $G(S_\varphi) = \operatorname{dom}\varphi_S$. This was introduced in [106] under the name of the *fitpatrification* of S. In general, S_φ is not monotone but it does, of course, have a convex graph. We note then that

$$D(S_\varphi) = \pi_1\operatorname{dom}\varphi_S \quad \text{and} \quad R(S_\varphi) = \pi_2\operatorname{dom}\varphi_S. \qquad (23.8)$$

The elementary results contained in Lemma 23.4 will be used explicitly in Theorem 27.1, Theorem 27.3, Theorem 27.5, Theorem 27.6, Lemma 28.1, Theorem 28.6, Theorem 32.2, Corollary 32.3, Theorem 33.1, Theorem 33.2, Corollary 34.4, Theorem 37.1 and Theorem 39.1.

Lemma 23.4. *Let E be a nonzero Banach space and $S\colon E \rightrightarrows E^*$ be nontrivial and monotone. Then*

$$G(S) \subset \operatorname{co} G(S) \subset \operatorname{dom} \varphi_S{}^@ \subset \operatorname{dom} \varphi_S = G(S_\varphi),$$

$$D(S) \subset \operatorname{co} D(S) \subset \pi_1\big(\operatorname{dom} \varphi_S{}^@\big) \subset D(S_\varphi)$$

and

$$R(S) \subset \operatorname{co} R(S) \subset \pi_2\big(\operatorname{dom} \varphi_S{}^@\big) \subset R(S_\varphi).$$

Proof. The first observation is immediate from (23.6) and the definition of S_φ, and the second and third follow from the first.   □

Definition 23.5. The *duality map* $J\colon E \rightrightarrows E^*$ is defined by:

$$x^* \in Jx \iff \tfrac{1}{2}\|x\|^2 + \tfrac{1}{2}\|x^*\|^2 = \langle x, x^* \rangle.$$

Since $J = \partial\big(\tfrac{1}{2}\| \cdot \|^2\big)$, Theorem 18.7 tells us that J is maximally monotone. Now let $(x, x^*) \in E \times E^*$ and $(s, s^*) \in G(J)$. Then

$$\langle s, x^* \rangle + \langle x, s^* \rangle - \langle s, s^* \rangle \leq \|s\|\|x^*\| + \|x\|\|s^*\| - \tfrac{1}{2}\|s\|^2 - \tfrac{1}{2}\|s^*\|^2$$

$$\leq \tfrac{1}{2}\|x\|^2 + \tfrac{1}{2}\|x^*\|^2.$$

Thus, from (23.3), $\varphi_J(x, x^*) \leq \tfrac{1}{2}\|x\|^2 + \tfrac{1}{2}\|x^*\|^2$ and so

$$G(J_\varphi) = \operatorname{dom} \varphi_J = E \times E^*. \tag{23.9}$$

So, even if S is maximally monotone, $G(S_\varphi)$ can be much larger than $G(S)$.

It is clear from (20.1) that if S is nontrivial and monotone then S is maximally monotone if, and only if,

$$b \in E \times E^* \text{ and } \varphi_S(b) \leq q(b) \implies b \in G(S). \tag{23.10}$$

Furthermore, if $\varphi_S \geq q$ on $E \times E^*$ then S has a unique maximally monotone extension. See Martínez-Legaz–Svaiter, [61, Lemma 38, p. 43–44]. Three descriptions of this extension can be found in [61, Proposition 36 and Lemma 37, pp. 42–43]. See also [108, Lemma 10.2, p. 313 and Theorem 10.5(c), p. 314].

We now consider how some of the above results can be strengthened for maximally monotone multifunctions. If S is maximally monotone then (20.2) and (20.4) imply that (23.4), (23.5) and part of (23.6) can be strengthened to

$$\varphi_S \text{ is a SBC–function} \quad \text{and} \quad \operatorname{pos} \varphi_S{}^@ = \operatorname{pos} \varphi_S = G(S) \tag{23.11}$$

(see Definition 20.2). (See [42, Corollary 3.9, p. 62].) Example 19.20 provides an example of a BC–function that is not the Fitzpatrick function of any maximally monotone multifunction.

Our next result is part of the folklore of the theory of maximal monotonicity. We recall that $T_{\mathcal{NW}}(E \times E^*)$ was defined in Example 19.2(a) to be the topology $T_{\| \, \|}(E) \times w(E^*, E)$ on $E \times E^*$.

Lemma 23.6 *Let E be a nonzero Banach space, $S: E \rightrightarrows E^*$ be maximally monotone, $\{s_\gamma\}$ be a bounded net of elements of $G(S)$, $b \in E \times E^*$ and $s_\gamma \to b$ in $T_{\mathcal{NW}}(E \times E^*)$. Then $b \in G(S)$.*

Proof. (23.11) implies that, for all γ, $\varphi_S(s_\gamma) = q(s_\gamma)$. It is easily seen that φ_S is $T_{\mathcal{NW}}(E \times E^*)$–lower semicontinuous and $q(s_\gamma) \to q(b)$. Thus, passing to the limit, $\varphi_S(b) \leq q(b)$. The result follows from (23.10). $\square$

The following result can be deduced from Theorem 20.3: *Let E be a nonzero Banach space. Then the mapping $S \mapsto \varphi_S$ is a bijection from the maximally monotone multifunctions $E: \rightrightarrows E^*$ onto the SBC–functions on $E \times E^*$. The inverse mapping is pos $(\cdot)$, via the identification of Lemma 23.1.* See also Fitzpatrick, [42, Theorem 3.8, p. 62] and Martínez-Legaz–Théra, [63, Theorem 2].

In fact, φ_S can be defined as in (23.3) for any nontrivial $S: E \rightrightarrows E^*$. It is then easy to see that S is monotone $\iff$ $\varphi_S \leq q$ on $G(S)$. This and many subtler criteria for monotonicity in terms of Fitzpatrick functions can be found in Penot, [66, Proposition 4, p. 860–861] and Martínez-Legaz–Svaiter, [61, Proposition 2, p. 25].

Notation 23.7. Let E and F be nonzero Banach spaces, $L: F \to E$ be continuous and linear, and $S: E \rightrightarrows E^*$ be monotone. We define $L^*SL: F \rightrightarrows F^*$ by $(L^*SL)\sigma := L^*\big(S(L\sigma)\big)$.

Our next result will be used in our work on the interaction between maximal monotonicity, continuous linear maps and closed subspaces in Theorem 28.9.

Lemma 23.8. *Let E and F be nonzero Banach spaces, $L: F \to E$ be continuous and linear, and $S: E \rightrightarrows E^*$ be monotone. Let $\Sigma := L^*SL: F \rightrightarrows F^*$. Then Σ is monotone. Now suppose that $\xi \in F$, $x^* \in E^*$ and $D(S) \subset L(F)$. Then*

$$\varphi_\Sigma(\xi, L^*x^*) = \varphi_S(L\xi, x^*). \qquad (23.12)$$

Proof. If $(\xi, \xi^*), (\eta, \eta^*) \in G(\Sigma)$ then we can choose $x^* \in S(L\xi)$ and $y^* \in S(L\eta)$ such that $\xi^* = L^*x^*$ and $\eta^* = L^*y^*$. Then the inequality

$$\langle \xi - \eta, \xi^* - \eta^* \rangle = \langle \xi - \eta, L^*x^* - L^*y^* \rangle = \langle L\xi - L\eta, x^* - y^* \rangle \geq 0$$

establishes the monotonicity of Σ.

Now let $\xi \in F$ and $x^* \in E^*$. If $(\sigma, \sigma^*) \in G(\Sigma)$ then there exists $s^* \in S(L\sigma)$ such that $\sigma^* = L^*s^*$, which implies that $\langle \xi, \sigma^* \rangle = \langle L\xi, s^* \rangle$ and $\langle \sigma, \sigma^* \rangle = \langle L\sigma, s^* \rangle$. Thus, using the fact that $(L\sigma, s^*) \in G(S)$,

$$\langle \xi, \sigma^* \rangle + \langle \sigma, L^*x^* \rangle - \langle \sigma, \sigma^* \rangle = \langle L\xi, s^* \rangle + \langle L\sigma, x^* \rangle - \langle L\sigma, s^* \rangle \leq \varphi_S(L\xi, x^*),$$

and the inequality "$\leq$" in (23.12) follows by taking the supremum over $(\sigma, \sigma^*) \in G(\Sigma)$.

Now suppose, in addition, that $D(S) \subset L(F)$. If $(s, s^*) \in G(S)$ then $s \in D(S)$, and so there exists $\sigma \in F$ such that $s = L\sigma$, which implies that $\langle s, x^* \rangle = \langle \sigma, L^*x^* \rangle$ and $\langle s, s^* \rangle = \langle \sigma, L^*s^* \rangle$. Thus, using the fact that $(\sigma, L^*s^*) \in G(\Sigma)$,

$$\langle L\xi, s^* \rangle + \langle s, x^* \rangle - \langle s, s^* \rangle = \langle \xi, L^*s^* \rangle + \langle \sigma, L^*x^* \rangle - \langle \sigma, L^*s^* \rangle \leq \varphi_\Sigma(\xi, L^*x^*),$$

and the inequality "$\geq$" in (23.12) follows by taking the supremum over $(s, s^*) \in G(S)$. This completes the proof of the lemma. □

The final result in this section is in preparation for our discussion of the sum problem in Theorem 24.1 and the Brezis–Haraux condition in Theorem 31.4. We recall that the binary operation $\oplus_2$ was defined in Definition 16.3.

Lemma 23.9. *Let E be a nonzero Banach space, $S, T \colon E \rightrightarrows E^*$ be monotone and $D(S) \cap D(T) \neq \emptyset$. Then*
(a) *$\varphi_S \oplus_2 \varphi_T \geq \varphi_{S+T}$ on $E \times E^*$.*
(b) *$R(S_\varphi + T_\varphi) \subset R((S+T)_\varphi)$.*

Proof. (a) Let $(x, x^*) \in E \times E^*$, $u^*, v^* \in E^*$ and $u^* + v^* = x^*$. Let (y, y^*) be an arbitrary element of $G(S+T)$. Then there exist $s^* \in Sy$ and $t^* \in Ty$ such that $s^* + t^* = y^*$. Thus we have

$$\varphi_S(x, u^*) + \varphi_T(x, v^*)$$
$$\geq \left[\langle x, s^* \rangle + \langle y, u^* \rangle - \langle y, s^* \rangle \right] + \left[\langle x, t^* \rangle + \langle y, v^* \rangle - \langle y, t^* \rangle \right]$$
$$= \langle x, s^* + t^* \rangle + \langle y, u^* + v^* \rangle - \langle y, s^* + t^* \rangle$$
$$= \langle x, y^* \rangle + \langle y, x^* \rangle - \langle y, y^* \rangle.$$

Taking the infimum over u^*, v^* and the supremum over (y, y^*) now gives us that $(\varphi_S \oplus_2 \varphi_T)(x, x^*) \geq \varphi_{S+T}(x, x^*)$. This completes the proof of (a).

(b) Let x^* be an arbitrary element of $R(S_\varphi + T_\varphi)$. Then there exist $x \in E$, $p^* \in S_\varphi(x)$ and $q^* \in T_\varphi(x)$ such that $p^* + q^* = x^*$. Consequently, $\varphi_S(x, p^*) < \infty$ and $\varphi_T(x, q^*) < \infty$. But then, using (a),

$$\varphi_{S+T}(x, x^*) \leq (\varphi_S \oplus_2 \varphi_T)(x, x^*) \leq \varphi_S(x, p^*) + \varphi_T(x, q^*) < \infty,$$

and so $x^* \in (S+T)_\varphi x \subset R((S+T)_\varphi)$. This completes the proof of (b). □

24 The maximal monotonicity of a sum

Let C and D be two closed disks in the plane that touch at the point p, as in the diagram below.

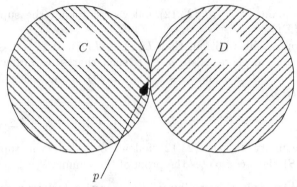

Then N_C and N_D $\big($see (17.2)$\big)$ are maximally monotone. By (17.1),

$$(N_C + N_D)(p) = N_C(p) + N_D(p),$$

which can be represented by the diagram

which is a proper subset of $\mathbb{R}^2$. Thus $G(N_C + N_D)$ is a proper subset of

$$G(N_{\{p\}}) = \{p\} \times \mathbb{R}^2,$$

and so $N_C + N_D$ is not maximally monotone. (See Phelps, [68, p. 54].) If $S\colon E \rightrightarrows E^*$ and $T\colon E \rightrightarrows E^*$ are nontrivial and monotone and $D(S) \cap D(T) \neq \emptyset$ then $S+T$ is obviously nontrivial and monotone. The above example shows that if S and T are maximally monotone and $D(S) \cap D(T) \neq \emptyset$ then $S + T$ is *not* necessarily maximally monotone. Determining conditions on S and T (normally called "constraint qualifications") that ensure that $S + T$ is maximally monotone is one of the fundamental questions in the theory of monotone multifunctions. The original milestone result due to Rockafellar (see [80, Theorem 1, p. 76]) was that *if E is reflexive, S, T are maximally monotone and*

$$D(S) \cap \operatorname{int} D(T) \neq \emptyset \qquad (24.1)$$

then $S+T$ is maximally monotone. We will give generalizations of this result in Theorem 24.1(a), Theorem 32.2 and Corollary 32.3. If E is nonreflexive the situation is very tantalizing — it is still unknown whether (24.1) suffices to ensure the maximal monotonicity of $S + T$. We will deal with this case in greater detail in Chapter VII. Theorem 24.1(b), which is based on an observation of Jonathan Borwein, will be the first step in our discussion of this case.

Theorem 24.1. *Let E be a nonzero Banach space, $S, T\colon E \rightrightarrows E^*$ be maximally monotone and $\bigcup_{\lambda>0} \lambda\big[D(S_\varphi) - D(T_\varphi)\big]$ be a closed subspace of E.*
(a) If E is reflexive then $S + T$ is maximally monotone.
(b) Even if E is not reflexive, if $\varphi_{S+T} \geq q$ on $E \times E^*$ then $S+T$ is maximally monotone.
(c) Even if E is not reflexive, if

$$b \in E \times E^* \text{ and } \varphi_{S+T}(b) \leq q(b) \quad \Longrightarrow \quad b_1 \in D(S+T) \qquad (24.2)$$

then $S + T$ is maximally monotone.

Proof. We know from (23.11) that φ_S and φ_T are BC–functions,

$$\operatorname{pos} \varphi_S^{@} = \operatorname{pos} \varphi_S = G(S) \quad \text{and} \quad \operatorname{pos} \varphi_T^{@} = \operatorname{pos} \varphi_T = G(T).$$

(23.8) implies that $D\big(S_\varphi\big) = \pi_1 \mathrm{dom}\, \varphi_S$ and $D\big(T_\varphi\big) = \pi_1 \mathrm{dom}\, \varphi_T$, and so Lemma 22.9 gives us that $\varphi_S \oplus_2 \varphi_T$ is a BC–function, and

$$b \in \operatorname{pos}\big(\varphi_S \oplus_2 \varphi_T\big)^{@} = \operatorname{pos}\big(\varphi_S \oplus_2 \varphi_T\big)$$

if, and only if, there exist $a \in G(S)$ and $c \in G(T)$ such that $a_1 = c_1 = b_1$ and $a_2 + c_2 = b_2$. In other words,

$$\operatorname{pos}\big(\varphi_S \oplus_2 \varphi_T\big)^{@} = G(S+T) = \operatorname{pos}\big(\varphi_S \oplus_2 \varphi_T\big). \qquad (24.3)$$

(a) If E is reflexive then Theorem 21.4(b), with $f := \varphi_S \oplus_2 \varphi_T$ implies that $G(S+T)$ is a maximally monotone subset of $E \times E^*$, and so the multifunction $S + T$ is maximally monotone.

(b) Even if E is not reflexive, (23.5) and the assumption that $\varphi_{S+T} \geq q$ on $E \times E^*$ imply that $D(S) \cap D(T) \neq \emptyset$ and φ_{S+T} is a BC–function. From Lemma 23.9(a), $\varphi_S \oplus_2 \varphi_T \geq \varphi_{S+T}$ on $E \times E^*$. Thus we have the string of inequalities

$$\varphi_{S+T}^{@} \geq \big(\varphi_S \oplus_2 \varphi_T\big)^{@} \geq \varphi_S \oplus_2 \varphi_T \geq \varphi_{S+T} \geq q \quad \text{on} \quad E \times E^*.$$

It is clear from this and (24.3) that

$$\operatorname{pos} \varphi_{S+T}^{@} \subset \operatorname{pos}\big(\varphi_S \oplus_2 \varphi_T\big)^{@} = G(S+T) = \operatorname{pos}\big(\varphi_S \oplus_2 \varphi_T\big) \subset \operatorname{pos} \varphi_{S+T}.$$

However, Lemma 19.12 implies that $\operatorname{pos} \varphi_{S+T}^{@} = \operatorname{pos} \varphi_{S+T}$, and so it follows that

$$\operatorname{pos} \varphi_{S+T} = G(S+T). \qquad (24.4)$$

If $b \in E \times E^*$ and $\varphi_{S+T}(b) \leq q(b)$, then the assumption that $\varphi_{S+T} \geq q$ on $E \times E^*$ gives $b \in \operatorname{pos} \varphi_{S+T}$ thus, from (24.4), $b \in G(S+T)$. (23.10) now implies that the multifunction $S + T$ is maximally monotone.

(c) Suppose that $b \in E \times E^*$ and $\varphi_{S+T}(b) \leq q(b)$. Then, by hypothesis, there exists $a \in G(S+T)$ such that $a_1 = b_1$. Consequently,

$$\varphi_{S+T}(b) \geq \langle b_1, a_2 \rangle + \langle a_1, b_2 \rangle - \langle a_1, a_2 \rangle = \langle b_1, a_2 \rangle + \langle b_1, b_2 \rangle - \langle b_1, a_2 \rangle = q(b).$$

Thus we have proved that $\varphi_{S+T} \geq q$ on $E \times E^*$, and the result follows from (b). □

IV Monotone multifunctions on general Banach spaces

25 Monotone multifunctions with bounded range

We now give a proof of the very useful result that if a maximally monotone multifunction has bounded range then it has full domain. Theorem 25.1 can also be established using the Debrunner–Flor extension theorem (which depends on Brouwer's fixed–point theorem, see Phelps, [69, Lemma 1.7, p. 4] and the comments preceding), or the Farkas Lemma (see Fitzpatrick–Phelps, [44, Lemma 2.4, pp. 580–581]), or a minimax theorem (see [99, Lemma 11.1, p. 41]). There will be a generalization of Theorem 25.1 to a special class of maximally monotone multifunctions in Corollary 41.2.

Theorem 25.1. Let E be a nonzero Banach space, $S\colon E \rightrightarrows E^*$ be maximally monotone and $R(S)$ be bounded. Then $D(S) = E$.

Proof. Choose $K \geq 0$ so that $s \in G(S) \implies \|s_2\| \leq K$. Then, for all $x \in E$ and $b \in E \times E^*$,

$$\begin{aligned}
\varphi_S(b) + K(\|x\| + \|b_1\|) &\geq \varphi_S(b) + K\|b_1 - x\| \\
&= \sup_{s\in G(S)} \left[\langle s_1, b_2 \rangle + \langle b_1, s_2 \rangle - \langle s_1, s_2 \rangle + K\|b_1 - x\| \right] \\
&\geq \sup_{s\in G(S)} \left[\langle s_1, b_2 \rangle + \langle x, s_2 \rangle - \langle s_1, s_2 \rangle \right] \\
&= \varphi_S(x, b_2) \geq \langle x, b_2 \rangle,
\end{aligned}$$

where the last inequality follows from (23.11). A second application of (23.11) and Theorem 22.8(b) now imply that $\pi_1 \mathrm{pos}\, \varphi_S{}^@ = E$. The result now follows from a third application of (23.11). $\square$

Remark 25.2. It was pointed out by H. Bauschke (personal communication) that the result "dual" to Theorem 25.1 fails. Suppose that E is not reflexive. Let C be the unit ball of E and $S := N_C$ (see Section 17). From James's theorem, Theorem 4.8, $R(S) \neq E^*$. On the other hand, $D(S) = C$ is bounded.

26 A general local boundedness theorem

Let $S\colon E \rightrightarrows E^*$ be nontrivial and $z \in E$. We say that S is *locally bounded* at z if there exist $\eta,\ K > 0$ such that

$$s \in G(S) \text{ and } \|s_1 - z\| < \eta \quad \Longrightarrow \quad \|s_2\| \le K.$$

Borwein–Fitzpatrick proved the following result in [21]: *Let S be nontrivial and monotone, and z be an absorbing point of $D(S)$. Then S is locally bounded at z.* This result is clearly extended by Theorem 26.1:

Theorem 26.1. *Let E be a nonzero Banach space, $S\colon E \rightrightarrows E^*$ be nontrivial and monotone and $z \in \operatorname{sur} D(S_\varphi)$. Then S is locally bounded at z.*

Proof. Let $f := \varphi_S{}^@$. From (23.7), $f \in \mathcal{PC}(E \times E^*)$ and $f^@ = \varphi_S$. From (23.8) and Theorem 22.5, $z \in \operatorname{sur} \pi_1 \operatorname{dom} f^@ = \operatorname{sur} \operatorname{dom} f_{\div}$. Lemma 22.7 now gives $K > 0$ and $\eta \in\ ]0,1[$ such that

$$b \in E \times E^*,\ \|b_1 - z\| \le \eta \text{ and } f(b) \le q(b) \quad \Longrightarrow \quad \|b_2\| \le K. \qquad (22.6)$$

Now if $b \in G(S)$ then (23.6) gives $f(b) = \varphi_S{}^@(b) = q(b)$, and so the result follows from (22.6). $\qquad \square$

Remark 26.2. Let $S\colon \mathbb{R}^2 \rightrightarrows \mathbb{R}^2$ be monotone and the four points $(\pm 1, \pm 1)$ lie in $D(S)$. Then Theorem 26.1 implies that S is locally bounded at 0 $\big($even if $0 \notin D(S)\big)$. We give this example in order to enable the reader to compare Theorem 26.1 with the result of Borwein–Fitzpatrick cited above.

27 The six set theorem and the nine set theorem

Define $f\colon \mathbb{R}^2 \to\ ]-\infty, \infty]$ by

$$f(x_1, x_2) := \begin{cases} |x_2| \vee \left(1 - \sqrt{1 - x_1{}^2}\right), & \text{if } |x_1| \vee |x_2| \le 1; \\ \infty, & \text{otherwise.} \end{cases}$$

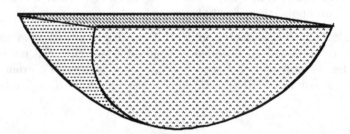

Despite the fact that ∂f is maximally monotone, $D(\partial f)$ is not convex (exercise!).

In this section, we investigate how close $D(S)$ is to being convex if S is a maximally monotone multifunction on a (possibly nonreflexive) Banach space. In the main result of this section, the "six set theorem", Theorem 27.1, we prove that the six sets $\operatorname{int} D(S)$, $\operatorname{int}(\operatorname{co} D(S))$, $\operatorname{int} D(S_\varphi)$, $\operatorname{sur} D(S)$, $\operatorname{sur}(\operatorname{co} D(S))$ and $\operatorname{sur} D(S_\varphi)$ coincide and, in its consequence, the "nine set theorem", Theorem 27.3, we prove that, if $\operatorname{sur} D(S_\varphi) \neq \emptyset$, then the nine sets $\overline{D(S)}$, $\overline{\operatorname{co} D(S)}$, $\overline{D(S_\varphi)}$, $\overline{\operatorname{int} D(S)}$, $\overline{\operatorname{int}(\operatorname{co} D(S))}$, $\overline{\operatorname{int} D(S_\varphi)}$, $\overline{\operatorname{sur} D(S)}$, $\overline{\operatorname{sur}(\operatorname{co} D(S))}$ and $\overline{\operatorname{sur} D(S_\varphi)}$ coincide. The six set theorem and the nine set theorem not only extend the results proved by Rockafellar in [78, Theorem 1, p. 398] (see also Phelps, [69, Theorem 1.9, p. 6]) that $D(S)$ is not far from being convex in the following sense: Let S be maximally monotone and $\operatorname{int}(\operatorname{co} D(S)) \neq \emptyset$. Then $\operatorname{int} D(S)$ is convex and $\overline{D(S)} = \overline{\operatorname{int} D(S)}$, so $\overline{D(S)}$ is also convex . Theorem 27.1 also answers in the affirmative a question raised by Phelps (see [68, p. 29] and [69, p. 8]), namely whether an absorbing point of $D(S)$ is necessarily an interior point.

(23.9) tells us that $G(S_\varphi)$ can be a much larger set than $G(S)$. It is this observation that makes Theorem 27.1, Theorem 27.3, Lemma 31.1, Theorem 31.2 and their consequences rather surprising.

Theorem 27.1. *Let E be a nonzero Banach space and $S\colon E \rightrightarrows E^*$ be maximally monotone. Then*

$$\operatorname{int} D(S) = \operatorname{int}(\operatorname{co} D(S)) = \operatorname{int} D(S_\varphi)$$
$$= \operatorname{sur} D(S) = \operatorname{sur}(\operatorname{co} D(S)) = \operatorname{sur} D(S_\varphi).$$

Consequently, $\operatorname{int} D(S)$ is convex and $\operatorname{sur} D(S_\varphi)$ is open.

Proof. Lemma 23.4 implies that $\operatorname{int} D(S) \subset \operatorname{int}(\operatorname{co} D(S)) \subset \operatorname{int} D(S_\varphi)$ and $\operatorname{sur} D(S) \subset \operatorname{sur}(\operatorname{co} D(S)) \subset \operatorname{sur} D(S_\varphi)$. Obviously $\operatorname{int}(\dots) \subset \operatorname{sur}(\dots)$. Now let $f := \varphi_S{}^{@}$. From (23.7), $f \in \mathcal{PC}(E \times E^*)$ and $f^{@} = \varphi_S$. Thus (23.8), (23.11), Theorem 22.8(a) and (23.11) give:

$$\operatorname{sur} D(S_\varphi) = \operatorname{sur} \pi_1 \operatorname{dom} \varphi_S = \operatorname{int} \pi_1 \operatorname{pos} \varphi_S = \operatorname{int} \pi_1 G(S) = \operatorname{int} D(S).$$

This completes the proof of Theorem 27.1.   □

Remark 27.2. Following on from the comments that we made in Remark 26.2, if $S\colon \mathbb{R}^2 \rightrightarrows \mathbb{R}^2$ is maximally monotone and the four points $(\pm 1, \pm 1)$ are in $D(S)$ then Theorem 27.1 implies that $]{-}1,1[\times]{-}1,1[\subset D(S)$ (even if we do not assume that $0 \in D(S)$).

Theorem 27.3. *Let E be a nonzero Banach space, $S\colon E \rightrightarrows E^*$ be maximally monotone and $\operatorname{sur} D(S_\varphi) \neq \emptyset$. Then*

$$\overline{D(S)} = \overline{\operatorname{co} D(S)} = \overline{D(S_\varphi)}$$
$$= \overline{\operatorname{int} D(S)} = \overline{\operatorname{int}(\operatorname{co} D(S))} = \overline{\operatorname{int} D(S_\varphi)}$$
$$= \overline{\operatorname{sur} D(S)} = \overline{\operatorname{sur}(\operatorname{co} D(S))} = \overline{\operatorname{sur} D(S_\varphi)}.$$

Proof. Lemma 23.4 implies that $\overline{D(S)} \subset \overline{\text{co}D(S)} \subset \overline{D(S_\varphi)}$, and Theorem 27.1 implies that int $D(S_\varphi) = \text{sur}\, D(S_\varphi) \neq \emptyset$ hence (see, for instance, Kelley–Namioka, [52, 13.1(i), pp. 110–111]), $\overline{D(S_\varphi)} = \overline{\text{int}\, D(S_\varphi)}$. Thus we have

$$\overline{\text{int}\, D(S)} \subset \overline{D(S)} \subset \overline{\text{co}D(S)} \subset \overline{D(S_\varphi)} = \overline{\text{int}D(S_\varphi)},$$

and the result now follows by combining this with Theorem 27.1.   □

Remark 27.4. It was observed in Simons–Zălinescu, [109, Remark 5.6, pp. 13–14] that, in the notation of [99, Definition 15.1, p. 53], $D(S_\varphi) = \text{dom}\,\chi_S$, so Theorem 27.1 and Theorem 27.3 are, in fact, identical with [99, Theorem 18.3 and Theorem 18.4, p. 67].

We now see what we can say about the first three sets in the statement of Theorem 27.3 without interiority conditions. Our result in this direction, Theorem 27.5, will be used explicitly in our analyses of maximally monotone multifunctions on a reflexive space in Theorem 31.2, and maximally monotone multifunctions of type (FPV) in Theorem 44.2.

Theorem 27.5. *Let E be a nonzero Banach space, $S: E \rightrightarrows E^*$ be monotone, and*

$$z \in E \setminus \overline{D(S)} \quad \Longrightarrow \quad \sup_{s \in G(S)} \frac{\langle z - s_1, s_2 \rangle}{\|z - s_1\|} = \infty. \tag{27.1}$$

Then:

(a)
$$E \setminus \overline{D(S)} \subset E \setminus D(S_\varphi).$$

(b)
$$\overline{D(S)} = \overline{\text{co}\, D(S)} = \overline{D(S_\varphi)}.$$

Proof. (a) If this were false then there would exist $b \in \text{dom}\,\varphi_S$ such that $b_1 \notin \overline{D(S)}$. Define $f: G(S) \to \mathbb{R}$ by $f(s) := \|b_1 - s_1\|$, so that $\inf_{G(S)} f > 0$. From Lemma 19.18, we would have

$$\inf_{s \in G(S)} \frac{\langle b_1 - s_1, b_2 - s_2 \rangle}{\|b_1 - s_1\|} > -\infty.$$

But this would contradict (27.1), since

$$\frac{\langle b_1 - s_1, b_2 - s_2 \rangle}{\|b_1 - s_1\|} = \frac{\langle b_1 - s_1, b_2 \rangle}{\|b_1 - s_1\|} - \frac{\langle b_1 - s_1, s_2 \rangle}{\|b_1 - s_1\|} \leq \|b_2\| - \frac{\langle b_1 - s_1, s_2 \rangle}{\|b_1 - s_1\|}.$$

(b) We derive from (a) that $D(S_\varphi) \subset \overline{D(S)}$, from which $\overline{D(S_\varphi)} \subset \overline{D(S)}$. The result now follows from Lemma 23.4.   □

We conclude this section with a result companion to Theorem 27.5, only for ranges instead of domains. Our result in this direction, Theorem 27.6, will be used explicitly in our analyses of maximally monotone multifunctions of type (FP) in Theorem 43.1, and maximally monotone multifunctions with surjective approximate resolvent in 43.4.

Theorem 27.6. *Let E be a nonzero Banach space, $S: E \rightrightarrows E^*$ be monotone, and*

$$z^* \in E^* \setminus \overline{R(S)} \quad \Longrightarrow \quad \sup_{s \in G(S)} \frac{\langle s_1, z^* - s_2 \rangle}{\|z^* - s_2\|} = \infty. \tag{27.2}$$

Then:

(a)
$$E^* \setminus \overline{R(S)} \subset E^* \setminus R(S_\varphi).$$

(b)
$$\overline{R(S)} = \overline{\text{co } R(S)} = \overline{R(S_\varphi)}.$$

Proof. (a) If this were false then there would exist $b \in \text{dom } \varphi_S$ such that $b_2 \notin \overline{R(S)}$. Define $f: G(S) \to \mathbb{R}$ by $f(s) := \|b_2 - s_2\|$, so that $\inf_{G(S)} f > 0$. From Lemma 19.18, we would have

$$\inf_{s \in G(S)} \frac{\langle b_1 - s_1, b_2 - s_2 \rangle}{\|b_2 - s_2\|} > -\infty.$$

But this would contradict (27.2), since

$$\frac{\langle b_1 - s_1, b_2 - s_2 \rangle}{\|b_2 - s_2\|} = \frac{\langle b_1, b_2 - s_2 \rangle}{\|b_2 - s_2\|} - \frac{\langle s_1, b_2 - s_2 \rangle}{\|b_2 - s_2\|} \le \|b_1\| - \frac{\langle s_1, b_2 - s_2 \rangle}{\|b_2 - s_2\|}.$$

(b) We derive from (a) that $R(S_\varphi) \subset \overline{R(S)}$, from which $\overline{R(S_\varphi)} \subset \overline{R(S)}$. The result now follows from Lemma 23.4. □

28 $D(S_\varphi)$ and various hulls

As a general matter of notation, we write "aff" for "affine hull" and "lin" for "linear hull". If F is a subspace of the nonzero Banach space E, we write

$$F^\perp := \{y^* \in E^*: \langle F, y^* \rangle = \{0\}\}.$$

In our analysis of constraint qualifications for pairs of maximally monotone multifunctions in Theorem 32.2, we will need to examine the closed subspaces $\overline{\text{lin}[D(S) - D(T)]}$ and $\overline{\text{lin}[D(S_\varphi) - D(T_\varphi)]}$. In this connection, Lemma 28.1(c) below will be important:

Lemma 28.1. *Let E be a nonzero Banach space and $S, T: E \rightrightarrows E^*$ be maximally monotone. Then:*

(a)
$$\overline{\text{aff}[D(S_\varphi) - D(T_\varphi)]} = \overline{\text{aff}[D(S) - D(T)]}.$$

(b)
$$\overline{\text{lin}[D(S_\varphi) - D(T_\varphi)]} = \overline{\text{lin}[D(S) - D(T)]}.$$

(c)
$$\bigcup_{\lambda > 0} \lambda[D(S_\varphi) - D(T_\varphi)] \subset \overline{\text{lin}[D(S) - D(T)]}.$$

Proof. (a) There exist $x \in E$ and a closed linear subspace F of E such that $\text{aff}\big[D(S) - D(T)\big] = x + F$. Let $y^* \in F^\perp$, and write $c := (0, y^*) \in E \times E^*$. Clearly, $q(c) = 0$. If $s = (s_1, s_2) \in G(S)$ and $t = (t_1, t_2) \in G(T)$ then $s_1 - t_1 \in D(S) - D(T)$ and so $\lfloor s - t, c \rfloor = \langle s_1 - t_1, y^* \rangle = \langle x, y^* \rangle$. Consequently, there exists $\mu \in \mathbb{R}$ such that $\lfloor G(S), c \rfloor = \{\mu + \langle x, y^* \rangle\}$ and $\lfloor G(T), c \rfloor = \{\mu\}$. Lemma 20.4(b) now implies that $\lfloor G(S_\varphi), c \rfloor = \{\mu + \langle x, y^* \rangle\}$ and $\lfloor G(T_\varphi), c \rfloor = \{\mu\}$, that is to say $\langle D(S_\varphi), y^* \rangle = \{\mu + \langle x, y^* \rangle\}$ and $\langle D(T_\varphi), y^* \rangle = \{\mu\}$. It follows from this that $\langle D(S_\varphi) - D(T_\varphi) - x, y^* \rangle = \{0\}$. Since this holds for all $y^* \in F^\perp$, and F is a closed subspace of E, the separation theorem of Theorem 4.4 now implies that $D(S_\varphi) - D(T_\varphi) - x \subset F$. Thus we have proved that

$$D(S_\varphi) - D(T_\varphi) \subset x + F = \overline{\text{aff}\big[D(S) - D(T)\big]}.$$

(a) now follows from this and Lemma 23.4.

(b) has a similar proof to that of (a), only taking $x = 0$, and (c) is immediate from (b). $\qquad\square$

Corollary 28.2. *Let E be a nonzero Banach space and $S \colon E \rightrightarrows E^*$ be maximally monotone. Then*

$$\overline{\text{aff}\, D(S_\varphi)} = \overline{\text{aff}\, D(S)} \quad \text{and} \quad \overline{\text{lin}\, D(S_\varphi)} = \overline{\text{lin}\, D(S)}.$$

Proof. Define T by $G(T) := \{0\} \times E^*$. It follows easily from the separation theorem, Theorem 4.4, that T is maximally monotone and $\text{dom}\, \varphi_T = G(T)$. Consequently, $D(T_\varphi) = D(T) = \{0\}$, and the result is immediate from Lemma 28.1. $\qquad\square$

Problem 28.3. Corollary 28.2 suggests the following question: let E be a nonzero Banach space and $S \colon E \rightrightarrows E^*$ be maximally monotone. Then is it necessarily true that

$$\overline{D(S_\varphi)} = \overline{\text{co}D(S)}?$$

We have seen in Theorem 27.3, and we shall see in Theorem 28.6, Theorem 31.2 and Theorem 44.2 that the answer is in the affirmative if $\text{sur}\, D(S_\varphi) \neq \emptyset$, $D(S)$ is closed and convex, E is reflexive or, more generally, S is "of type (FPV)". It is clear from the analysis in Section 44 that it could be a hard problem to find a counterexample.

In Theorem 28.6, we will describe the solution to Problem 28.3 when $D(S)$ is closed and convex. Lemmas 28.4 and 28.5 will be used explicitly in Theorem 28.6, and in our proof of Voisei's theorem, Theorem 51.1. In addition, Lemma 28.4 will be used explicitly in Theorem 52.1 and Theorem 52.2. Theorem 28.6 will be used explicitly in our proof of Voisei's theorem and Theorem 52.2. We recall from (17.2) that if C is a closed convex subset of E then the normality multifunction $N_C \colon E \rightrightarrows E^*$ is defined by

$$(x, x^*) \in G(N_C) \iff x \in C \text{ and } \langle x, x^* \rangle = \max{}_C\, x^*.$$

Lemma 28.4. *Let C be a nonempty closed convex subset of a Banach space E, $U\colon E \rightrightarrows E^*$ be monotone and $D(U) \supset C$. Then*

$$D\big((U + N_C)_\varphi\big) = \pi_1 \mathrm{dom}\, \varphi_{U+N_C} = C.$$

Proof. It is clear from Lemma 23.4 that $\pi_1 \mathrm{dom}\, \varphi_{U+N_C} = D\big((U+N_C)_\varphi\big) \supset D(U + N_C) = D(U) \cap D(N_C) = D(U) \cap C = C$. Conversely, let $b \in \mathrm{dom}\, \varphi_{U+N_C}$. We shall prove that

$$a \in G(N_C) \quad \Longrightarrow \quad \langle b_1, a_2 \rangle \le \langle a_1, a_2 \rangle. \tag{28.1}$$

It will then follow from Theorem 18.10 that $b_1 \in C$, which gives the required result. So let a be an arbitrary element of $G(N_C)$. Since $a_1 \in C \subset D(U)$, we can fix $x^* \in Ua_1$. If $n \ge 1$ then $(a_1, na_2) \in G(N_C)$ and so $(a_1, x^* + na_2) \in G(U + N_C)$. Thus $\langle b_1, x^* + na_2 \rangle + \langle a_1, b_2 \rangle - \langle a_1, x^* + na_2 \rangle \le \varphi_{U+N_C}(b)$, and so $n\big[\langle b_1, a_2 \rangle - \langle a_1, a_2 \rangle\big] \le \varphi_{U+N_C}(b) - \langle b_1, x^* \rangle - \langle a_1, b_2 \rangle + \langle a_1, x^* \rangle < \infty$. Letting $n \to \infty$, we have established (28.1), which completes the proof of the Lemma. $\square$

Lemma 28.5. *Let E be a nonzero Banach space, $S\colon E \rightrightarrows E^*$ be maximally monotone and $D(S)$ be closed and convex. Then*

$$S = S + N_{D(S)}.$$

Proof. Since $G(S) \subset G\big(S + N_{D(S)}\big)$ and $S + N_{D(S)}$ is monotone, this is immediate from the maximal monotonicity of S. $\square$

Theorem 28.6. *Let E be a nonzero Banach space, $S\colon E \rightrightarrows E^*$ be maximally monotone and $D(S)$ be closed and convex. Then*

$$D(S) = \mathrm{co}\, D(S) = D(S_\varphi).$$

Proof. Lemma 28.4 with $U := S$ and $C := D(S)$ and Lemma 28.5 give

$$D(S_\varphi) = D\big((S + N_{D(S)})_\varphi\big) = D(S).$$

The required result now follows from Lemma 23.4. $\square$

We close this section by analyzing how maximal monotonicity interacts with continuous linear maps and closed subspaces. Lemma 28.8(a) below contains a result companion to Corollary 28.2, only without the assumption of maximality. We first make a preliminary definition.

Definition 28.7. Let F be a subspace of a Banach space E, and $S\colon E \rightrightarrows E^*$ be nontrivial and monotone. We say that S is *F-saturated* if

$$x \in D(S) \quad \Longrightarrow \quad Sx + F^\perp = Sx.$$

By permuting quantifiers, this can also be written:

$$y^* \in F^\perp \quad \Longrightarrow \quad G(S) + (0, y^*) \subset G(S). \tag{28.2}$$

Lemma 28.8. *Let E be a nonzero Banach space, $S\colon E \rightrightarrows E^*$ be nontrivial and monotone, F be a subspace of E, $y \in E$, and $D(S) \subset F + y$.*
(a) If F is closed and S is F-saturated then $D(S_\varphi) \subset F + y$.
(b) If S is maximally monotone then S is F-saturated.

Proof. (a) Let y^* be an arbitrary element of $F^\perp$, and write $c := (0, y^*) \in E \times E^*$. Clearly, $q(c) = 0$. Furthermore,

$$\lfloor G(S), c \rfloor = \langle D(S), y^* \rangle \subset \langle F + y, y^* \rangle = \{\langle y, y^* \rangle\}$$

and

$$\lfloor \operatorname{dom} \Phi_{G(S)}, c \rfloor = \lfloor \operatorname{dom} \varphi_S, c \rfloor = \lfloor G(S_\varphi), c \rfloor = \langle D(S_\varphi), y^* \rangle.$$

From (28.2), $G(S) + \mathbb{R}c \subset G(S)$ and so Lemma 20.4(a) with $A := G(S)$ implies that $\langle D(S_\varphi), y^* \rangle = \{\langle y, y^* \rangle\}$, that is to say, $\langle D(S_\varphi) - y, y^* \rangle = \{0\}$. Since F is closed, it follows from the separation theorem, Theorem 4.4, that $D(S_\varphi) - y \subset F$. This completes the proof of (a)
  (b) Let y^* be an arbitrary element of $F^\perp$, and write $c := (0, y^*) \in E \times E^*$. Clearly, $q(c) = 0$. If $a, b \in G(S)$ then $a_1, b_1 \in D(S)$, and so $\lfloor a - b, c \rfloor = \langle a_1 - b_1, y^* \rangle = 0$. Consequently, there exists $\mu \in \mathbb{R}$ such that $\lfloor G(S), c \rfloor = \{\mu\}$. From Lemma 20.4(b) with $A := G(S)$, $G(S) + \mathbb{R}c \subset G(S)$. (b) now follows from (28.2). $\qquad \square$

  We recall that L^*SL was defined in Notation 23.7.

Theorem 28.9. *Let E and F be nonzero Banach spaces, $L\colon F \to E$ be continuous and linear with $L(F)$ closed and $L^*(E^*) = F^*$, and $S\colon E \rightrightarrows E^*$ be nontrivial and monotone with $D(S) \subset L(F)$. Then S is maximally monotone $\iff S$ is $L(F)$-saturated and L^*SL is maximally monotone.*

Proof. Let $\Sigma := L^*SL$, as in Lemma 23.8.
  We first assume that S is $L(F)$-saturated and Σ is maximally monotone, and prove that S is maximally monotone. Let

$$(x, x^*) \in E \times E^* \quad \text{and} \quad \varphi_S(x, x^*) \le \langle x, x^* \rangle. \tag{28.3}$$

Then $x \in \pi_1 \operatorname{dom} \varphi_S = D(S_\varphi)$, and Lemma 28.8(a) (with $y := 0$) implies that $x \in L(F)$. Fix $\xi \in F$ such that $L\xi = x$. From Lemma 23.8,

$$\varphi_\Sigma(\xi, L^*x^*) = \varphi_S(L\xi, x^*) = \varphi_S(x, x^*) \le \langle x, x^* \rangle = \langle L\xi, x^* \rangle = \langle \xi, L^*x^* \rangle,$$

and so (23.10) (applied to Σ) now implies that

$$(\xi, L^*x^*) \in G(\Sigma).$$

Consequently, there exists $s^* \in S(L\xi)$ such that $L^*s^* = L^*x^*$. Let $y^* := x^* - s^*$, so that $L^*y^* = 0$, from which $y^* \in L(F)^\perp$. But then

$$(x, x^*) = (L\xi, x^*) = (L\xi, s^*) + (0, y^*).$$

Now $(L\xi, s^*) \in G(S)$, so (28.2) implies that $(x, x^*) \in G(S)$. Since this holds for all $(x, x^*) \in E \times E^*$ satisfying (28.3), the maximal monotonicity of S is now clear from (23.10).

Suppose, conversely, that S is maximally monotone. Since it follows from Lemma 28.8(b) (with $y := 0$) that S is $L(F)$-saturated, it only remains to prove that Σ is maximally monotone. Let

$$(\xi, \xi^*) \in F \times F^* \quad \text{and} \quad \varphi_\Sigma(\xi, \xi^*) \le \langle \xi, \xi^* \rangle. \tag{28.4}$$

Since $L^*(E^*) = F^*$, there exists $x^* \in E^*$ such that $L^* x^* = \xi^*$. From Lemma 23.8,

$$\varphi_S(L\xi, x^*) = \varphi_\Sigma(\xi, L^* x^*) = \varphi_\Sigma(\xi, \xi^*) \le \langle \xi, \xi^* \rangle = \langle \xi, L^* x^* \rangle = \langle L\xi, x^* \rangle,$$

and so (23.10) gives $(L\xi, x^*) \in G(S)$, from which

$$(\xi, \xi^*) = (\xi, L^* x^*) \in G(\Sigma).$$

Since this holds for all $(\xi, \xi^*) \in F \times F^*$ satisfying (28.4), the maximal monotonicity of Σ is now clear from (23.10). $\qquad \square$

Corollary 28.10. *Let F be a nonzero closed subspace of a Banach space E, $S \colon E \rightrightarrows E^*$ be nontrivial and monotone and $D(S) \subset F$. Define $T \colon F \rightrightarrows F^*$ by*

$$G(T) := \{(s_1, s_2|_F) \colon s \in G(S)\}.$$

Then S is maximally monotone $\iff S$ is F-saturated and T is maximally monotone.

Proof. Let L be the inclusion map from F into E. The extension form of the Hahn–Banach theorem, Corollary 2.2, gives us that $L^*(E^*) = F^*$, and the result follows from Theorem 28.9. $\qquad \square$

V Monotone multifunctions on reflexive Banach spaces

29 Criteria for maximality, and Rockafellar's surjectivity theorem

We now turn our attention to the SSDB space considered in Remark 17.1 and Example 21.2(a). As in Section 22, let E be a nonzero Banach space and E^* be its topological dual space. The difference with Section 22 is that we now suppose that E is *reflexive*. Let $B := E \times E^*$ normed by $\|b\| := \sqrt{\|b_1\|^2 + \|b_2\|^2}$ $(b = (b_1, b_2) \in B)$. Then B is a SSDB space with $\lfloor b, c \rceil := \langle b_1, c_2 \rangle + \langle c_1, b_2 \rangle$ $(b = (b_1, b_2), c = (c_1, c_2) \in B)$ and $q(b) = \langle b_1, b_2 \rangle$ $(b = (b_1, b_2) \in B)$ and $g_0(b) := \frac{1}{2}(\|b_1\|^2 + \|b_2\|^2)$. Furthermore,

$$\text{neg}\, g_0 = \left\{ b \in E \times E^* \colon \tfrac{1}{2}\|b_1\|^2 + \tfrac{1}{2}\|b_2\|^2 + \langle b_1, b_2 \rangle = 0 \right\}.$$

If $b \in \text{neg}\, g_0$ then $\frac{1}{2}\|b_1\|^2 + \frac{1}{2}\|b_2\|^2 - \|b_1\|\|b_2\| \le 0$, from which $\|b_1\| = \|b_2\|$. Thus

$$b \in \text{neg}\, g_0 \quad \Longrightarrow \quad \langle b_1, b_2 \rangle = -\|b_1\|\|b_2\| = -\|b_1\|^2. \tag{29.1}$$

We now indicate how $\text{neg}\, g_0$ can be expressed in terms of more familiar concepts.

Lemma 29.1. $\text{neg}\, g_0 = G(-J)$, where $-J \colon E \rightrightarrows E^*$ is defined by: $(-J)x := -Jx$. (We recall that the multifunction J was defined in Definition 23.5.)

Proof. Exercise! □

Now let $S \colon E \rightrightarrows E^*$ be nontrivial. Then, as in Section 23, S is monotone exactly when $G(S)$ is q–positive, and S is maximally monotone exactly when $G(S)$ is maximally q–positive.

We now come to the main result of this section, Theorem 29.2. We will deduce from this in Theorem 29.5 one direction of Rockafellar's "surjectivity theorem". We point out Theorem 29.5 does not assume that E has been renormed to have any special properties.

Theorem 29.2. *Let E be a nonzero reflexive Banach space and $S \colon E \rightrightarrows E^*$ be nontrivial and monotone. Then S is maximally monotone $\iff$ for all $b \in E \times E^*$, there exists $s \in G(S)$ such that*

$$\|b_1 - s_1\|^2 + \|b_2 - s_2\|^2 + 2q(b - s) = 0. \tag{29.2}$$

Proof. This is a restatement of Theorem 21.7. □

We next deduce from Theorem 29.2 the "negative alignment" criterion for maximality.

Corollary 29.3. *Let E be a nonzero reflexive Banach space and $S: E \rightrightarrows E^*$ be nontrivial and monotone. Then S is maximally monotone $\iff$ for all $b \in E \times E^* \setminus G(S)$, there exists $s \in G(S)$ such that*

$$s_1 \neq b_1, \ s_2 \neq b_2 \quad \text{and} \quad q(b - s) = -\|b_1 - s_1\|\|b_2 - s_2\|. \tag{29.3}$$

Proof. ($\Longrightarrow$) Let $b \in E \times E^* \setminus G(S)$. Choose $s \in G(S)$ as in (29.2). Then either $s_1 \neq b_1$ or $s_2 \neq b_2$ (or both!). It is clear from (29.3) that

$$\|b_1 - s_1\|^2 + \|b_2 - s_2\|^2 - 2\|b_1 - s_1\|\|b_2 - s_2\| = 0,$$

from which $\|b_1 - s_1\| = \|b_2 - s_2\|$. So, in fact $s_1 \neq b_1$ and $s_2 \neq b_2$. Using (29.2) again,

$$
\begin{aligned}
0 &= \|b_1 - s_1\|^2 + \|b_2 - s_2\|^2 + 2q(b - s) \\
&\geq \|b_1 - s_1\|^2 + \|b_2 - s_2\|^2 - 2\|b_1 - s_1\|\|b_2 - s_2\| = 0,
\end{aligned}
$$

from which the rest of (29.3) now follows easily.

($\Longleftarrow$) Since (29.3) implies that $q(b - s) < 0$, this is immediate from the definition of maximality. $\qquad\square$

Remark 29.4. There is no hope of getting a result analogous to Corollary 29.3 if E is not reflexive. Let E be a nonzero Banach space and $S: E \rightrightarrows E^*$ be defined by $S := 0$. S is clearly maximally monotone. Let x^* be an arbitrary nonzero element of E^*, and $b := (0, x^*) \in E \times E^*$. If (29.3) is true for this choice of b then there would exist $s = (s_1, 0) \in G(S)$ such that $s_1 \neq 0$ and $\langle s_1, x^* \rangle = \|s_1\|\|x^*\|$. Setting $x := s_1/\|s_1\|$, we would have $\|x\| = 1$ and $\langle x, x^* \rangle = \|x^*\|$. From James's theorem, Theorem 4.8, E is reflexive. It is these considerations that gave rise to the definition of maximally monotone multifunctions of "type (ANA)" in Definition 36.11.

Now, it was proved by Minty that *if E is a Hilbert space and $S: E \rightrightarrows E^*$ is monotone then S is maximally monotone $\iff R(S+J) = E^*$*. Rockafellar showed in [80, Proposition 1, p. 77] that Minty's result can be extended to the case where E is reflexive and J and J^{-1} are single–valued. Further, it was proved by Asplund that any reflexive Banach space can be renormed so that J and J^{-1} are single–valued. (Of course, renorming does not affect monotonicity or maximality). This renorming theorem is not easy. We shall show in Theorem 29.5 that the implication ($\Longrightarrow$), known as "Rockafellar's surjectivity theorem", is true even without the renorming (and also without the implicit use of Brouwer's fixed–point theorem). The proof given here is a simplification of that given in [109, Theorem 3.1(b), p. 8]. We will give a nontrivial generalization of Theorem 29.5 in Theorem 30.2.

Theorem 29.5. *Let E be a nonzero reflexive Banach space and $S\colon E \rightrightarrows E^*$ be maximally monotone. Then*

$$R(S + J) = E^*.$$

Proof. Let $w^* \in E^*$. From Lemma 29.1 and Theorem 21.7,

$$(0, w^*) \in G(S) + G(-J).$$

Thus there exist $x \in E$, $x^* \in Sx$ and $y^* \in (-J)(-x)$ such that $x^* + y^* = w^*$. But then $y^* \in Jx$, hence

$$w^* = x^* + y^* \in Sx + Jx \subset R(S + J). \qquad \square$$

The proof of the exact value of $\min\{\|x\|\colon x \in E,\ (S + J)x \ni 0\}$ in terms of φ_S given in Theorem 29.6 below is a simplification of that given in Simons–Zălinescu, [109, Theorem 3.1(b), p. 8]. There is, in fact, a continuum of formulae for this minimum. In particular, if $\|b\|_1 := \|b_1\| + \|b_2\|$ and $\|b\|_\infty := \|b_1\| \vee \|b_2\|$ then

$$\min\{\|x\|\colon x \in E,\ (S + J)x \ni 0\}$$

$$= \tfrac{1}{2} \sup_{b \in E \times E^*} \left[\|b\|_1 - \sqrt{4\varphi_S(b) + \|b\|_1{}^2}\right] \vee 0$$

$$= \sup_{b \in E \times E^*} \left[\|b\|_\infty - \sqrt{\varphi_S(b) + \|b\|_\infty{}^2}\right] \vee 0.$$

These formulae might be more convenient for computation. See [109, Section 7, pp. 18–21] for more details.

Theorem 29.6. *Let E be a nonzero reflexive Banach space, $S\colon E \rightrightarrows E^*$ be maximally monotone and*

$$N := \sup_{b \in E \times E^*} \left[\|b\| - \sqrt{2\varphi_S(b) + \|b\|^2}\right] \vee 0.$$

Then

$$\min\{\|x\|\colon x \in E,\ (S + J)x \ni 0\} = \tfrac{1}{\sqrt{2}} N$$

and

$$\sup\{\|x\|\colon x \in E,\ (S + J)x \ni 0\} \le \sqrt{2} \inf_{b \in G(S)} \left[\|b_1\| + \|b_2\|\right].$$

Proof. (23.11), Lemma 29.1 and Theorem 21.4(c) with $f := \varphi_S$ and $c := 0$ give us that

$$\min\{\|a\|\colon a \in G(-J) \cap G(S)\} = N. \qquad (29.4)$$

Suppose first that $x \in E$ and $(S + J)x \ni 0$. Then there exists $x^* \in Sx$ such that $-x^* \in Jx$. But then $(x, x^*) \in G(-J) \cap G(S)$, so (29.4) implies that $\|(x, x^*)\| \ge N$. Since $\|(x, x^*)\| = \sqrt{2}\|x\|$, it follows that $\|x\| \ge \tfrac{1}{\sqrt{2}}N$. Conversely, (29.4) provides us with $a \in G(-J) \cap G(S)$ such that $\|a\| \le N$. Let

$a = (x, x^*)$. Then $x^* \in Sx$ and $-x^* \in Jx$, from which $0 = x^* - x^* \in (S+J)x$. Further, $\|x\| = \frac{1}{\sqrt{2}}\|(x, x^*)\| = \frac{1}{\sqrt{2}}\|a\| \leq \frac{1}{\sqrt{2}}N$.

Arguing exactly as above, we can also deduce that if $x \in E$ and $(S+J)x \ni 0$ then

$$\|x\| \leq \frac{1}{\sqrt{2}} \inf_{b \in E \times E^*} \left[\|b\| + \sqrt{2\varphi_S(b) + \|b\|^2} \right].$$

But then we have

$$\|x\| \leq \frac{1}{\sqrt{2}} \inf_{b \in G(S)} \left[\|b\| + \sqrt{2\langle b_1, b_2 \rangle + \|b\|^2} \right]$$

$$\leq \frac{1}{\sqrt{2}} \inf_{b \in G(S)} \left[\sqrt{\|b_1\|^2 + \|b_2\|^2} + \sqrt{2\|b_1\|\|b_2\| + \|b_1\|^2 + \|b_2\|^2} \right]$$

$$\leq \sqrt{2} \inf_{b \in G(S)} \left[\|b_1\| + \|b_2\| \right]. \qquad \square$$

Remark 29.7. We outline a proof of Rockafellar's result that if E is a nonzero reflexive Banach space and J and J^{-1} are single–valued then

$$R(S + J) = E^* \implies S \text{ is maximally monotone.} \qquad (29.5)$$

Suppose that $b \in E \times E^*$ is monotonically related to $G(S)$. Since $R(S+J) = E^*$, we can choose $s \in G(S)$ so that $s_2 + Js_1 = b_2 + Jb_1$, from which

$$Js_1 - Jb_1 = -(s_2 - b_2). \qquad (29.6)$$

(Remember that, for all $x \in E$, Jx is now a *singleton*.) Let $s^J := (s_1, Js_1)$ and $b^J := (b_1, Jb_1)$. Now $s^J, b^J \in G(J) = \text{pos } g_0$, from which

$$g_0(s_1, Jb_1) + g_0(b_1, Js_1) = g_0(s^J) + g_0(b^J) = q(s^J) + q(b^J). \qquad (29.7)$$

Since b is monotonically related to $G(S)$, $q(s - b) \geq 0$, and so it follows from (29.6) and the bilinearity of q on $E \times E^*$ that $q(s^J - b^J) \leq 0$, or equivalently $q(s^J) + q(b^J) \leq q(s_1, Jb_1) + q(b_1, Js_1)$. Combining this with (29.7),

$$g_0(s_1, Jb_1) + g_0(b_1, Js_1) \leq q(s_1, Jb_1) + q(b_1, Js_1).$$

Using the fact that $g_0 \geq q$ on $E \times E^*$, we derive that $g_0(s_1, Jb_1) = q(s_1, Jb_1)$, that is to say, $(s_1, Jb_1) \in G(J)$. Thus we have proved that

$$Jb_1 = Js_1. \qquad (29.8)$$

Substituting this in (29.6), $b_2 = s_2$. It also follows from (29.8) and the fact that J^{-1} is single–valued that $b_1 = s_1$. Thus $b = s \in G(S)$. Since this holds whenever b is monotonically related to $G(S)$, it follows that S is maximally monotone.

It was pointed out by S. Fitzpatrick (personal communication) that *if J is not single–valued then (29.5) does not follow*. Here was his reasoning: if J is not single–valued then there exist $x \in E$ and distinct elements y^* and z^* of Jx. Let $x^* := \frac{1}{2}(y^* + z^*)$, and define S by setting $G(S) := G(J) \setminus \{(x, x^*)\}$. Now let $w^* \in E^*$. Since E is reflexive, there exists $w \in E$ such that $\frac{1}{2}w^* \in Jw$. If $(w, \frac{1}{2}w^*) \neq (x, x^*)$ then $\frac{1}{2}w^* \in Sw$ hence

$$w^* = \tfrac{1}{2}w^* + \tfrac{1}{2}w^* \in Sw + Jw = (S + J)w.$$

If, on the other hand, $(w, \frac{1}{2}w^*) = (x, x^*)$ then $w^* = 2x^* = y^* + z^*$. Since $y^* \in Jx$ and $y^* \neq x^*$, $y^* \in Sx$. Consequently,

$$w^* = y^* + z^* \in Sx + Jx = (S + J)x.$$

Thus we have proved that

$$w^* \in E^* \quad \Longrightarrow \quad w^* \in R(S + J),$$

that is to say, $R(S+J) = E^*$. On the other hand, S is obviously not maximally monotone.

It was pointed out by H. Bauschke (personal communication) that *if J is single–valued and J^{-1} is not single–valued then, again, (29.5) does not follow*. Here is his reasoning: there exist $z^* \in E^*$ and distinct elements x and y of $J^{-1}z^*$, and define S by setting $G(S) := (E \setminus \{x\}) \times \{0\}$. Now let $w^* \in E^*$. If $w^* = z^*$ then

$$w^* = 0 + z^* \in (S + J)y.$$

If, on the other hand, $w^* \neq z^*$ then, since E is reflexive, there exists $w \in E$ such that $w^* \in Jw$. Since J is single–valued, $z^* \in Jx$ and $w^* \neq z^*$, $w^* \notin Jx$. It follows that $w \neq x$, and so

$$w^* = 0 + w^* \in (S + J)w.$$

Thus we have proved that

$$w^* \in E^* \quad \Longrightarrow \quad w^* \in R(S + J),$$

that is to say, $R(S + J) = E^*$. On the other hand, S is, again, obviously not maximally monotone.

If E is reflexive and $S \colon E \rightrightarrows E^*$ is maximally monotone, the existence of an autoconjugate function (see Definition 21.8) $h \in \mathcal{PC}(E \times E^*)$ such that $\varphi_S \leq h \leq \varphi_S^{@}$ on $E \times E^*$ was established using Zorn's Lemma by Penot in [65, Corollary 5] and [66, Theorem 10, p. 865]. A similar result was proved by Svaiter in [110, Theorem 2.4, p. 3854] without any mention of reflexivity. Penot–Zălinescu give in [67, Proposition 4.2] an *explicit* formula for an autoconjugate function h such that $\text{pos}\, h = G(S)$ in the case where E is reflexive, $S \colon E \rightrightarrows E^*$ is maximally monotone and $\text{aff}\, D(S)$ is closed.

The autoconjugate h in Theorem 29.8 below is a (very) special case of a "kernel average" — such functions have been considered in great detail in Bauschke–Wang [13]. An allied question is to find an expression for a maximally monotone extension of a given nontrivial monotone multifunction on a Banach space (which always exists by Zorn's lemma). In Theorem 29.8(c), we give a somewhat more explicit description of such an extension in the reflexive case. In fact, Theorem 29.8(a–b) follow from [13, Theorem 5.7]. In a recent paper, [47], Ghoussoub has proved similar results for what he calls "Self-dual lagrangians".

Theorem 29.8. *Let E be a nonzero reflexive Banach space and $S: E \rightrightarrows E^*$ be nontrivial and monotone. For all $b \in E \times E^*$, let*

$$h(b) := \inf_{c \in E \times E^*} \left[\tfrac{1}{2} \varphi_S(b + c) + \tfrac{1}{2} \varphi_S{}^@(b - c) + g_0(c) \right].$$

Then:
(a) h is autoconjugate.
(b) $\varphi_S \vee q \le h \le \varphi_S{}^@$ on $E \times E^*$ and pos h is a maximally monotone superset of $G(S)$. If S is maximally monotone then pos $h = G(S)$.
(c) $b \in$ pos h if, and only if,

$$\text{there exists } d \in G(J) \text{ such that } (b - d, b + d) \in G(\partial \varphi_S).$$

Proof. This is immediate from Theorem 21.11.   □

We end this section by giving a result motivated by the Brøndsted–Rockafellar theorem for subdifferentials (see Corollary 18.5 and Theorem 18.6) which was proved by Torralba in [111, Proposition 6.17] in connection with problems of scale–change:

Theorem 29.9. *Let E be a nonzero reflexive Banach space, $T: E \rightrightarrows E^*$ be maximally monotone, $b \in E \times E^*$, α, $\beta > 0$ and $\inf_{t \in G(T)} q(t - b) \ge -\alpha\beta$. Then there exists $t \in G(T)$ such that $\|t_1 - b_1\| \le \alpha$ and $\|t_2 - b_2\| \le \beta$.*

Proof. Define the (invertible) linear map $\Delta: E \times E^* \to E \times E^*$ by $\Delta(c_1, c_2) := (c_1/\alpha, c_2/\beta)$ $(c = (c_1, c_2) \in E \times E^*)$. Let $S: E \rightrightarrows E^*$ be defined by $G(S) = \Delta(G(T))$. S is easily seen to be maximally monotone. Theorem 21.7 provides us an element s of $G(S)$ such that $s - \Delta(b) \in \text{neg } g_0$. So $\|s_1 - b_1/\alpha\|^2 = \|s_2 - b_2/\beta\|^2 = -q(s - \Delta(b))$. Choose $t \in G(T)$ such that $s = \Delta(t)$. Then we have $\|t_1 - b_1\|^2/\alpha^2 = \|t_2 - b_2\|^2/\beta^2 = -q(t - b)/\alpha\beta \le 1$. This gives the required result.   □

30 Surjectivity and an abstract Hammerstein theorem

In this section, we prove various surjectivity results, including a generalization of Rockafellar's surjectivity theorem (Theorem 30.2) and an abstract Hammerstein theorem (Theorem 30.4).

We suppose that E is a nonzero reflexive Banach space and define the "reflection" $\rho_2 \colon E \times E^* \to E \times E^*$ by $\rho_2(b) := (b_1, -b_2)$.

Theorem 30.1. *Let E be a nonzero reflexive Banach space, $w^* \in E^*$ and $f, g \in \mathcal{PCLSC}(E \times E^*)$ be BC–functions such that $E \times \{w^*\} \subset \mathrm{dom}\, f$ and $\pi_2 \, \mathrm{dom}\, g = E^*$. Then:*
(a) $\mathrm{pos}\, f - \rho_2 \, \mathrm{pos}\, g = E \times E^*$ *and* $\mathrm{pos}\, f + \rho_2 \, \mathrm{pos}\, g = E \times E^*$.
(b) *If $x \in E$ then there exist $(y, y^*) \in \mathrm{pos}\, f$ and $(z, y^*) \in \mathrm{pos}\, g$ such that $y + z = x$.*
(c) *If $x^* \in E^*$ then there exist $(y, y^*) \in \mathrm{pos}\, f$ and $(y, z^*) \in \mathrm{pos}\, g$ such that $y^* + z^* = x^*$.*

Proof. (a) Let (x, x^*) be an arbitrary element of $E \times E^*$. Since $\pi_2 \, \mathrm{dom}\, g = E^*$, there exists $y \in E$ such that $(y, x^* - w^*) \in \mathrm{dom}\, g$. But $(x + y, w^*) \in \mathrm{dom}\, f$, hence

$$(x, x^*) = (x + y, w^*) - \rho_2(y, x^* - w^*) \in \mathrm{dom}\, f - \rho_2 \, \mathrm{dom}\, g.$$

Thus we have proved that $\mathrm{dom}\, f - \rho_2 \, \mathrm{dom}\, g = E \times E^*$. It is easy to see by direct computation that $g \circ \rho_2 \in \mathcal{PCLSC}(E \times E^*)$, $g \circ \rho_2$ is a TBC–function, $\rho_2 \mathrm{dom}\, g = \mathrm{dom}(g \circ \rho_2)$ and $\mathrm{neg}(g \circ \rho_2) = \rho_2 \mathrm{pos}\, g$. Thus $\mathrm{dom}\, f - \mathrm{dom}(g \circ \rho_2) = E \times E^*$, and so the local transversality theorem, Theorem 21.12, gives $\mathrm{pos}\, f - \mathrm{neg}(g \circ \rho_2) = E \times E^*$, from which

$$\mathrm{pos}\, f - \rho_2 \mathrm{pos}\, g = E \times E^*,$$

as required. If we now replace g by the function $b \mapsto g(-b)$, we can also obtain that $\mathrm{pos}\, f + \rho_2 \mathrm{pos}\, g = E \times E^*$.

(b) It follows from (a) that there exist $(y, y^*) \in \mathrm{pos}\, f$ and $(z, z^*) \in \mathrm{pos}\, g$ such that $(y, y^*) + (z, -z^*) = (x, 0)$. But then $z^* = y^*$ and $y + z = x$.

(c) It follows from (a) that there exist $(y, y^*) \in \mathrm{pos}\, f$ and $(z, z^*) \in \mathrm{pos}\, g$ such that $(y, y^*) - (z, -z^*) = (0, x^*)$. But then $z = y$ and $y^* + z^* = x^*$. $\square$

Since $G(J_\varphi) = E \times E^*$, the following result generalizes Theorem 29.5. We note that it ultimately uses Theorem 15.1, which is a much harder result than Corollary 8.6. It can also be deduced from [121, Theorem 3 and Corollary 4].

Theorem 30.2. *Let E be a nonzero reflexive Banach space, $S, T \colon E \rightrightarrows E^*$ be maximally monotone and $G(S_\varphi) - \rho_2 \, G(T_\varphi) = E \times E^*$. Then $R(S + T) = E^*$.*

Proof. Let $f := \varphi_S$ and $g := \varphi_T$. Definition 23.3 gives us the equality $\mathrm{dom}\, \varphi_S - \rho_2 \, \mathrm{dom}\, \varphi_T = E \times E^*$, and the second half of the argument of Theorem 30.1(a) implies that $\mathrm{pos}\, \varphi_S - \rho_2 \, \mathrm{pos}\, \varphi_T = E \times E^*$. The result follows from (23.11) and the argument of Theorem 30.1(c). $\square$

We now reverse the direction of T.

Theorem 30.3. *Let E be a nonzero reflexive Banach space and $S\colon E \rightrightarrows E^*$ and $T\colon E^* \rightrightarrows E$ be maximally monotone. Suppose that $D(T_\varphi) = E^*$ and $\bigcap_{x \in E} S_\varphi(x) \neq \emptyset$. Then:*
(a) *If I_E is the identity map on E, $(I_E + TS)(E) = E$.*
(b) *If I_{E^*} is the identity map on E^*, $(I_{E^*} + ST)(E^*) = E^*$.*

Proof. Let $f := \varphi_S$ and $g := \varphi_{T^{-1}}$, so that $\pi_2 \operatorname{dom} g = E^*$.

(a) Let x be an arbitrary element of E. From Theorem 30.1(b), there exist $(y, y^*) \in \operatorname{pos} f$ and $(z, y^*) \in \operatorname{pos} g$ such that $y + z = x$. From (23.11), $(y, y^*) \in G(S)$ and $(y^*, z) \in G(T)$, so $z \in Ty^* \subset TSy$ and $x = y + z \in (I_E + TS)(y) \subset (I_E + TS)(E)$.

(b) Let x^* be an arbitrary element of E^*. From Theorem 30.1(c), there exist $(y, y^*) \in \operatorname{pos} f$ and $(y, z^*) \in \operatorname{pos} g$ such that $y^* + z^* = x^*$. From (23.11), $(y, y^*) \in G(S)$ and $(z^*, y) \in G(T)$, so $y^* \in Sy \subset STz^*$ and $x^* = z^* + y^* \in (I_{E^*} + ST)(z^*) \subset (I_{E^*} + ST)(E^*)$.   □

The next result is a considerable generalization of [122, Theorem 32.O, p. 909] which, in turn, was applied to Hammerstein integral equations. See Remark 30.5 below.

Theorem 30.4. *Let E be a nonzero reflexive Banach space, $S\colon E \rightrightarrows E^*$ and $T\colon E^* \rightrightarrows E$ be maximally monotone. Suppose that **either** $D(T_\varphi) = E^*$ and $\bigcap_{x \in E} S_\varphi(x) \neq \emptyset$ **or** $D(S_\varphi) = E$ and $\bigcap_{x^* \in E^*} T_\varphi(x^*) \neq \emptyset$. Then $(I_E + TS)(E) = E$.*

Proof. The first case has already been established in Theorem 30.3(a), while the second case follows from Theorem 30.3(b), with E replaced by E^* and the roles of S and T interchanged.   □

Remark 30.5. We now compare Theorem 30.4 and [122, Theorem 32.O]. [122] assumes that E is a Hilbert space, while Theorem 30.4 assumes that E is a reflexive Banach space. [122] assumes that $\bigcap_{x \in E} S_\varphi(x) \supset R(S)$ ($\bigcap_{x^* \in E^*} T_\varphi(x^*) \supset R(T)$, respectively) while Theorem 30.4 makes the weaker assumption that $\bigcap_{x \in E} S_\varphi(x) \neq \emptyset$ ($\bigcap_{x^* \in E^*} T_\varphi(x^*) \neq \emptyset$, respectively). However, Theorem 27.1 tells us that the assumptions $D(T) = E^*$ ($D(S) = E$, respectively) of [122] are, in fact, equivalent to the formally weaker assumptions $D(T_\varphi) = E^*$ ($D(S_\varphi) = E$, respectively) of Theorem 30.4.

We give the following result for completeness, though it is probably less interesting than Theorem 30.1.

Theorem 30.6. *Let E be a nonzero reflexive Banach space, $f \in \mathcal{PC}(E \times E^*)$ be a BC–function and $g \in \mathcal{PC}(E \times E^*)$ be a continuous real–valued BC–function. Then:*
(a) $\operatorname{pos} f - \rho_2 \operatorname{pos} g = E \times E^*$ *and* $\operatorname{pos} f + \rho_2 \operatorname{pos} g = E \times E^*$.
(b) *If $x \in E$ then there exist $(y, y^*) \in \operatorname{pos} f$ and $(z, y^*) \in \operatorname{pos} g$ such that $y + z = x$.*

(c) If $x^* \in E^*$ then there exist $(y, y^*) \in \text{pos } f$ and $(y, z^*) \in \overset{\circ}{\ } \text{pos } g$ such that $y^* + z^* = x^*$.

Proof. This proceeds exactly as in Theorem 30.1, except that it appeals to the transversality theorem, Theorem 19.16, instead of the the local transversality theorem, Theorem 21.12. □

31 The Brezis–Haraux condition

In this section, we characterize the interior and the closure of the range of a maximally monotone multifunction on a reflexive space in terms of its fitzpatrification, and deduce in Theorem 31.4(c) a strengthening of the result of Brezis–Haraux on the approximation of the range of a sum by the sum of the ranges.

Lemma 31.1 below goes back to [99, Theorem 18.3, p. 67] and Simons–Zălinescu, [109, Remark 5.6, pp. 13–14].

Lemma 31.1. *Let E be a nonzero reflexive Banach space and $U\colon E \rightrightarrows E^*$ be maximally monotone. Then*

$$\text{int } R(U) = \text{int } (\text{co } R(U)) = \text{int } R(U_\varphi)$$
$$= \text{sur } R(U) = \text{sur } (\text{co } R(U)) = \text{sur } R(U_\varphi).$$

Proof. This follows from Theorem 27.1, applied to $S := U^{-1}$. □

Theorem 31.2 extends the result proved in Rockafellar, [79, Theorem 2, p. 89] that if E is reflexive and $S\colon E \rightrightarrows E^*$ is maximally monotone then $\overline{D(S)}$ is convex.

Theorem 31.2. *Let E be a nonzero reflexive Banach space and $S\colon E \rightrightarrows E^*$ be maximally monotone. Then*

$$\overline{D(S)} = \overline{\text{co } D(S)} = \overline{D(S_\varphi)} \quad \text{and} \quad \overline{R(S)} = \overline{\text{co } R(S)} = \overline{R(S_\varphi)}.$$

Consequently, $\overline{D(S)}$ and $\overline{R(S)}$ are both convex.

Proof. Let $z \in E \setminus \overline{D(S)}$ and $\alpha := \text{dist}\big(z, D(S)\big) > 0$. Let $n \geq 1$. Since S/n is maximally monotone, Theorem 21.7 gives

$$G(S/n) - \text{neg } g_0 \ni (z, 0),$$

from which there exists $s \in G(S)$ such that $(s_1, s_2/n) - (z, 0) \in \text{neg } g_0$. It follows from (29.1) that $\langle z - s_1, s_2/n \rangle = \|z - s_1\|^2$. Since $s_1 \in D(S)$, this implies in turn that $\langle z - s_1, s_2/n \rangle \geq \alpha \|z - s_1\|$. Thus

$$\frac{\langle z - s_1, s_2 \rangle}{\|z - s_1\|} \geq n\alpha,$$

and the first set of equalities follows from Theorem 27.5(b). The second set follows similarly either from Theorem 27.6(b), or by applying the result already proved to S^{-1}. □

Theorem 27.3 and Theorem 31.2 suggest the following problem:

Problem 31.3. Is $\overline{D(S)}$ necessarily convex when E is not reflexive, S is maximally monotone and   sur $D(S_\varphi) = \emptyset$? (See Section 44 for more on this problem.)

Theorem 31.4(b) should be compared with Theorem 30.2, and Theorem 31.4(c) will be applied in Corollary 31.6.

Theorem 31.4. *Let E be a nonzero reflexive Banach space, $S, T: E \rightrightarrows E^*$ be monotone, and $S + T$ be maximally monotone. Then:*
(a) sur $R(S_\varphi + T_\varphi)$ = int $R(S + T)$.
(b) *If* sur $R(S_\varphi + T_\varphi) = E^*$ *then* $R(S + T) = E^*$.
(c) *If*

$$R(S) + R(T) \subset R(S_\varphi + T_\varphi) \tag{31.1}$$

then

$$\text{int}\left[R(S) + R(T)\right] = \text{int}\left[R(S_\varphi + T_\varphi)\right] = \text{int } R((S+T)_\varphi) = \text{int } R(S+T)$$

$$= \text{sur}\left[R(S) + R(T)\right] = \text{sur}\left[R(S_\varphi + T_\varphi)\right] = \text{sur } R((S+T)_\varphi) = \text{sur } R(S+T).$$

and

$$\overline{R(S) + R(T)} = \overline{R(S_\varphi + T_\varphi)} = \overline{R((S+T)_\varphi)} = \overline{R(S+T)}.$$

Proof. (a) The inclusion "$\supset$" is immediate since $G(S_\varphi) \supset G(S)$ and $G(T_\varphi) \supset G(T)$, and the inclusion "$\subset$" follows from Lemma 23.9(b) and Lemma 31.1.
   (b) is immediate from (a).
   (c) (31.1) and Lemma 23.9(b) imply that

$$R(S) + R(T) \subset R(S_\varphi + T_\varphi) \subset R((S+T)_\varphi).$$

Thus Theorem 31.2 and Lemma 31.1 give

$$\overline{R(S) + R(T)} \subset \overline{R(S_\varphi + T_\varphi)} \subset \overline{R((S+T)_\varphi)} = \overline{R(S+T)},$$

$$\text{int}\left[R(S) + R(T)\right] \subset \text{int}\left[R(S_\varphi + T_\varphi)\right] \subset \text{int } R((S+T)_\varphi) = \text{int } R(S+T),$$

and

$$\text{sur}\left[R(S) + R(T)\right] \subset \text{sur}\left[R(S_\varphi + T_\varphi)\right] \subset \text{sur } R((S+T)_\varphi) = \text{sur } R(S+T).$$

(c) now follows easily from Lemma 31.1 since $R(S+T) \subset R(S) + R(T)$. $\square$

Definition 31.5. We say that a monotone multifunction $S\colon E \rightrightarrows E^*$ is *rectangular* if
$$D(S) \times R(S) \subset G(S_\varphi).$$
It is easily seen that S is rectangular $\iff$ S is "3*–monotone" in the sense of [122, Definition 32.40(c), p. 901] $\iff$ S satisfies property "(**)" of [30, p. 166] (when E is a Hilbert space). It follows from [122, Proposition 32.41, p. 902] that if S is monotone with bounded range, or monotone and strongly coercive, or there exists a proper convex function $f\colon E \to \,]{-\infty}, \infty]$ such that $S = \partial f$, then S is rectangular. This last observation can also be seen directly from the result that appears in [12, Proposition 2.1] that dom $f \times$ dom $f^@ \subset$ dom $\varphi_{\partial f}$. Now if E is reflexive and S is maximally monotone then, from Theorem 27.1, Lemma 31.1 and Theorem 31.2,

$$\operatorname{int} G(S_\varphi) \subset \operatorname{int} D(S_\varphi) \times \operatorname{int} R(S_\varphi) = \operatorname{int} D(S) \times \operatorname{int} R(S) \subset \operatorname{int}\bigl[D(S) \times (R(S))\bigr]$$

and

$$\overline{G(S_\varphi)} \subset \overline{D(S_\varphi)} \times \overline{R(S_\varphi)} = \overline{D(S)} \times \overline{R(S)} = \overline{D(S) \times R(S)}.$$

So, in this case, if S is rectangular then $G(S_\varphi)$ is almost a rectangle.

The fact that either (31.2) or (31.3) implies (31.4) in Corollary 31.6 below was proved by Brezis and Haraux in Hilbert spaces in [30, Théorème 3, pp. 173–174] and [30, Théorème 4, pp. 174–175], with applications to Hammerstein integral equations, partial differential equations with nonlinear boundary conditions, and nonlinear periodic equations of evolution. These results were extended by Reich in [72, Theorem 2.2, p. 315] to the case where E is a nonzero reflexive Banach space in which J and J^{-1} are single–valued. In Corollary 31.6, we do not need to use an Asplund renorming, in contrast to the proofs of the results quoted above.

Corollary 31.6. *Let E be a nonzero reflexive Banach space, $S, T\colon E \rightrightarrows E^*$ be monotone, and $S + T$ be maximally monotone. If either*

$$S \text{ and } T \text{ are both rectangular}, \tag{31.2}$$

or

$$D(S) \subset D(T) \quad \text{and} \quad T \text{ is rectangular}, \tag{31.3}$$

then the Brezis–Haraux condition

$$\operatorname{int} R(S + T) = \operatorname{int}\bigl[R(S) + R(T)\bigr] \quad \text{and} \quad \overline{R(S + T)} = \overline{R(S) + R(T)} \tag{31.4}$$

is satisfied.

Proof. We first prove that (31.1) is satisfied. To this end, let x^* be an arbitrary element of $R(S) + R(T)$. Then there exist $s^* \in R(S)$ and $t^* \in R(T)$ such that $s^* + t^* = x^*$. If (31.2) is satisfied then, since $S + T$ is maximally monotone, there exists $x \in D(S) \cap D(T)$, from which $(x, s^*) \in D(S) \times R(S) \subset G(S_\varphi)$ and $(x, t^*) \in D(T) \times R(T) \subset G(T_\varphi)$, so

$$x^* = s^* + t^* \in (S_\varphi + T_\varphi)(x) \subset R(S_\varphi + T_\varphi).$$

If (31.3) is satisfied then we choose $x \in E$ so that $(x, s^*) \in G(S) \subset G(S_\varphi)$, from which $x \in D(S) \subset D(T)$, and so $(x, t^*) \in D(T) \times R(T) \subset G(T_\varphi)$, giving $x^* \in R(S_\varphi + T_\varphi)$ again. This completes the proof of (31.1). The result now follows from Theorem 31.4(c).   □

Remark 31.7. There is a another condition which implies (31.4). This condition, due to Pazy, appears in the appendix of [30] for Hilbert spaces. Pazy's results were generalized (with applications) by Reich in [72, Proposition 2.3, pp. 315–316] to the case where E is a reflexive Banach spaces in which J and J^{-1} are single–valued. By analogy with the proof above, one might hope that $R(S) + R(T) \subset R(S_\varphi + T_\varphi)$ in the situation considered there. However, we do not know if this is the case.

32 Bootstrapping the sum theorem

The main result of this section is the "sandwiched closed subspace theorem", Theorem 32.2. We shall show in Corollary 32.3 how different choices for F lead to known sufficient conditions for $S + T$ to be maximally monotone.

We start off with a result that is purely algebraic in character. In fact, Lemma 32.1 is equivalent to the known fact that if C is convex then $a \in C$ and $b \in \mathrm{icr}\, C \Longrightarrow]a, b] \subset \mathrm{icr}\, C$. (See Zălinescu, [119, p. 3].)

Lemma 32.1. Let C be a convex subset of a vector space E, and $F :=$ $\bigcup_{\lambda > 0} \lambda C$ be a subspace of E. Let $x \in C$ and $\alpha \in]0, 1[$. Then

$$\bigcup_{\lambda > 0} \lambda [C - \alpha x] = F. \tag{32.1}$$

Proof. $C - \alpha x \subset F - F = F$, which gives the inclusion "$\subset$" in (32.1). Now let $y \in F$. Then there exist $\mu > 0$ and $z \in C$ such that $y = \mu z$. Thus

$$(1 - \alpha)z = [(1 - \alpha)z + \alpha x] - \alpha x \in C - \alpha x,$$

and so

$$y = \mu z \in \frac{\mu}{1 - \alpha}[C - \alpha x] \subset \bigcup_{\lambda > 0} \lambda [C - \alpha x],$$

which gives the inclusion "$\supset$" in (32.1), and thus completes the proof of Lemma 32.1.   □

Theorem 32.2 first appeared in Simons–Zălinescu, [109, Theorem 5.5, p. 13].

Theorem 32.2. *Let E be a nonzero reflexive Banach space, $S, T\colon E \rightrightarrows E^*$ be maximally monotone, and suppose that there exists a closed subspace F of E such that*

$$D(S) - D(T) \subset F \subset \bigcup_{\lambda>0} \lambda \big[D(S_\varphi) - D(T_\varphi)\big]. \tag{32.2}$$

Then $S + T$ is maximally monotone. Furthermore, for all $\varepsilon > 0$,

$$D(S) - D(T) \subset D(S_\varphi) - D(T_\varphi) \subset (1+\varepsilon)\big(D(S) - D(T)\big), \tag{32.3}$$

(that is to say, $D(S_\varphi) - D(T_\varphi)$ and $D(S) - D(T)$ are almost identical) and

$$\bigcup_{\lambda>0} \lambda\big[D(S_\varphi) - D(T_\varphi)\big] = \bigcup_{\lambda>0} \lambda\big[D(S) - D(T)\big]. \tag{32.4}$$

Proof. (32.2) gives $\overline{\mathrm{lin}}\big(D(S) - D(T)\big) \subset F$. We then obtain from Lemma 28.1(c) that $\bigcup_{\lambda>0} \lambda\big[D(S_\varphi) - D(T_\varphi)\big] \subset F$, and another application of (32.2) implies that

$$\bigcup_{\lambda>0} \lambda\big[D(S_\varphi) - D(T_\varphi)\big] = F, \tag{32.5}$$

so the maximal monotonicity of $S + T$ follows from Theorem 24.1(a). Let $\varepsilon > 0$ and $\alpha := 1/(1+\varepsilon) \in\,]0,1[$. Let $d = (d_1, d_2) \in G(S_\varphi) - G(T_\varphi)$ and define the maximally monotone $U\colon E \rightrightarrows E^*$ so that $G(U) = G(S) - \alpha d$. We now obtain from (19.11), (32.5), and Lemma 32.1 with $C = D(S_\varphi) - D(T_\varphi)$ and $x = d_1$, that

$$\bigcup_{\lambda>0} \lambda\big[D(U_\varphi) - D(T_\varphi)\big] = \bigcup_{\lambda>0} \lambda\big[D(S_\varphi) - \alpha d_1 - D(T_\varphi)\big]$$

$$= \bigcup_{\lambda>0} \lambda\big[D(S_\varphi) - D(T_\varphi) - \alpha d_1\big]$$

$$= \bigcup_{\lambda>0} \lambda\big[D(S_\varphi) - D(T_\varphi)\big] = F,$$

and so Theorem 24.1(a) (with S replaced by U) implies that $U + T$ is maximally monotone and, in particular, $D(U) \cap D(T) \neq \emptyset$, that is to say

$$\big(D(S) - \alpha d_1\big) \cap D(T) \neq \emptyset,$$

from which $\alpha d_1 \in D(S) - D(T)$, and so $d_1 \in (1+\varepsilon)\big(D(S) - D(T)\big)$. Since this holds for any $d \in G(S_\varphi) - G(T_\varphi)$, we have proved that

$$D(S_\varphi) - D(T_\varphi) \subset (1+\varepsilon)\big(D(S) - D(T)\big).$$

(32.3) now follows from Lemma 23.4, and (32.4) is an immediate consequence of (32.3). □

Corollary 32.3. *Let E be a nonzero reflexive Banach space, $S, T: E \rightrightarrows E^*$ be maximally monotone, and suppose that one of the conditions (32.6)–(32.8) below is satisfied:*

$$\bigcup_{\lambda > 0} \lambda[D(S) - D(T)] \text{ is a closed subspace of } E. \tag{32.6}$$

$$\bigcup_{\lambda > 0} \lambda[\text{co}\, D(S) - \text{co}\, D(T)] \text{ is a closed subspace of } E. \tag{32.7}$$

$$\bigcup_{\lambda > 0} \lambda[\pi_1 \text{dom}\, \varphi_S{}^@ - \pi_1 \text{dom}\, \varphi_T{}^@] \text{ is a closed subspace of } E. \tag{32.8}$$

Then $S + T$ is maximally monotone. Furthermore, in any of these cases,

$$\bigcup_{\lambda > 0} \lambda[D(S) - D(T)] = \bigcup_{\lambda > 0} \lambda[\text{co}\, D(S) - \text{co}\, D(T)]$$

$$= \bigcup_{\lambda > 0} \lambda[\pi_1 \text{dom}\, \varphi_S{}^@ - \pi_1 \text{dom}\, \varphi_T{}^@]$$

$$= \bigcup_{\lambda > 0} \lambda[D(S_\varphi) - D(T_\varphi)].$$

Proof. These results follows from Lemma 23.4, and Theorem 32.2 with either $F = \bigcup_{\lambda > 0} \lambda[D(S) - D(T)]$, or $F = \bigcup_{\lambda > 0} \lambda[\text{co}\, D(S) - \text{co}\, D(T)]$, or $F = \bigcup_{\lambda > 0} \lambda[\pi_1 \text{dom}\, \varphi_S{}^@ - \pi_1 \text{dom}\, \varphi_T{}^@]$, as the case may be. □

Remark 32.4. In case (32.6), we obtain Attouch–Riahi–Théra, [2, Corollaire 1], in case (32.7), we obtain Chu, [35, Corollary 3.5]. Either of these cases can be used to establish Penot, [66, Theorem 15]. In case (32.8), we obtain Zălinescu, [120, Corollary 4]. Of course, there are also many valid "hybrid" choices, such as $F = \bigcup_{\lambda > 0} \lambda[D(S) - D(T_\varphi)]$. Clearly Rockafellar's condition (24.1) implies (32.6). Finally, we point out that our analysis does not use an Asplund renorming of E (see the remarks preceding Theorem 29.5).

Sufficient conditions (which are not subsumed by the results of Theorem 24.1 and Corollary 32.3) for the sum of maximally monotone multifunctions on a reflexive Banach space to be maximally monotone using the so–called *closedness–type regularity conditions* have been introduced recently by Boţ–Csetnek–Wanka and Boţ–Grad–Wanka in [25] and [26].

33 The > six set and the > nine set theorems for pairs of multifunctions

Let E be reflexive and $S: E \rightrightarrows E^*$ and $T: E \rightrightarrows E^*$ be maximally monotone. Our main result here is the "> 6 set theorem for pairs", Theorem 33.1, in which we prove the identity of $\text{int}[D(S) - D(T)]$ with at least five other sets. (We can certainly add in the sets $\text{int}[\pi_1 \text{dom}\, \varphi_S{}^@ - \pi_1 \text{dom}\, \varphi_T{}^@]$ and $\text{sur}[\pi_1 \text{dom}\, \varphi_S{}^@ - \pi_1 \text{dom}\, \varphi_T{}^@]$, and many hybrid sets of the kind mentioned in Remark 32.4.) In the "> 9 set theorem for pairs", Theorem 33.2, we prove the identity of $\overline{D(S) - D(T)}$ with eight other sets if $D(S) - D(T)$ is sufficiently fat. The results in this section are extensions of results proved in [98].

Theorem 33.1. *Let E be a nonzero reflexive Banach space and the multifunctions $S, T \colon E \rightrightarrows E^*$ be maximally monotone. Then:*

$$\text{int}\big[D(S) - D(T)\big] = \text{int}\big[\text{co}D(S) - \text{co}D(T)\big] = \text{int}\big[D(S_\varphi) - D(T_\varphi)\big]$$
$$= \text{sur}\big[D(S) - D(T)\big] = \text{sur}\big[\text{co}D(S) - \text{co}D(T)\big] = \text{sur}\big[D(S_\varphi) - D(T_\varphi)\big].$$

Consequently, $\text{int}\big[D(S) - D(T)\big]$ *is convex.*

Proof. We first prove that

$$\text{sur}\big[D(S_\varphi) - D(T_\varphi)\big] \subset D(S) - D(T). \tag{33.1}$$

To this end, let $x \in \text{sur}\big[D(S_\varphi) - D(T_\varphi)\big]$. In particular, $x \in D(S_\varphi) - D(T_\varphi)$, so there exists $d \in G(S_\varphi) - G(T_\varphi)$ such that $d_1 = x$. Define the maximally monotone $U \colon E \rightrightarrows E^*$ so that $G(U) = G(S) - d$. We now obtain from (19.11) that

$$\bigcup_{\lambda > 0} \lambda\big[D(U_\varphi) - D(T_\varphi)\big] = \bigcup_{\lambda > 0} \lambda\big[D(S_\varphi) - x - D(T_\varphi)\big]$$
$$= \bigcup_{\lambda > 0} \lambda\big[(D(S_\varphi) - D(T_\varphi)) - x\big] = E,$$

and so Theorem 24.1(a) (with S replaced by U) implies that $U + T$ is maximally monotone and, in particular, $D(U) \cap D(T) \neq \emptyset$, that is to say

$$\big(D(S) - x\big) \cap D(T) \neq \emptyset,$$

from which $x \in D(S) - D(T)$. This completes the proof of (33.1).

Let $f := \varphi_S{}^@$ and $g := \varphi_T{}^@$. Then $f, g \in \mathcal{PC}(E \times E^*)$ and, from (19.9), $f^@ = \varphi_S$ and $g^@ = \varphi_T$. Thus, from Corollary 22.6, $\text{sur}\big[\pi_1 \text{dom}\, \varphi_S - \pi_1 \text{dom}\, \varphi_T\big]$ is open. Since

$$\text{sur}\big[\pi_1 \text{dom}\, \varphi_S - \pi_1 \text{dom}\, \varphi_T\big] = \text{sur}\big[D(S_\varphi) - D(T_\varphi)\big],$$

(33.1) gives

$$\text{sur}\big[D(S_\varphi) - D(T_\varphi)\big] \subset \text{int}\big[D(S) - D(T)\big]. \tag{33.2}$$

From Lemma 23.4,

$$\text{int}\big[D(S) - D(T)\big] \subset \text{int}\big[\text{co}D(S) - \text{co}D(T)\big] \subset \text{int}\big[D(S_\varphi) - D(T_\varphi)\big]$$

and

$$\text{sur}\big[D(S) - D(T)\big] \subset \text{sur}\big[\text{co}D(S) - \text{co}D(T)\big] \subset \text{sur}\big[D(S_\varphi) - D(T_\varphi)\big].$$

Since $\text{int}(\ldots) \subset \text{sur}(\ldots)$, the result now follows from (33.2). $\qquad\square$

Theorem 33.2. *Let E be a nonzero reflexive Banach space, $S, T\colon E \rightrightarrows E^*$ be maximally monotone and $\mathrm{sur}\,[D(S_\varphi) - D(T_\varphi)] \neq \emptyset$. Then:*

$$\overline{D(S) - D(T)} = \overline{\mathrm{co}D(S) - \mathrm{co}D(T)} = \overline{D(S_\varphi) - D(T_\varphi)}$$
$$= \overline{\mathrm{int}\,[D(S) - D(T)]} = \overline{\mathrm{int}\,[\mathrm{co}D(S) - \mathrm{co}D(T)]} = \overline{\mathrm{int}\,[D(S_\varphi) - D(T_\varphi)]}$$
$$= \overline{\mathrm{sur}\,[D(S) - D(T)]} = \overline{\mathrm{sur}\,[\mathrm{co}D(S) - \mathrm{co}D(T)]} = \overline{\mathrm{sur}\,[D(S_\varphi) - D(T_\varphi)]}.$$

Consequently, $\overline{D(S) - D(T)}$ is convex.

Proof. Clearly, $\overline{\mathrm{int}\,[D(S) - D(T)]} \subset \overline{D(S) - D(T)}$, and from Lemma 23.4,

$$\overline{D(S) - D(T)} \subset \overline{\mathrm{co}D(S) - \mathrm{co}D(T)} \subset \overline{D(S_\varphi) - D(T_\varphi)}.$$

From Theorem 33.1, $\mathrm{int}\,[D(S_\varphi) - D(T_\varphi)] \neq \emptyset$ hence (see, for instance, Kelley–Namioka, [52, 13.1(i), pp. 110–111])

$$\overline{D(S_\varphi) - D(T_\varphi)} = \overline{\mathrm{int}\,[D(S_\varphi) - D(T_\varphi)]}.$$

The result now follows from Theorem 33.1.   □

34 The Brezis–Crandall–Pazy condition

In this section, we investigate sufficient conditions for $S + T$ to be maximally monotone of a kind different from those considered in previous sections, and which do not depend on Baire's theorem. The most general result in this section is Theorem 34.3, which is generalization of [99, Theorem 24.3, p. 94]. We show in Corollary 34.5 how to deduce from this the result of Brezis, Crandall and Pazy. The Brezis–Crandall–Pazy condition can be thought of as a *perturbation condition* and has found applications to partial differential equations. We refer the reader to their original paper, [29], for more details.

As in Section 29, E is a nonzero reflexive Banach space and $B := E \times E^*$. Then B is a SSDB space with $\lfloor b, c \rfloor := \langle b_1, c_2 \rangle + \langle c_1, b_2 \rangle$ and $\|b\| := \sqrt{\|b_1\|^2 + \|b_2\|^2}$, and $q(b) = \langle b_1, b_2 \rangle$ $\left(b = (b_1, b_2), c = (c_1, c_2) \in B\right)$.

Let $B^2 = B \times B = E \times E^* \times E \times E^*$. We consider B^2 as a SSDB space under the norm

$$\|b\| := \sqrt{\|b_1\|^2 + \|b_2\|^2 + \|b_3\|^2 + \|b_4\|^2}$$

and the bilinear form

$$\lceil b, c \rceil := \langle b_1, c_2 \rangle + \langle c_1, b_2 \rangle + \langle b_3, c_4 \rangle + \langle c_3, b_4 \rangle,$$

for all $b = (b_1, b_2, b_3.b_4)$ and $c = (c_1, c_2, c_3, c_4) \in B^2$. We then define

$$\overline{g_0}(b) := \tfrac{1}{2}\|b\|^2 = \tfrac{1}{2}\big(\|b_1\|^2 + \|b_2\|^2 + \|b_3\|^2 + \|b_4\|^2\big) = g_0(b_1, b_2) + g_0(b_3, b_4)$$

and

$$\overline{q}(b) := \tfrac{1}{2}\lceil b, b \rceil = \langle b_1, b_2 \rangle + \langle b_3, b_4 \rangle = q(b_1, b_2) + q(b_3, b_4).$$

Thus, since $g_0 \geq -q$ on B,

$$b \in \operatorname{neg}\overline{g_0} \iff g_0(b_1, b_2) + g_0(b_3, b_4) = -q(b_1, b_2) - q(b_3, b_4)$$
$$\iff g_0(b_1, b_2) = -q(b_1, b_2) \text{ and } g_0(b_3, b_4) = -q(b_3, b_4),$$

and so

$$\operatorname{neg}\overline{g_0} = \operatorname{neg} g_0 \times \operatorname{neg} g_0. \tag{34.1}$$

Let $\pi_{12}, \pi_{34}\colon B^2 \to B$ be the projection maps, defined by $\pi_{12}(b) := (b_1, b_2)$ and $\pi_{34}(b) := (b_3, b_4)$. If $n \geq 1$, define a linear map $L^n\colon B^2 \to B^2$ by

$$L^n(b) := (b_1, b_2 + nb_4, b_1 - b_3/n, -nb_4),$$

and $k^n \in \mathcal{PC}(B^2)$ by

$$k^n := f \circ \pi_{12} \circ L^n + g \circ \pi_{34} \circ L^n. \tag{34.2}$$

Lemma 34.1. *Let $f, g \in \mathcal{PC}(B)$ be BC–functions and $n \geq 1$. Then*

$$b, c \in B^2 \implies \lceil L^n(b), L^n(c) \rceil = \lceil b, c \rceil, \tag{34.3}$$

$$c, d \in B^2 \implies \lceil (L^n)^{-1}d, c \rceil = \lceil d, L^n(c) \rceil \tag{34.4}$$

and

$$d \in B^2 \implies (L^n)^{-1}(d) = (d_1, d_2 + d_4, nd_1 - nd_3, -d_4/n). \tag{34.5}$$

Furthermore,

$$(k^n)^@ = f^@ \circ \pi_{12} \circ L^n + g^@ \circ \pi_{34} \circ L^n, \tag{34.6}$$

k^n *is a BC–function, and* $L^n(\operatorname{pos} k^n) = \operatorname{pos} f \times \operatorname{pos} g$.

Proof. (34.3) follows since

$$\langle b_1, c_2 + nc_4 \rangle + \langle c_1, b_2 + nb_4 \rangle + \langle b_3 - nb_1, c_4 \rangle + \langle c_3 - nc_1, b_4 \rangle$$
$$= \langle b_1, c_2 \rangle + \langle c_1, b_2 \rangle + \langle b_3, c_4 \rangle + \langle c_3, b_4 \rangle = \lceil b, c \rceil,$$

and (34.4) is a direct consequence of (34.3). (34.5) follows from a simple computation. Furthermore, using (34.4), for all $c \in B^2$,

$$(k^n)^@(c) = \sup_{b \in B^2} \big[\lceil b, c \rceil - f \circ \pi_{12} \circ L^n(b) - g \circ \pi_{34} \circ L^n(b)\big]$$
$$= \sup_{d \in B^2} \big[\lceil (L^n)^{-1}d, c \rceil - f \circ \pi_{12}(d) - g \circ \pi_{34}(d)\big]$$
$$= \sup_{d \in B^2} \big[\lceil d, L^n(c) \rceil - f(d_1, d_2) - g(d_3, d_4)\big]$$
$$= \sup_{d \in B^2} \big[\lfloor (d_1, d_2), \pi_{12} \circ L^n(c) \rfloor - f(d_1, d_2)$$
$$+ \lfloor (d_3, d_4), \pi_{34} \circ L^n(c) \rfloor - g(d_3, d_4)\big]$$
$$= f^@ \circ \pi_{12} \circ L^n(c) + g^@ \circ \pi_{34} \circ L^n(c),$$

which establishes (34.6). Since $f^@ \geq f$ and $g^@ \geq g$ on B, it follows by comparing this with (34.2) that $(k^n)^@ \geq k^n$ on B^2. Since $f \geq q$ and $g \geq q$ on B and, from (34.3), $\bar{q} \circ L^n = \bar{q}$,

$$k^n = f \circ \pi_{12} \circ L^n + g \circ \pi_{34} \circ L^n \geq q \circ \pi_{12} \circ L^n + q \circ \pi_{34} \circ L^n = \bar{q} \circ L^n = \bar{q}, \quad (34.7)$$

thus k^n is a BC–function, as required. (34.7) also implies that $c \in \text{pos } k^n$ if, and only if,

$$f \circ \pi_{12} \circ L^n(c) = q \circ \pi_{12} \circ L^n(c) \quad \text{and} \quad g \circ \pi_{34} \circ L^n(c) = q \circ \pi_{34} \circ L^n(c),$$

i.e., $L^n(c) \in \text{pos } f \times \text{pos } g$. Thus $L^n(\text{pos } k^n) = \text{pos } f \times \text{pos } g$.   □

Lemma 34.2. Let $f, g \in \mathcal{PC}(B)$ be BC–functions, $\pi_1 \text{dom } f \cap \pi_1 \text{dom } g \neq \emptyset$ and $y \in B$. Then there exists $R \geq 0$ (independent of n) such that: for all $n \geq 1$, there exists $b^n \in \text{pos } k^n \cap (\text{neg } \overline{g_0} + (y, 0))$ with $\|b^n\| \leq R$.

Proof. Since $\pi_1 \text{dom } f \cap \pi_1 \text{dom } g \neq \emptyset$, we can choose $d \in \text{dom } f \times \text{dom } g$ such that that $d_1 = d_3$. We first define some quantities that depend only on y and d: Let

$$P := f(d_1, d_2) + g(d_3, d_4),$$

$$Q := \sqrt{\|d_1\|^2 + \|d_2 + d_4\|^2 + \|d_4\|^2},$$

$$a := (y_1, y_2, y_1, 0) \in B^2,$$

$$c := (y_1, y_2, 0, 0) = (y, 0) \in B^2,$$

and

$$R := Q + 2\|c\| + \sqrt{2P - 2\lceil d, a \rceil + 2\bar{q}(c) + (Q + \|c\|)^2}.$$

Let $n \geq 1$, and write $t^n := (L^n)^{-1}(d) \in B^2$. We note from (34.2) and (34.5) that

$$k^n(t^n) = P \quad \text{and} \quad \|t^n\| = \sqrt{\|d_1\|^2 + \|d_2 + d_4\|^2 + \|d_4\|^2/n^2} \leq Q.$$

Now $a := L^n(c)$, and so (34.4) implies that $\lceil t^n, c \rceil = \lceil d, a \rceil$. Since k^n is a BC–function, Theorem 21.4 and Remark 21.5 give us $b^n \in \text{pos } k^n \cap (\text{neg } \overline{g_0} + c)$ such that

$$\|b^n - c\| \leq \|t^n - c\| + \sqrt{2k^n(t^n) - 2\lceil t^n, c \rceil + 2\bar{q}(c) + \|t^n - c\|^2}$$

$$= \|t^n - c\| + \sqrt{2P - 2\lceil d, a \rceil + 2\bar{q}(c) + \|t^n - c\|^2}$$

$$\leq Q + \|c\| + \sqrt{2P - 2\lceil d, a \rceil + 2\bar{q}(c) + (Q + \|c\|)^2}.$$

It is immediate from this that $\|b^n\| \leq R$.   □

Theorem 34.3. *Let* $f, g \in \mathcal{PCLSC}(B)$ *be BC–functions, and suppose that* $\pi_1 \mathrm{dom}\, f \cap \pi_1 \mathrm{dom}\, g \neq \emptyset$ *and that there exists an increasing function* $j \colon [0, \infty[\to [0, \infty[$ *such that*

$$\left. \begin{array}{l} s \in \mathrm{pos}\, f \times \mathrm{pos}\, g, \ s_1 \neq s_3 \ \text{and} \ \langle s_1 - s_3, s_4 \rangle = \|s_1 - s_3\| \|s_4\| \\ \implies \quad \|s_4\| \leq j\big(\|s_1\| + \|s_2 + s_4\| + \|s_3\| + \|s_1 - s_3\| \|s_4\|\big). \end{array} \right\} \quad (34.8)$$

Then $M := \{(x, s^* + t^*) \colon (x, s^*) \in \mathrm{pos}\, f, \ (x, t^*) \in \mathrm{pos}\, g\}$ *is a maximally q–positive subset of* B.

Proof. Let $y = (y_1, y_2)$ be an arbitrary element of B. We will first prove that there exists $s \in B^2$ such that $s_1 = s_3$ and

$$f(s_1, s_2) + g(s_3, s_4) \leq \lfloor (s_1, s_2 + s_4), y \rfloor - q(y) - g_0\big((s_1, s_2 + s_4) - y\big). \quad (34.9)$$

Lemma 34.1 and Lemma 34.2 give $R \geq 0$ and, for all $n \geq 1$, $b^n \in \mathrm{neg}\, \overline{g_0} + (y, 0)$ such that

$$\|b^n\| \leq R \quad (34.10)$$

and $L^n(b^n) \in \mathrm{pos}\, f \times \mathrm{pos}\, g$. Then, from (34.1),

$$q\big((b_1^n, b_2^n) - y\big) = -g_0\big((b_1^n, b_2^n) - y\big) \quad \text{and} \quad q(b_3^n, b_4^n) = -g_0(b_3^n, b_4^n) \leq 0. \quad (34.11)$$

Let

$$s^n := L^n(b^n) \in \mathrm{pos}\, f \times \mathrm{pos}\, g. \quad (34.12)$$

Thus, from (34.11) and (34.12),

$$\left. \begin{array}{rl} f(s_1^n, s_2^n) + g(s_3^n, s_4^n) & = q(s_1^n, s_2^n) + q(s_3^n, s_4^n) \\ & = q(b_1^n, b_2^n) + q(b_3^n, b_4^n) \\ & \leq q(b_1^n, b_2^n) \\ & = \lfloor (b_1^n, b_2^n), y \rfloor - q(y) + q\big((b_1^n, b_2^n) - y\big) \\ & = \lfloor (b_1^n, b_2^n), y \rfloor - q(y) - g_0\big((b_1^n, b_2^n) - y\big). \end{array} \right\} \quad (34.13)$$

If there exists $n \geq 1$ such that $s_1^n = s_3^n$ then, using (34.5), this gives (34.9) with $s := s^n$. So we can and will assume that, for all $n \geq 1$, $s_1^n \neq s_3^n$. Using (34.5), (34.11) and (29.1), for all $n \geq 1$,

$$\left. \begin{array}{rl} \langle s_1^n - s_3^n, s_4^n \rangle & = \langle b_3^n / n, -n b_4^n \rangle = -\langle b_3^n, b_4^n \rangle \\ & = \|b_3^n\| \|b_4^n\| = \|b_3^n / n\| \|-n b_4^n\| = \|s_1^n - s_3^n\| \|s_4^n\|. \end{array} \right\} \quad (34.14)$$

Thus, from (34.12) and (34.8),

$$\|s_4^n\| \leq j\big(\|s_1^n\| + \|s_2^n + s_4^n\| + \|s_3^n\| + \|s_1^n - s_3^n\| \|s_4^n\|\big). \quad (34.15)$$

Using (34.5) and (34.10), $\|s_1^n\| = \|b_1^n\| \leq R$, $\|s_2^n + s_4^n\| = \|b_2^n\| \leq R$ and $\|s_1^n - s_3^n\| = \|b_3^n / n\| \leq R/n \leq R$, from which $\|s_3^n\| \leq 2R$. Combining with (34.14), we also have $\|s_1^n - s_3^n\| \|s_4^n\| \leq R^2$. Substituting into (34.15):

$$\|s_4^n\| \leq j(R + R + 2R + R^2) = j(4R + R^2).$$

Further, $\|s_2^n\| = \|b_2^n - s_4^n\| \leq R + j(4R + R^2)$ and $\|s_1^n - s_3^n\| \leq R/n$. Thus, by passing to a subnet, we can suppose that $s^\alpha \rightharpoonup s \in B$ with $s_1 = s_3$. We now obtain (34.9) by passing to the limit in (34.13), and using the weak lower semicontinuity of f, g and g_0. Let $w := (s_1, s_2 + s_4) \in B$. From (34.9),

$$\left.\begin{aligned}
f(s_1, s_2) + g(s_1, s_4) &\leq \lfloor w, y \rfloor - q(y) - g_0(w - y) \\
&= q(w) - q(w - y) - g_0(w - y) \\
&\leq q(w) = q(s_1, s_2) + q(s_1, s_4).
\end{aligned}\right\} \tag{34.16}$$

Since $f \geq q$ and $g \geq q$ on B, $(s_1, s_2) \in \mathrm{pos}\, f$ and $(s_1, s_4) \in \mathrm{pos}\, g$, from which $w \in M$. We also deduce from (34.16) that $q(w - y) + g_0(w - y) = 0$, from which $w - y \in \mathrm{neg}\, g_0$. Thus $y = w - (w - y) \in M - \mathrm{neg}\, g_0$. Since this construction can be carried out for all $y \in E \times E^*$, Lemma 21.3 implies that M is maximally q–positive. □

In what follows, if $U \colon E \rightrightarrows E^*$ and $x \in E$, we write $|Ux| = \inf \|Ux\|$. The next result is an implicit version of the Brezis–Crandall–Pazy theorem on the perturbation of multifunctions ("implicit" because the quantity $|Tx|$ appears on both sides of the inequality in (34.17)). The original explicit version will appear in Corollary 34.5, and a new explicit version in Corollary 34.6.

Corollary 34.4. *Let E be a nonzero reflexive Banach space, $S \colon E \rightrightarrows E^*$ and $T \colon E \rightrightarrows E^*$ be maximally monotone, $D(S) \subset D(T)$, and suppose that there exists an increasing function $j \colon [0, \infty[\to [0, \infty[$ such that,*

$$x \in D(S) \implies |Tx| \leq j(\|x\| + (|Sx| - |Tx|) \vee 0). \tag{34.17}$$

Then $S + T$ is maximally monotone.

Proof. We first note from Lemma 23.4 that $D(S_\varphi) \cap D(T_\varphi) \supset D(S) \neq \emptyset$, that is to say $\pi_1 \mathrm{dom}\, \varphi_S \cap \pi_1 \mathrm{dom}\, \varphi_T \neq \emptyset$. We now show that (34.8) is satisfied with $f := \varphi_S$ and $g := \varphi_T$. To this end, suppose that

$$s \in G(S) \times G(T), \quad s_1 \neq s_3 \quad \text{and} \quad \langle s_1 - s_3, s_4 \rangle = \|s_1 - s_3\| \|s_4\|.$$

This clearly implies that $s_1 \in D(S) \subset D(T)$. Now let t^* be an arbitrary element of Ts_1. Since T is monotone,

$$\langle s_1 - s_3, s_4 \rangle \leq \langle s_1 - s_3, t^* \rangle,$$

and so

$$\|s_1 - s_3\| \|s_4\| \leq \|s_1 - s_3\| \|t^*\|.$$

Dividing by $\|s_1 - s_3\|$ and taking the infimum over t^*, we obtain $\|s_4\| \leq |Ts_1|$. Since $(s_1, s_2) \in G(S)$, we also have

$$|Ss_1| \leq \|s_2\| \leq \|s_2 + s_4\| + \|s_4\| \leq \|s_2 + s_4\| + |Ts_1|,$$

and so $|Ss_1| - |Ts_1| \leq \|s_2 + s_4\|$, from which $(|Ss_1| - |Ts_1|) \vee 0 \leq \|s_2 + s_4\|$. Thus (34.17) implies that

$$\|s_4\| \leq |Ts_1| \leq j(\|s_1\| + \|s_2 + s_4\|),$$

and it now follows from Theorem 34.3 that $S + T$ is maximally monotone. $\square$

Corollary 34.5. *Let E be a nonzero reflexive Banach space, $S\colon E \rightrightarrows E^*$ and $T\colon E \rightrightarrows E^*$ be maximally monotone, $D(S) \subset D(T)$, and suppose that there exist increasing functions $k\colon [0,\infty[\to [0,1[$ and $C\colon [0,\infty[\to [0,\infty[$ such that*

$$x \in D(S) \implies |Tx| \leq k(\|x\|)|Sx| + C(\|x\|). \tag{34.18}$$

Then $S + T$ is maximally monotone.

Proof. Let $x \in D(S)$. From (34.18),

$$\begin{aligned}
(1 - k(\|x\|))|Tx| &\leq k(\|x\|)(|Sx| - |Tx|) + C(\|x\|) \\
&\leq k(\|x\|)(|Sx| - |Tx|) \vee 0 + C(\|x\|),
\end{aligned}$$

and the result now follows from Corollary 34.4 with

$$j(\rho) := \frac{k(\rho)\rho + C(\rho)}{1 - k(\rho)}. \qquad \square$$

In our final result, we allow k to take values bigger than 1, but we replace $|Sx|$ by $|Sx|^p$ in the statement of Corollary 34.5.

Corollary 34.6. *Let E be a nonzero reflexive Banach space, $S\colon E \rightrightarrows E^*$ and $T\colon E \rightrightarrows E^*$ be maximally monotone, $D(S) \subset D(T)$, and suppose that $0 < p < 1$ and there exist increasing functions $k\colon [0,\infty[\to [0,\infty[$ and $C\colon [0,\infty[\to [0,\infty[$ such that,*

$$x \in D(S) \implies |Tx| \leq k(\|x\|)|Sx|^p + C(\|x\|). \tag{34.19}$$

Then $S + T$ is maximally monotone.

Proof. Let $x \in D(S)$. From (34.19) and the fact that

$$\lambda, \mu \geq 0 \implies (\lambda + \mu)^p \leq \lambda^p + \mu^p,$$

we have

$$\begin{aligned}
|Tx| &\leq k(\|x\|)(|Tx| \vee |Sx|)^p + C(\|x\|) \\
&= k(\|x\|)(|Tx| + (|Sx| - |Tx|) \vee 0)^p + C(\|x\|) \\
&\leq k(\|x\|)|Tx|^p + k(\|x\|)((|Sx| - |Tx|) \vee 0)^p + C(\|x\|).
\end{aligned}$$

Now if $k(\|x\|)|Tx|^p \leq \frac{1}{2}|Tx|$ then this gives

$$|Tx| \leq 2k(\|x\|)((|Sx| - |Tx|) \vee 0)^p + 2C(\|x\|),$$

while if $\frac{1}{2}|Tx| < k(\|x\|)|Tx|^p$ then, of course, $|Tx| < (2k(\|x\|))^{1/(1-p)}$. Thus the result follows from Corollary 34.4, with

$$j(\rho) := [2k(\rho)\rho^p + 2C(\rho)] \vee (2k(\rho))^{1/(1-p)}. \qquad \square$$

Problem 34.7. Let $f, g \in \mathcal{PCLSC}(B)$ be BC–functions. Suppose that $\pi_1 \mathrm{dom}\, f \cap \pi_1 \mathrm{dom}\, g \neq \emptyset$ and that there exists an increasing function $j\colon [0, \infty[\to [0, \infty[$ such that

$$\left.\begin{array}{l} s \in \mathrm{pos}\, f \times \mathrm{pos}\, g, \; s_1 \neq s_3 \text{ and } \langle s_1 - s_3, s_4 \rangle = \|s_1 - s_3\| \|s_4\| \\[2mm] \implies \quad \|s_4\| \leq j\big(\|s_1\| + \|s_2 + s_4\| + \|s_3\| + \|s_1 - s_3\| \|s_4\|\big). \end{array}\right\} \quad (34.8)$$

Then is it true that, for all $b \in B$, there exist $a, c \in B$ such that $a_1 = c_1 = b_1$, $a_2 + c_2 = b_2$ and $f^@(a) + g^@(c) \leq (f \oplus_2 g)^@(a)$? If the answer to this question is "yes", then the results of this section can be more completely integrated into the theory of BC–functions.

Remark 34.8. We emphasize that, unlike in the analysis in [29], we do not use any renorming or fixed–point theorems in any of the results in this section.

VI Special maximally monotone multifunctions

35 The norm–dual of the space $E \times E^*$ and $\widetilde{BC}$–functions

We now develop the machinery that we will use for the discussion of the various subclasses of the maximally monotone multifunctions that we will consider in the following sections. The main result of this section is Theorem 35.8 which, together with its two corollaries, Corollary 35.9 and Corollary 35.10, will have direct applications in Sections 37, 39, 40 and 41. So let E be a nonzero Banach space, E^* be its topological dual space and $B := E \times E^*$. In Section 22, we considered B as a SSD space. We recall that if $b = (b_1, b_2) \in B$ then $q(b) = \langle b_1, b_2 \rangle$. Of course, B is also a normed space under the norm $\|(b_1, b_2)\| = \sqrt{\|b_1\|^2 + \|b_2\|^2}$. As in Definition 21.1, we write $g_0 := \frac{1}{2}\| \cdot \|^2$ on B. Even though B is a SSD space and a Banach space, it is not a SSDB space if E is not reflexive, since the topology $\mathcal{T}_{\| \cdot \|}(B)$ is not B–compatible. Following Notation 16.1, the norm–dual of B is $B^* = E^{**} \times E^*$ under the pairing

$$\langle b, v \rangle := \langle b_1, v_2 \rangle + \langle b_2, v_1 \rangle \quad (b = (b_1, b_2) \in B, \ v = (v_1, v_2) \in B^*),$$

and the dual norm of B^* is given by $\|(v_1, v_2)\| = \sqrt{\|v_1\|^2 + \|v_2\|^2}$. B^* is also a SSD space under the bilinear form

$$[u, v] := \langle v_2, u_1 \rangle + \langle u_2, v_1 \rangle \quad (u = (u_1, u_2) \in B^*, \ v = (v_1, v_2) \in B^*).$$

We define $\widetilde{q}\colon B^* \to \mathbb{R}$ by $\widetilde{q}(v) := \frac{1}{2}[v, v]$. $\widetilde{q}$ is a quadratic form on B^* and, since $\widetilde{q}(v_1, v_2) = \langle v_2, v_1 \rangle$, bilinear on $E^{**} \times E^*$. We write ι for the linear isometry of B into B^* defined by $\iota(b_1, b_2) := (\widehat{b_1}, b_2)$. Clearly, $\widetilde{q} \circ \iota = q$. Furthermore,

$$v \in B^* \implies |\widetilde{q}(v)| \leq \tfrac{1}{2}\|v\|^2. \tag{35.1}$$

We note for future reference that we proved in (22.8) that if $f \in \mathcal{PC}(B)$ then

$$f^{@} = f^* \circ \iota \quad \text{on} \quad B.$$

Let $\mathcal{T}_{\mathcal{WN}}(B^*)$ be the topology $w(E^{**}, E^*) \times \mathcal{T}_{\| \cdot \|}(E^*)$ on B^*. Then $\mathcal{T}_{\mathcal{WN}}(B^*)$ is B^*–compatible. We will use frequently the easily verifiable result that if $\{v_\gamma\}$ is a *bounded* net in B^* and $v \in B^*$ then

$$v_\gamma \to v \text{ in } \mathcal{T}_{\mathcal{WN}}(B^*) \quad \Longrightarrow \quad \widetilde{q}(v_\gamma) \to \widetilde{q}(v). \tag{35.2}$$

We say that $h\colon B \to]-\infty,\infty]$ is a $\widetilde{BC}$–function if h is a BC–function (see (23.1)) and $h^* \geq \widetilde{q}$ on B^*. (If E is reflexive then this is equivalent to saying that h is a BC–function.) We say that $h\colon B \to]-\infty,\infty]$ is a $\widetilde{TBC}$–function if h is a TBC–function (see Definition 19.14) and $h^* \geq -\widetilde{q}$ on B^*. (If E is reflexive then this is equivalent to saying that h is a TBC–function.) We note from (35.1) that g_0 is a $\widetilde{BC}$–function and a $\widetilde{TBC}$–function.

We now give a useful example of a $\widetilde{BC}$–function. We remark that "k^{**}" in the next result stands for the *full biconjugate* of k (defined on E^{**}) — this is discussed in greater detail in Section 38.

Lemma 35.1. *Let E be a nonzero Banach space, $k \in \mathcal{PCLSC}(E)$ and $g(c) := k(c_1) + k^*(c_2)$ $(c = (c_1, c_2) \in E \times E^*)$. Then g is a $\widetilde{BC}$–function.*

Proof. Since $g^*(v) = k^{**}(v_1) + k^*(v_2)$ $(v = (v_1, v_2) \in E^{**} \times E^*)$, this is immediate from the Fenchel–Young inequality, (8.2), and the Fenchel–Moreau theorem, Corollary 12.4.   □

We next give an example of a maximally monotone (single-valued and linear) multifunction, S, such that φ_S is *not* a $\widetilde{BC}$-function. (In this connection, see Theorem 36.3.)

Example 35.2. Let E be a nonzero Banach space and $S\colon E \to E^*$ be continuous, linear and skew, that is to say, for all $x \in E$, $\langle x, Sx \rangle = 0$ or, equivalently, for all $x, y \in E$, $\langle x, Sy \rangle = -\langle y, Sx \rangle$. Then, considering S as a single–valued monotone multifunction, for all $b = (b_1, b_2) \in E \times E^*$,

$$\varphi_S(b) = \sup_{s \in E} \left[\langle s, b_2 \rangle + \langle b_1, Ss \rangle - \langle s, Ss \rangle \right] = \sup_{s \in E} \langle s, b_2 - Sb_1 \rangle,$$

and so $\varphi_S = \mathbb{I}_{G(S)}$. It follows that, for all $v = (v_1, v_2) \in E^{**} \times E^*$,

$$\varphi_S{}^*(v) = \sup_{b \in E \times E^*} \left[\langle b, v \rangle - \mathbb{I}_{G(S)}(b) \right] = \sup_{a \in G(S)} \langle a, v \rangle$$

$$= \sup_{x \in E} \left[\langle x, v_2 \rangle + \langle Sx, v_1 \rangle \right] = \sup_{x \in E} \langle x, v_2 + S^* v_1 \rangle,$$

and so $\varphi_S{}^* = \mathbb{I}_{G(-S^*)}$. We are going to give an example where there exists $(x^{**}, x^*) \in G(S^*)$ such that $\langle x^*, x^{**} \rangle < 0$. Then $\varphi_S{}^*(x^{**}, -x^*) = 0$ but $\widetilde{q}(x^{**}, -x^*) = -\langle x^*, x^{**} \rangle > 0$, and so φ_S is not a $\widetilde{BC}$–function.

Our example was originally due to Gossez, and the fact that it has the required properties follows from Bauschke–Borwein, [9, Theorem 4.1, pp. 10–12] — see also [9, Example 5.2, p. 15]. Here, we give a direct proof. Let $E := \ell^1$, and $S\colon \ell^1 \to E^* = \ell^\infty$ be defined by

$$(Sx)_n = -\sum_{k<n} x_k + \sum_{k>n} x_k \quad (x \in \ell^1).$$

S is skew. Let $e := (1, 1, 1, \ldots) \in \ell^\infty$ and, for all $p \geq 1$, $e^{(p)}$ be the pth basic vector of ℓ^1. Then $\left(Se^{(p)}\right)_n = 1$ if $n < p$ and $\left(Se^{(p)}\right)_n = -1$ if $n > p$. Since c_0

is a closed subspace of ℓ^∞ and $e \notin c_0$, it follows from Theorem 4.4 that there exists $x^{**} \in E^{**}$ that vanishes on c_0 with $\langle e, x^{**} \rangle = 1$, for instance a "Banach limit". It is clear from the above that $\langle e^{(p)}, S^* x^{**} \rangle = \langle S e^{(p)}, x^{**} \rangle = -1$, and so $S^* x^{**} = -e$. The result follows with $x^* := -e$ since $\langle -e, x^{**} \rangle = -1 < 0$.

We start off our analysis with a useful analog of Lemma 19.13. For this, we will need some notation: If $\psi \in \mathcal{PC}(B^*)$ and $\psi \geq \widetilde{q}$ on B^*, let

$$\widetilde{\text{pos}}\, \psi = \{v \in B^*: \psi(v) = \widetilde{q}(v)\}.$$

Lemma 19.8 implies that if $\widetilde{\text{pos}}\, \psi \neq \emptyset$ then $\widetilde{\text{pos}}\, \psi$ is a $\widetilde{q}$–positive subset of B^*. Similarly, if $\psi \in \mathcal{PC}(B^*)$ and $\psi \geq -\widetilde{q}$ on B^* let

$$\widetilde{\text{neg}}\, \psi = \{v \in B^*: \psi(v) = -\widetilde{q}(v)\}.$$

As a special case of this, if $v \in \widetilde{\text{neg}}\, g_0^*$ then, using the arguments of the first paragraph of Section 29,

$$\|v_1\|^2 = \|v_2\|^2 = -\widetilde{q}(v). \tag{35.3}$$

We will use the following simple result many times:

Lemma 35.3. *Let E be a nonzero Banach space, $B := E \times E^*$, $b \in B$, $f \in \mathcal{PC}(B)$ be a $\widetilde{BC}$–function, and $\iota(b) \in \widetilde{\text{pos}}\, f^*$. Then $b \in \text{pos}\, f$.*

Proof. Since $\widetilde{q} \circ \iota(b) = f^* \circ \iota(b) = f^{@}(b) \geq f(b) \geq q(b) = \widetilde{q} \circ \iota(b) = \widetilde{q}(w)$, it follows that $f(b) = q(b)$. $\qquad\square$

Lemma 35.4. *Let E be a nonzero Banach space, $B := E \times E^*$, $f \in \mathcal{PC}(B)$, $f \geq q$ on B, $f^* \geq \widetilde{q}$ on B^* and $d \in B$. Then f_d (see Lemma 19.13) $\geq q$ on B,*

$${f_d}^* = f^*\big(\cdot + \iota(d)\big) - \langle d, \cdot \rangle - q(d) \geq \widetilde{q} \text{ on } B^*$$

and

$$\widetilde{\text{pos}}\, {f_d}^* = \widetilde{\text{pos}}\, f^* - \iota(d).$$

Proof. For all $b \in B$,

$$f_d(b) = f(b + d) - \lfloor b, d \rfloor - q(d) \geq q(b + d) - \lfloor b, d \rfloor - q(d) = q(b).$$

Further, for all $v \in B^*$,

$$\begin{aligned}
{f_d}^*(v) &= \sup_{b \in B}\big[\langle b, v \rangle + \lfloor b, d \rfloor + q(d) - f(b + d)\big] \\
&= \sup_{e \in B}\big[\langle e - d, v \rangle + \lfloor e - d, d \rfloor + q(d) - f(e)\big] \\
&= \sup_{e \in B}\big[\langle e, v \rangle + \langle e, \iota(d) \rangle - f(e)\big] - \langle d, v \rangle - q(d) \\
&= \sup_{e \in B}\big[\langle e, v + \iota(d) \rangle - f(e)\big] - \langle d, v \rangle - q(d) \\
&= f^*\big(v + \iota(d)\big) - \langle d, v \rangle - q(d) \geq \widetilde{q}\big(v + \iota(d)\big) - \langle d, v \rangle - q(d) = \widetilde{q}(v).
\end{aligned}$$

Finally, since

$$\begin{aligned}
v \in \widetilde{\text{pos}}\, {f_d}^* &\iff f^*\big(v + \iota(d)\big) - \langle d, v \rangle - q(d) = \widetilde{q}(v) \\
&\iff f^*\big(v + \iota(d)\big) = \widetilde{q}\big(v + \iota(d)\big) \iff v + \iota(d) \in \widetilde{\text{pos}}\, f^*,
\end{aligned}$$

we have $\widetilde{\text{pos}}\, {f_d}^* = \widetilde{\text{pos}}\, f^* - \iota(d)$, as required. $\qquad\square$

Our next "transversality" result is an analog of Theorem 19.16, which will be used explicitly in Lemma 35.6 and Theorem 36.3.

Lemma 35.5. *Let E be a nonzero Banach space, $B := E \times E^*$, $f \in \mathcal{PC}(B)$, $f \geq q$ on B, $f^* \geq \tilde{q}$ on B^*, $g \colon B \to \mathbb{R}$ be a $\mathcal{T}_{\| \, \|}(B)$–continuous convex function, $g \geq -q$ on B and $g^* \geq -\tilde{q}$ on B^*. Then*

$$\widetilde{\mathrm{pos}}\, f^* + \widetilde{\mathrm{neg}}\, g^* \supset \iota(B).$$

In particular, since $0 \in B$,

$$\widetilde{\mathrm{pos}}\, f^* \cap \left(-\widetilde{\mathrm{neg}}\, g^*\right) \neq \emptyset.$$

Proof. Let d be an arbitrary element of B. As in Theorem 19.16,

$$b \in B \quad \Longrightarrow \quad f_d(b) + g(b) \geq q(b) - q(b) = 0.$$

Thus Rockafellar's version of the Fenchel duality theorem, Corollary 8.6, gives $v \in B^*$ such that $f_d^*(v) + g^*(-v) \leq 0$. However,

$$f_d^*(v) \geq \tilde{q}(v), \quad g^*(-v) \geq -\tilde{q}(-v) = -\tilde{q}(v) \quad \text{and} \quad \tilde{q}(v) - \tilde{q}(v) = 0.$$

Consequently,

$$f_d^*(v) = \tilde{q}(v) \quad \text{and} \quad g^*(-v) = -\tilde{q}(-v),$$

that is to say, using Lemma 35.4, $v \in \widetilde{\mathrm{pos}}\, f_d^* = \widetilde{\mathrm{pos}}\, f^* - \iota(d)$, and also $-v \in \widetilde{\mathrm{neg}}\, g^*$. But then $\iota(d) = (\iota(d) + v) - v \in \widetilde{\mathrm{pos}}\, f^* + \widetilde{\mathrm{neg}}\, g^*$.   □

Our next preliminary result is an analog of Theorem 21.4(b).

Lemma 35.6. *Let E be a nonzero Banach space, $B := E \times E^*$, $f \in \mathcal{PC}(B)$, $f \geq q$ on B, $f^* \geq \tilde{q}$ on B^*, $d \in B$, and*

$$v \in \widetilde{\mathrm{pos}}\, f^* \quad \Longrightarrow \quad \tilde{q}(v - \iota(d)) \geq 0.$$

Then $\iota(d) \in \widetilde{\mathrm{pos}}\, f^*$.

Proof. Lemma 35.5 with $g := g_0$ provides us with $v \in \widetilde{\mathrm{pos}}\, f^*$ such that $g_0(\iota(d) - v) = -\tilde{q}(\iota(d) - v) = -\tilde{q}(v - \iota(d)) \leq 0$. Consequently, $v = \iota(d)$, which gives the required result.   □

Our final preliminary result is an analog of Lemma 22.9.

Lemma 35.7. *Let E be a nonzero Banach space, $B := E \times E^*$ and $f, g \in \mathcal{PCLSC}(B)$ be $\widetilde{BC}$–functions.*
(a) *If*

$$\bigcup_{\lambda > 0} \lambda\left[\pi_1 \mathrm{dom}\, f - \pi_1 \mathrm{dom}\, g\right] \text{ is a closed subspace of } E$$

then $f \oplus_2 g$ is a $\widetilde{BC}$–function. Furthermore, $v \in \widetilde{\mathrm{pos}}\, (f \oplus_2 g)^*$ if, and only if, there exist $u \in \widetilde{\mathrm{pos}}\, f^*$ and $w \in \widetilde{\mathrm{pos}}\, g^*$ such that

$$u_1 = w_1 = v_1 \quad \text{and} \quad u_2 + w_2 = v_2.$$

(b) *If*

$$\bigcup_{\lambda > 0} \lambda \left[\pi_2 \operatorname{dom} f - \pi_2 \operatorname{dom} g \right] \text{ is a closed subspace of } E^*$$

then $f \oplus_1 g \geq q$ *on* B *and* $(f \oplus_1 g)^* \geq \tilde{q}$ *on* B^*. *Furthermore,* $v \in \widetilde{\operatorname{pos}} \, (f \oplus_1 g)^*$ *if, and only if, there exist* $u \in \widetilde{\operatorname{pos}} \, f^*$ *and* $w \in \widetilde{\operatorname{pos}} \, g^*$ *such that*

$$u_1 + w_1 = v_1 \quad \text{and} \quad u_2 = w_2 = v_2.$$

(For the reasons explained in Problem 22.12, we cannot assert that $f \oplus_1 g$ is a $\widetilde{BC}$–function.)

Proof. In case (a), Lemma 22.9 implies that $f \oplus_2 g$ is a BC–function. From Remark 22.10, for all $v = (v_1, v_2) \in B^*$,

$$(f \oplus_2 g)^*(v)$$
$$= \min \left\{ f^*(u) + g^*(w) \colon u, w \in B^*, \ u_1 = w_1 = v_1, \ u_2 + w_2 = v_2 \right\}$$
$$\geq \inf \left\{ \tilde{q}(u) + \tilde{q}(w) \colon u, w \in B^*, \ u_1 = w_1 = v_1, \ u_2 + w_2 = v_2 \right\} = \tilde{q}(v),$$

thus $f \oplus_2 g$ is a $\widetilde{BC}$–function, and we obtain the required characterization of $\widetilde{\operatorname{pos}} \, (f \oplus_2 g)^*$.

Case (b) is proved similarly, using Remark 22.11. $\qquad\square$

We now come to the main result of this section, which will be used (via Corollary 35.9 and Corollary 35.10) in Theorem 37.1, Theorem 39.1 and Theorem 40.1.

Theorem 35.8. *Let E be a nonzero Banach space, $B := E \times E^*$, $d \in B$, and $f, g \in \mathcal{PCLSC}(B)$ be $\widetilde{BC}$–functions.*
(a) *Let*

$$\bigcup_{\lambda > 0} \lambda \left[\pi_1 \operatorname{dom} f - \pi_1 \operatorname{dom} g \right] \text{ be a closed subspace of } E$$

and

$$u \in \widetilde{\operatorname{pos}} \, f^*, \ w \in \widetilde{\operatorname{pos}} \, g^* \text{ and } u_1 = w_1 \Longrightarrow \langle u_2 + w_2 - d_2, u_1 - \widehat{d_1} \rangle \geq 0. \quad (35.4)$$

Then there exist $a \in \operatorname{pos} f$ and $c \in \operatorname{pos} g$ such that

$$a_1 = c_1 = d_1 \quad \text{and} \quad a_2 + c_2 = d_2.$$

(b) *Let*

$$\bigcup_{\lambda > 0} \lambda \left[\pi_2 \operatorname{dom} f - \pi_2 \operatorname{dom} g \right] \text{ be a closed subspace of } E^*,$$

$$u \in \widetilde{\operatorname{pos}} \, f^*, \ w \in \widetilde{\operatorname{pos}} \, g^* \text{ and } u_2 = w_2 \Longrightarrow \langle u_2 - d_2, u_1 + w_1 - \widehat{d_1} \rangle \geq 0, \quad (35.5)$$

and

$$(x^{**}, d_2) \in \widetilde{\operatorname{pos}} \, g^* \quad \Longrightarrow \quad x^{**} \in \widehat{E}. \quad (35.6)$$

Then there exist $a \in \operatorname{pos} f$ and $c \in \operatorname{pos} g$ such that

$$a_1 + c_1 = d_1 \quad \text{and} \quad a_2 = c_2 = d_2.$$

Proof. (a) Lemma 35.7(a) implies that $f \oplus_2 g$ is a $\widetilde{BC}$–function Let $v \in \widetilde{\mathrm{pos}}\,(f \oplus_2 g)^*$. Then, from Lemma 35.7(a) again, there exist $u \in \widetilde{\mathrm{pos}}\,f^*$ and $w \in \widetilde{\mathrm{pos}}\,g^*$ such that

$$u_1 = w_1 = v_1 \quad \text{and} \quad u_2 + w_2 = v_2.$$

But then, using (35.4),

$$\tilde{q}(v - \iota(d)) = \langle v_2 - d_2, v_1 - \hat{d}_1 \rangle = \langle u_2 + w_2 - d_2, u_1 - \hat{d}_1 \rangle \geq 0,$$

and so Lemma 35.6 implies that $\iota(d) \in \widetilde{\mathrm{pos}}\,(f \oplus_2 g)^*$. As above, there exist $u \in \widetilde{\mathrm{pos}}\,f^*$ and $w \in \widetilde{\mathrm{pos}}\,g^*$ such that

$$u_1 = w_1 = \hat{d}_1 \quad \text{and} \quad u_2 + w_2 = d_2.$$

Let $c := (d_1, w_2) \in B$, so that $w = \iota(c)$. Since $w \in \widetilde{\mathrm{pos}}\,g^*$, Lemma 35.3 implies that $(d_1, w_2) = c \in \mathrm{pos}\,g$. We can prove similarly that $a := (d_1, u_2) \in \mathrm{pos}\,f$, and (a) now follows.

(b) Arguing exactly as in (a), only using Lemma 35.7(b) and (35.5), there exist $u \in \widetilde{\mathrm{pos}}\,f^*$ and $w \in \widetilde{\mathrm{pos}}\,g^*$ such that

$$u_1 + w_1 = \hat{d}_1 \quad \text{and} \quad u_2 = w_2 = d_2.$$

(35.6) now implies that there exists $x \in E$ such that $w_1 = \hat{x}$, and so $u_1 = \hat{d}_1 - \hat{x}$. Let $c := (x, d_2) \in B$, so that $w = \iota(c)$. Since $w \in \widetilde{\mathrm{pos}}\,g^*$, Lemma 35.3 implies that $(x, d_2) = c \in \mathrm{pos}\,g$. We can prove similarly that $a := (d_1 - x, d_2) \in \mathrm{pos}\,f$, and (b) now follows. $\square$

Corollary 35.9 below will be used explicitly in Theorem 37.1 and Theorem 40.1.

Corollary 35.9. *Let E be a nonzero Banach space and D be a nonempty $w(E^*, E)$–compact convex subset of E^*. Let $B := E \times E^*$.*
(a) *Suppose that $y \in E$, $h \in \mathcal{PCLSC}(B)$ is a $\widetilde{BC}$–function and*

$$v \in \widetilde{\mathrm{pos}}\,h^* \implies \langle v_2, v_1 - \hat{y} \rangle + \sup \langle D, v_1 - \hat{y} \rangle \geq 0. \tag{35.7}$$

Then there exists $x^* \in -D$ such that $(y, x^*) \in \mathrm{pos}\,h$.
(b) *Suppose that $d \in E \times \mathrm{int}\,D$, $f \in \mathcal{PCLSC}(B)$ is a $\widetilde{BC}$–function, $\pi_2 \mathrm{dom}\,f \cap \mathrm{int}\,D \neq \emptyset$ and*

$$u \in \widetilde{\mathrm{pos}}\,f^* \text{ and } u_2 \in D \implies \tilde{q}(u - \iota(d)) \geq 0. \tag{35.8}$$

Then $d \in \mathrm{pos}\,f$.

Proof. Define $k \colon E \to \mathbb{R}$ by $k := \sup\langle \cdot, D \rangle$. Then $k^* = \mathbb{I}_D$ and $k^{**} = \sup\langle D, \cdot \rangle$. Define $g \in \mathcal{PCLSC}(B)$ by

$$c = (c_1, c_2) \in B \quad \Longrightarrow \quad g(c) := k(c_1) + k^*(c_2).$$

From Lemma 35.1, g is a $\widetilde{BC}$–function and

$$w = (w_1, w_2) \in B^* \quad \Longrightarrow \quad g^*(w) = k^{**}(w_1) + k^*(w_2).$$

Thus we have

$$c \in \text{pos}\, g \iff c_2 \in D \text{ and } \sup\langle c_1, D \rangle = \langle c_1, c_2 \rangle \tag{35.9}$$

and

$$w \in \widetilde{\text{pos}}\, g^* \iff w_2 \in D \text{ and } \langle w_2, w_1 \rangle = \sup\langle D, w_1 \rangle, \tag{35.10}$$

from which

$$w \in \widetilde{\text{pos}}\, g^* \text{ and } d \in E \times D \quad \Longrightarrow \quad w_2 \in D \text{ and } \langle w_2 - d_2, w_1 \rangle \ge 0. \tag{35.11}$$

(a) Let $f := h_{(y,0)}$. Lemma 19.13 and Lemma 35.4 imply that f is a $\widetilde{BC}$–function. Since $\pi_1 \text{dom}\, g = E$, $\bigcup_{\lambda > 0} \lambda \big[\pi_1 \text{dom}\, f - \pi_1 \text{dom}\, g \big] = E$. We now verify that (35.4) is satisfied with $d := 0$. To this end, let $u \in \widetilde{\text{pos}}\, f^*$, $w \in \widetilde{\text{pos}}\, g^*$ and $u_1 = w_1$. From Lemma 35.4 again, $u = (v_1 - \widehat{y}, v_2)$ for some $v \in \widetilde{\text{pos}}\, h^*$. (35.10) gives

$$\langle w_2, u_1 \rangle = \langle w_2, w_1 \rangle = \sup\langle D, w_1 \rangle = \sup\langle D, u_1 \rangle = \sup\langle D, v_1 - \widehat{y} \rangle,$$

and combining this with (35.7) gives

$$\langle u_2 + w_2, u_1 \rangle = \langle v_2, v_1 - \widehat{y} \rangle + \sup\langle D, v_1 - \widehat{y} \rangle \ge 0.$$

Thus (35.4) is satisfied with $d = 0$, and (a) is now immediate from Lemma 19.13, Theorem 35.8(a) and (35.9).

(b) Since $\pi_2 \text{dom}\, g = D$ and $\pi_2 \text{dom}\, f \cap \text{int}\, D \ne \emptyset$, it follows that $\bigcup_{\lambda > 0} \lambda \big[\pi_2 \text{dom}\, f - \pi_2 \text{dom}\, g \big] = E^*$. We now verify that (35.5) is satisfied. To this end, let $u \in \widetilde{\text{pos}}\, f^*$, $w \in \widetilde{\text{pos}}\, g^*$ and $u_2 = w_2$. (35.10) and (35.8) imply that $u_2 \in D$ and thus $\widetilde{q}(u - \iota(d)) \ge 0$. Since $u_2 = w_2$, (35.11) gives $\langle u_2 - d_2, w_1 \rangle \ge 0$. Adding up these two inequalities, we have

$$\langle u_2 - d_2, u_1 + w_1 - \widehat{d_1} \rangle \ge 0,$$

and so (35.5) is satisfied. (35.6) is clear since if $(x^{**}, d_2) \in \widetilde{\text{pos}}\, g^*$ then, from (35.10), $\langle d_2, x^{**} \rangle = \sup\langle D, x^{**} \rangle$ and, since $d_2 \in \text{int}\, D$, $x^{**} = 0 \in \widehat{E}$. Theorem 35.8(b) now gives $a \in \text{pos}\, f$ and $c \in \text{pos}\, g$ such that

$$a_1 + c_1 = d_1 \quad \text{and} \quad a_2 = c_2 = d_2.$$

From (35.9), $\sup\langle c_1, D \rangle = \langle c_1, d_2 \rangle$. Since $\text{int}\, D \ni d_2$, this implies that $c_1 = 0$, from which $d = a \in \text{pos}\, f$. This completes the proof of (b). $\square$

Corollary 35.10 below will be used explicitly in Theorem 39.1 and Theorem 40.1.

Corollary 35.10. *Let E be a nonzero Banach space, D be a nonempty bounded convex subset of E and $\ddot{D}$ be the $w(E^{**}, E^*)$–closure of $\widehat{D}$ in E^{**}. Let $B := E \times E^*$.*
(a) Suppose that, $d \in \text{int } D \times E^*$, $f \in \mathcal{PCLSC}(B)$ is a $\widetilde{BC}$–function, $\pi_1 \text{dom } f \cap \text{int } D \neq \emptyset$ and

$$u \in \widetilde{\text{pos}}\, f^* \text{ and } u_1 \in \ddot{D} \implies \tilde{q}(u - \iota(d)) \geq 0. \qquad (35.12)$$

Then $d \in \text{pos } f$.
(b) Suppose that $y^* \in E^*$, $h \in \mathcal{PCLSC}(B)$ is a $\widetilde{BC}$–function,

$$v \in \widetilde{\text{pos}}\, h^* \implies \langle v_2 - y^*, v_1 \rangle + \sup\langle D, v_2 - y^* \rangle \geq 0, \qquad (35.13)$$

and D is $w(E, E^*)$–compact. Then there exists $x \in -D$ such that $(x, y^*) \in \text{pos } h$.

Proof. Define $k\colon E \to [0, \infty]$ by $k := \mathbb{I}_D$. Then $k^* = \sup\langle D, \cdot \rangle$ and $k^{**} = \mathbb{I}_{\ddot{D}}$. Define $g \in \mathcal{PCLSC}(B)$ by:

$$c = (c_1, c_2) \in B \implies g(c) := k(c_1) + k^*(c_2).$$

From Lemma 35.1, g is a $\widetilde{BC}$–function and

$$w = (w_1, w_2) \in B^* \implies g^*(w) = k^{**}(w_1) + k^*(w_2).$$

Thus we have

$$c \in \text{pos } g \iff c_1 \in D \text{ and } \sup\langle D, c_2 \rangle = \langle c_1, c_2 \rangle \qquad (35.14)$$

and

$$w \in \widetilde{\text{pos}}\, g^* \iff w_1 \in \ddot{D} \text{ and } \langle w_2, w_1 \rangle = \sup\langle D, w_2 \rangle, \qquad (35.15)$$

from which

$$w \in \widetilde{\text{pos}}\, g^* \text{ and } d \in D \times E^* \implies w_1 \in \ddot{D} \text{ and } \langle w_2, w_1 - \hat{d}_1 \rangle \geq 0. \ (35.16)$$

(a) Since $\pi_1 \text{dom } g = D$, and $\pi_1 \text{dom } f \cap \text{int } D \neq \emptyset$, it follows that $\bigcup_{\lambda > 0} \lambda [\pi_1 \text{dom } f - \pi_1 \text{dom } g] = E$. We now verify that (35.4) is satisfied. To this end, let $u \in \widetilde{\text{pos}}\, f^*$, $w \in \widetilde{\text{pos}}\, g^*$ and $u_1 = w_1$. (35.15) and (35.12) imply that $u_1 \in \ddot{D}$, and thus $\tilde{q}(u - \iota(d)) \geq 0$. Since $u_1 = w_1$, (35.16) also gives $\langle w_2, u_1 - \hat{d}_1 \rangle \geq 0$. Adding up these two inequalities, we have

$$\langle u_2 + w_2 - d_2, u_1 - \hat{d}_1 \rangle \geq 0,$$

and so (35.4) is satisfied. Theorem 35.8(a) now gives $a \in \text{pos } f$ and $c \in \text{pos } g$ such that

$$a_1 = c_1 = d_1 \quad \text{and} \quad a_2 + c_2 = d_2.$$

From (35.14), $\sup\langle D, c_2\rangle = \langle d_1, c_2\rangle$. Since int $D \ni d_1$, this implies that $c_2 = 0$, from which $d = a \in \text{pos } f$. This completes the proof of (a).

(b) Let $f := h_{(0,y^*)}$. Lemma 19.13 and Lemma 35.4 imply that f is a $\widetilde{\text{BC}}$–function. Since $\pi_2 \text{ dom } g = E^*$, $\bigcup_{\lambda>0} \lambda[\pi_2 \text{ dom } f - \pi_2 \text{ dom } g] = E^*$. We now verify that (35.5) is satisfied with $d := 0$. To this end, let $u \in \widetilde{\text{pos }} f^*$, $w \in \widetilde{\text{pos }} g^*$ and $u_2 = w_2$. From Lemma 35.4 again, $u = (v_1, v_2 - y^*)$ for some $v \in \widetilde{\text{pos }} h^*$. (35.15) gives

$$\langle u_2, w_1\rangle = \langle w_2, w_1\rangle = \sup\langle D, w_2\rangle = \sup\langle D, u_2\rangle = \sup\langle D, v_2 - y^*\rangle,$$

and combining this with (35.13) gives

$$\langle u_2, u_1 + w_1\rangle = \langle v_2 - y^*, v_1\rangle + \sup\langle D, v_2 - y^*\rangle \geq 0.$$

Thus (35.5) is satisfied with $d = 0$. Finally, (35.6) is satisfied with $d = 0$ because if $(x^{**}, 0) \in \widetilde{\text{pos }} g^*$ then (35.15) gives $x^{**} \in \ddot{D}$, and the $w(E, E^*)$–compactness of D implies that $\ddot{D} = \hat{D} \subset \hat{E}$. (b) now follows from Lemma 19.13, Theorem 35.8(b) and (35.14).   □

36 Subclasses of the maximally monotone multifunctions

In recent years, many subclasses of the class of maximally monotone multifunctions have been introduced. In this section we introduce those that are "of type (D)", those that are "of type (NI)", those that are "of type (FP)", those that are "of type (FPV)", those that are "strongly maximally monotone", those that are "of type (ANA)", and those that are "of type (BR)". There is an eighth subclass of the class of maximally monotone multifunctions which has a very interesting theory, those that are "of type (ED)". The definition of these requires more preliminary work, and so it will be postponed until Definition 38.3. We will also discuss briefly the "ultramaximally monotone" multifunctions.

We first define multifunctions of type (D). These were essentially introduced by Gossez in [48, Lemme 2.1, p. 375] — see Phelps, [69, Section 3] for an exposition. Gossez considered this kind of multifunction in order to generalize to nonreflexive spaces some of the results previously known for reflexive spaces. In order to make this definition, we must introduce another concept due to Gossez: if $S\colon E \rightrightarrows E^*$, $\overline{S}\colon E^{**} \rightrightarrows E^*$ was originally defined by: $v = (v_1, v_2) \in G(\overline{S}) \iff \inf_{s \in G(S)}\langle s_2 - v_2, \hat{s}_1 - v_1\rangle \geq 0$. In terms of our notation, this can be rewritten

$$v \in G(\overline{S}) \iff \inf_{s \in G(S)} \tilde{q}(\iota(s) - v) \geq 0. \tag{36.1}$$

Using (23.4), this is equivalent to

$$v \in G(\overline{S}) \iff \sup_{s \in G(S)} \left[\langle s, v \rangle - \varphi_S(s) \right] \leq \widetilde{q}(v).$$

It follows that

$$v \in E^{**} \times E^* \text{ and } \varphi_S{}^*(v) = \widetilde{q}(v) \implies v \in G(\overline{S}). \qquad (36.2)$$

Definition 36.1. Let E be a nonzero Banach space, $B := E \times E^*$ and $B^* = E^{**} \times E^*$. S is said to be *maximally monotone of type (D)* if S is maximally monotone and, for all $v \in G(\overline{S})$, there exists a bounded net $\{s_\gamma\}$ of elements of $G(S)$ such that $\iota(s_\gamma) \to v$ in $\mathcal{T}_{\mathcal{WN}}(B^*)$.

If E is reflexive then *every maximally monotone multifunction is of type (D)* and, even if E is not reflexive, if $f \in \mathcal{PCLSC}(E)$ then ∂f is of type (D). Gossez proved this latter result in [48] by adapting one of Rockafellar's proof of the maximal monotonicity theorem. This proof involves some rather delicate functional analysis. We will prove a significant sharpening of Gossez's result in Theorem 48.4(b), using the formula for the biconjugate of a maximum that we develop in Theorem 45.3(b).

We introduced multifunctions of type (NI) in [96, Definition 10, p. 183], motivated by some questions about the range of maximally monotone operators in nonreflexive spaces. Here is the definition — "NI" stands for "negative infimum".

Definition 36.2. Let E be a nonzero Banach space and $B^* = E^{**} \times E^*$. Let $S \colon E \rightrightarrows E^*$ be maximally monotone. According to the original definition, S is said to be *of type (NI)* if

$$v = (v_1, v_2) \in B^* \implies \inf_{s \in G(S)} \langle s_2 - v_2, \widehat{s_1} - v_1 \rangle \leq 0.$$

In terms of our notation, this can be rewritten

$$v \in B^* \implies \inf_{s \in G(S)} \widetilde{q}(\iota(s) - v) \leq 0. \qquad (36.3)$$

Using (23.4), this is equivalent to

$$v \in B^* \implies \sup_{s \in G(S)} \left[\langle s, v \rangle - \varphi_S(s) \right] \geq \widetilde{q}(v). \qquad (36.4)$$

The importance of maximally monotone multifunctions of type (NI) stems from the following result. Part of Theorem 36.3(b) was proved in [96, Theorem 12(a), p. 184]. Theorem 36.3 will be used explicitly in Theorem 37.1, Theorem 39.1, Theorem 40.1 and Lemma 42.4.

Theorem 36.3. Let E be a nonzero Banach space, $B^* = E^{**} \times E^*$ and $S \colon E \rightrightarrows E^*$ be maximally monotone.
(a) Let S be of type (D). Then S is of type (NI).
(b) Let S be of type (NI). Then φ_S is a $\widetilde{BC}$–function, $\widetilde{\text{pos}}\,\varphi_S{}^* \subset G(\overline{S})$, and there exists $v \in G(\overline{S})$ such that $\|v_1\|^2 = \|v_2\|^2 = -\widetilde{q}(v)$.

Proof. (a) If S is maximally monotone of type (D) but not of type (NI) then (36.3) gives $\delta > 0$ and $v \in B^*$ such that $\inf_{s \in G(S)} \widetilde{q}(\iota(s) - v) \geq \delta$. But then $v \in G(\overline{S})$ and so Definition 36.1 provides a bounded net $\{s_\gamma\}$ of elements of $G(S)$ such that $\iota(s_\gamma) \to v$ in $\mathcal{T}_{\mathcal{WN}}(B^*)$. Then, for all γ, $\widetilde{q}(\iota(s_\gamma) - v) \geq \delta$. This is impossible since (35.2) implies that $\widetilde{q}(\iota(s_\gamma) - v) \to \widetilde{q}(v - v) = 0$.

(b) It is immediate from (23.11) and (36.4) that φ_S is a $\widehat{\text{BC}}$–function, and from (36.2) that $\widetilde{\text{pos}}\,\varphi_S{}^* \subset G(\overline{S})$. The rest follows from Lemma 35.5 with $g := g_0$ and (35.3). $\qquad\square$

Problem 36.4. If S is maximally monotone of type (NI) then does it necessarily follow that S is maximally monotone of type (D)? It will be proved in Theorem 46.1(c) that the answer to this question is in the affirmative if $G(S)$ is convex.

It follows from Theorem 36.3(a) that maximally monotone multifunctions of type (NI) share the structural properties of type (D) multifunctions discussed above: if E is reflexive then *every maximally monotone multifunction on E is of type (NI)* (see also (23.11)) and, even if E is not reflexive, *subdifferentials are of type (NI)*.

The multifunctions of type (FP) were introduced by Fitzpatrick–Phelps in [44, Section 3]. Fitzpatrick–Phelps called these multifunctions "locally maximally monotone". We have renamed them since many people have found the original terminology confusing. The motivation for their introduction was as follows. If E is reflexive then every maximally monotone operator on E can be approximated by "nicer" maximally monotone operators using the Moreau-Yosida approximation. If E is nonreflexive then every subdifferential can also be approximated by "nicer" subdifferentials by using the operation of inf–convolution. So the question arises whether a general maximally monotone operators on a nonreflexive space can also be approximated by "nicer" maximally monotone operators in some appropriate sense. Fitzpatrick–Phelps defined an appropriate sense of approximation in [44], and showed that the multifunctions of type (FP) can be approximated by "nicer" maximally monotone operators in their sense.

Definition 36.5. Let E be a nonzero Banach space. We say that a monotone multifunction $S \colon E \rightrightarrows E^*$ is *of type (FP)* if, for any open convex subset U of E^* and $d \in E \times U$ such that $U \cap R(S) \neq \emptyset$ and

$$s \in G(S) \text{ and } s_2 \in U \quad \Longrightarrow \quad q(s - d) \geq 0, \qquad (36.5)$$

then we have $d \in G(S)$. (If we take $U := E^*$, we see that every multifunction of type (FP) is maximally monotone.)

We will prove in Theorem 37.1 that *every multifunction of type (D) is of type (FP)*. Thus, from what we already know about multifunctions of type (D), if E is reflexive then *every maximally monotone multifunction on E is of type (FP)* (see Fitzpatrick–Phelps, [44, Proposition 3.3, p. 585]) and, even if E is not reflexive, *if* $f \in \mathcal{PCLSC}(E)$ *then* ∂f *is of type (FP)* (see [92]). Finally, it was proved in Fitzpatrick–Phelps, [45, Theorem 3.7, p. 67] that

S *is maximally monotone and* $R(S) = E^*$ $\Longrightarrow$ S *is of type (FP)*.

Remark 36.6. The continuous linear skew operator already considered in Example 35.2 is not of type (FP). If we only ask that S be *positive* rather than *skew*, it is possible to give the much simpler example below, known as the *tail operator*, which is taken from Phelps–Simons, [70, Remark 6.10, p. 324]. Let $E := \ell^1$, and $S\colon \ell^1 \to E^* = \ell^\infty$ be defined by

$$(Sx)_n = \sum_{k \geq n} x_k \quad (x \in \ell^1).$$

Then S is positive (exercise!). Let $e := (1,1,1,\ldots)$ and $e^{(1)} = (1,0,0,\ldots) \in \ell^\infty$. Since c_0 is a closed subspace of ℓ^∞ and $e \notin c_0$, it follows from Theorem 4.4 that there exists $x^{**} \in E^{**}$ that vanishes on c_0 with $\langle e, x^{**} \rangle = 2$. Now let

$$U := \{x^* \in E^*\colon x^*_1 < \langle x^*, x^{**} \rangle\}.$$

U is a convex open subset of E^*. Suppose that $x \in E$. Then, by direct computation, $(Sx)_1 = \langle x, e \rangle$. Further, since $Sx \in c_0$, $\langle Sx, x^{**} \rangle = 0$. So if also $Sx \in U$ then $\langle x, e \rangle < 0$ and, consequently,

$$\left.\begin{array}{l} x \in E \text{ and } Sx \in U \implies \\ q\big((x, Sx) - (0, e)\big) = \langle x - 0, Sx - e \rangle = \langle x, Sx \rangle - \langle x, e \rangle > 0. \end{array}\right\} \quad (36.6)$$

Now $\big[S\big(-e^{(1)}\big)\big]_1 = \big(-e^{(1)}\big)_1 = -1$ and $\big\langle S\big(-e^{(1)}\big), x^{**} \big\rangle = \big\langle -e^{(1)}, x^{**} \big\rangle = 0$. Since $-1 < 0$, $S\big(-e^{(1)}\big) \in U$, and so $U \cap R(S) \neq \emptyset$. Thus if S were of type (FP), it would follow from (36.6) that $(0, e) \in G(S)$, which is obviously impossible.

The multifunctions of type (FPV) were introduced by Fitzpatrick–Phelps in [45, p. 65] and Verona–Verona, [113, p. 268] by dualizing Definition 36.5.

Definition 36.7. Let E be a nonzero Banach space. We say that a monotone multifunction $S\colon E \rightrightarrows E^*$ is of type (FPV) if, for any open convex subset U of E and $d \in U \times E^*$ such that $U \cap D(S) \neq \emptyset$ and

$$s \in G(S) \text{ and } s_1 \in U \implies q(s - d) \geq 0, \quad (36.7)$$

then we have $d \in G(S)$. (If we take $U := E$, we see that every multifunction of type (FPV) is maximally monotone.)

In Definition 38.3, we will define maximally monotone multifunctions
of type (ED); we will observe in Theorem 38.4 that if E is reflexive then
every maximally monotone multifunction on E is of type (ED); we will prove
in Theorem 39.1 that *every maximally monotone multifunction of type (ED)
is of type (FVP)*; finally, we will prove in Theorem 48.4(b) that, even if E
is not reflexive, if $f \in \mathcal{PCLSC}(E)$ then ∂f is of type (ED). Consequently, if
E is reflexive then *every maximally monotone multifunction on E is of type
(FPV)* (see Fitzpatrick–Phelps, [44, Proposition 3.3, p. 585]). It also follows
that, even if E is not reflexive, if $f \in \mathcal{PCLSC}(E)$ then ∂f is of type (FPV).
This was first proved by Fitzpatrick–Phelps in [45, Corollary 3.4, p. 66] and
Verona–Verona in [113, Theorem 3, p. 269]. In addition, we will prove in
Theorem 46.1(b) that *every maximally monotone multifunction with convex
graph is of type (FPV)*. Finally, it was noted in [45, Theorem 3.10, p. 68]
that

$$S \text{ is maximally monotone and } D(S) = E \implies S \text{ is of type (FPV)}.$$

These observations lead naturally to the following problem. It will become
clear in Theorem 44.1 that this could be a hard problem. The significance of
multifunctions of type (FPV) is, to some extent, explained by Section 44.

Problem 36.8. Is every maximally monotone multifunction of type (FPV)?

Definition 36.9. Let E be a nonzero Banach space. We say that a multifunction $S\colon E \rightrightarrows E^*$ is *strongly maximally monotone* if S is monotone and
whenever C is a nonempty $w(E, E^*)$–compact convex subset of E, $y^* \in E^*$
and,

$$\text{for all } s \in G(S), \quad \text{there exists } b \in C \times \{y^*\} \text{ such that } \quad q(s - b) \geq 0, \quad (36.8)$$

then

$$G(S) \cap \left(C \times \{y^*\} \right) \neq \emptyset, \tag{36.9}$$

and, further, whenever C is a nonempty $w(E^*, E)$–compact convex subset of
E^*, $y \in E$ and,

$$\text{for all } s \in G(S), \quad \text{there exists } b \in \{y\} \times C \text{ such that } \quad q(s - b) \geq 0, \quad (36.10)$$

then

$$G(S) \cap \left(\{y\} \times C \right) \neq \emptyset. \tag{36.11}$$

Obviously, every strongly maximally monotone multifunction is maximal
monotone. As we have already noted, if E is reflexive then *every maximally
monotone multifunction on E is of type (ED)* and, even if E is not reflexive, if $f \in \mathcal{PCLSC}(E)$ then ∂f is of type (ED). We will prove in Theorem
40.1 that *every maximally monotone multifunction of type (ED) is strongly
maximally monotone*. Consequently, if E is reflexive then *every maximally
monotone multifunction on E is strongly maximal*. It also follows that, even if
E is not reflexive, if $f \in \mathcal{PCLSC}(E)$ then ∂f is strongly maximal — this was

first proved in [94, Theorem 6.1 and Theorem 6.2, p. 1386], using the properties of sublinear functionals and directional derivatives. We will also prove in Theorem 46.1(a) that *every maximally monotone multifunction with convex graph is strongly maximally monotone.*

These observations lead naturally to the following problem:

Problem 36.10. Is every maximally monotone multifunction strongly maximally monotone?

We now discuss a class of multifunctions that have a property of a more metric character. In the following definition, "ANA" stands for "almost negative alignment".

Definition 36.11. Let E be a nonzero Banach space and $S\colon E \rightrightarrows E^*$ be maximally monotone. We say that S is *maximally monotone of type (ANA)* if, whenever $b \in E \times E^* \setminus G(S)$ then, for all $n \geq 1$, there exists $s_n = (s_{n1}, s_{n2}) \in G(S)$ such that $s_{n1} \neq b_1$, $s_{n2} \neq b_2$ and

$$\frac{q(s_n - b)}{\|s_{n1} - b_1\|\|s_{n2} - b_2\|} \to -1 \quad \text{as } n \to \infty.$$

It is clear from Corollary 29.3 that if E is reflexive then *every maximally monotone multifunction on E is of type (ANA)* — in fact the pathology exhibited in Remark 29.4 was the motivation for this definition. We proved in [97] that, even if E is not reflexive, if $f \in \mathcal{PCLSC}(E)$ then ∂f is maximally monotone of type (ANA). Both of these result are subsumed by the result to be proved in Theorem 42.6(b) that *every maximally monotone multifunction of type (ED) is of type (ANA).* On the other hand, it was proved in [11] that *every continuous positive linear operator is maximally monotone of type (ANA)* (see Theorem 47.7). This result is *not* subsumed by Theorem 42.6(b) — as can be seen by considering Example 35.2, or the example in Remark 36.6.

These observations lead naturally to the following problem:

Problem 36.12. Is every maximally monotone multifunction of type (ANA)? (We do not even know what the situation is for *discontinuous* positive linear operators.)

We next recall Torralba's theorem, Theorem 29.9: *Let E be a nonzero reflexive Banach space, $T\colon E \rightrightarrows E^*$ be maximally monotone, $b \in E \times E^*$, α, $\beta > 0$ and $\inf_{t \in G(T)} q(t - b) \geq -\alpha\beta$. Then there exists $t \in G(T)$ such that $\|t_1 - b_1\| \leq \alpha$ and $\|t_2 - b_2\| \leq \beta$.* We first observe that there is no hope of proving a similar result in a nonreflexive space. To see this, let T be the tail operator of Remark 36.6, and take $b := \left(-e^{(1)}, e\right) \in \ell^1 \times \ell^\infty$. We note that, for all $x \in \ell^1$, $\langle e^{(1)}, Tx \rangle = \langle x, e \rangle$, and so

$$\inf_{t \in G(T)} q(t - b) = \inf_{x \in \ell^1} \left\langle x + e^{(1)}, Tx - e \right\rangle = \inf_{x \in \ell^1} \langle x, Tx \rangle - 1 = -1$$

but, for all $t \in G(T)$, $t_2 \in c_0$ and so we cannot have $\|t_2 - b_2\| \leq 1/2$.

As we have already pointed out, Torralba's theorem was motivated by the Brøndsted–Rockafellar theorem for subdifferentials. We will, in fact, give a generalization of Torralba's theorem to the nonreflexive case in Theorem 42.10, however it holds for the extremely restricted subclass of the maximally monotone multifunctions, which we call *ultramaximal* (see Definition 42.9). So it makes sense to ask if, by modifying the question slightly, we can obtain a useful result of a similar character that is true for a significant subclass of the maximally monotone multifunctions in the nonreflexive case. With this in mind, we make the following definition:

Definition 36.13. Let E be a nonzero Banach space and $S\colon E \rightrightarrows E^*$ be maximally monotone. We say that S is *maximally monotone of type (BR)* if, whenever $b \in E \times E^*$, $\alpha, \beta > 0$ and $\inf_{s \in G(S)} q(s - b) > -\alpha\beta$ then there exists $s \in G(S)$ such that $\|s_1 - b_1\| < \alpha$ and $\|s_2 - b_2\| < \beta$.

We will see in Theorem 42.6(c) that *every maximally monotone multifunction of type (ED) (see Definition 38.3) is of type (BR)*.

37 First application of Theorem 35.8: type (D) implies type (FP)

The result of Theorem 37.1 below was first established in [101, Theorem 17, pp. 405–406], using the "free convexification" of a multifunction and a minimax theorem. The proof given here using the Fitzpatrick function and Theorem 35.8 provide an enormous simplification.

Theorem 37.1. *Let E be a nonzero Banach space, and $S\colon E \rightrightarrows E^*$ be maximally monotone of type (D). Then S is of type (FP).*

Proof. Let U be an open convex subset of E^*, $d \in E \times U$, $U \cap R(S) \neq \emptyset$ and

$$s \in G(S) \text{ and } s_2 \in U \implies q(s - d) \geq 0. \tag{36.5}$$

Now let $f := \varphi_S$. Then Theorem 36.3 implies that

$$f \text{ is a } \widetilde{\text{BC}}\text{–function} \quad \text{and} \quad \widetilde{\text{pos}}\, f^* \subset G(\overline{S}). \tag{37.1}$$

By hypothesis, there exists $y^* \in U \cap R(S)$. Since the segment $[d_2, y^*]$ is a compact subset of the open set U, we can choose $\varepsilon > 0$ so that

$$D := [d_2, y^*] + \{x^* \in E^*\colon \|x^*\| \leq \varepsilon\} \subset U.$$

We now show that the conditions of Corollary 35.9(b) are satisfied. Clearly $d \in E \times \operatorname{int} D$. Lemma 23.4 gives us that $\pi_2 \operatorname{dom} f \supset R(S) \ni y^*$, from which $\pi_2\operatorname{dom} f \cap \operatorname{int} D \neq \emptyset$. Now let $u \in \widetilde{\text{pos}}\, f^*$ and $u_2 \in D$. Since S is of type (D), (37.1) and Definition 36.1 give a bounded net $\{s_\gamma\}$ of elements of $G(S)$ such that $\iota(s_\gamma) \to u$ in $\mathcal{T}_{\mathcal{WN}}(E^{**} \times E^*)$. Since $u_2 \in D \subset U$, eventually,

$\pi_2(s_\gamma) \in U$, and so, from (36.5), eventually $\widetilde{q}(\iota(s_\gamma) - \iota(d)) = q(s_\gamma - d) \geq 0$. Passing to the limit and using (35.2), we have $\widetilde{q}(u - \iota(d)) \geq 0$. Thus (35.8) is also satisfied.

Corollary 35.9(b) now implies that $d \in \operatorname{pos} f$. From (23.11), $d \in G(S)$, which completes the proof that S is of type (FP). $\qquad\square$

Remark 37.2. Theorem 37.1 can be strengthened in two ways without a change of proof. Firstly, the assumption that S is maximally monotone of type (D) can be relaxed to the assumption that S is maximally monotone, $\varphi_S{}^* \geq \widetilde{q}$ on $E^{**} \times E^*$ and, whenever $u \in \widehat{\operatorname{pos}} f^*$, there exists a bounded net $\{s_\gamma\}$ of elements of $G(S)$ such that $\iota(s_\gamma) \to u$ in $\mathcal{T}_{\mathcal{WN}}(E^{**} \times E^*)$. Secondly, in fact, $d \in G(S)$ whenever U is an open convex subset of E^* such that $U \cap R(S_\varphi) \neq \emptyset$ and $d \in E \times U$ satisfies (36.5).

38 $\mathcal{T}_{\mathcal{CLB}}(E^{**})$, $\mathcal{T}_{\mathcal{CLBN}}(B^*)$ and type (ED)

In this section, we suppose that E is a nonzero Banach space, and we write $B := E \times E^*$ and $B^* = E^{**} \times E^*$. It is true (and was realized by Gossez) that it is advantageous to replace the topology $\mathcal{T}_{\mathcal{WN}}(B^*)$ in Definition 36.1 by a stronger one. In this section, we will define $\mathcal{T}_{\mathcal{CLBN}}(B^*)$, which is such a replacement, and produces a subclass of the maximally monotone multi-functions that has a number of extremely attractive properties. $\mathcal{T}_{\mathcal{CLBN}}(B^*)$ is defined in terms of a topology, $\mathcal{T}_{\mathcal{CLB}}(E^{**})$, on E^{**}.

Before embarking on the details, we make some general comments. Suppose that $\mathcal{T}$ is a topology on E^{**} such that $(E^{**}, \mathcal{T})$ is a topological vector space, $\widehat{E}$ is dense in $(E^{**}, \mathcal{T})$ and the function $\|\cdot\|$ is continuous on $(E^{**}, \mathcal{T})$. Let x^{**} be an arbitrary element of E^{**}. Then, by hypothesis, there exists a net $\{x_\gamma\}$ of elements of E such that $\widehat{x_\gamma} \to x^{**}$ in $\mathcal{T}$. Since $(E^{**}, \mathcal{T})$ is a topological vector space, $\widehat{x_\gamma} - x^{**} \to 0$ in $\mathcal{T}$, and so $\|\widehat{x_\gamma} - x^{**}\| \to 0$. Since $\widehat{}$ is a norm–isometry from E into E^{**}, it follows that $x^{**} \in \widehat{E}$. Thus E is reflexive. Put another way, if E is not reflexive (which is the interesting case), $\widehat{E}$ is dense in $(E^{**}, \mathcal{T})$ and the function $\|\cdot\|$ is continuous on $(E^{**}, \mathcal{T})$ then $(E^{**}, \mathcal{T})$ *cannot* be a topological vector space. Consequently, the topology $\mathcal{T}_{\mathcal{CLB}}(E^{**})$ on E^{**} that we will introduce in Definition 38.1 will fail to make E^{**} a topological vector space in the interesting situations. However, $\mathcal{T}_{\mathcal{CLB}}(E^{**})$ and $\mathcal{T}_{\mathcal{CLBN}}(B^*)$ have a number of nice properties, which are collected together in Lemma 38.2. Subtler properties of these topologies will be considered in Section 45.

The corresponding class of maximally monotone multifunctions, those that are of "type (ED)", will be introduced in Definiton 38.3.

Now for the details. We write $\mathcal{PCPC}(E)$ for the set of all those $f \in \mathcal{PC}(E)$ such that $\operatorname{dom} f^* \neq \emptyset$. (The extra "$\mathcal{PC}$" stands for "proper conjugate".) It is clear from the Fenchel–Moreau theorem, Corollary 12.4, that

$$f \in \mathcal{PCLSC}(E) \implies f \in \mathcal{PCPC}(E) \implies f^* \in \mathcal{PCLSC}(E^*).$$

Suppose now that $f \in \mathcal{PCPC}(E)$. Then the *(full) biconjugate*, f^{**}, of f is the function from the bidual, E^{**}, of E into $]-\infty, \infty]$ defined by

$$f^{**}(x^{**}) := (f^*)^*(x^{**}) = \sup\nolimits_{x^* \in E^*} \left[\langle x^*, x^{**}\rangle - f^*(x^*)\right] \quad (x^{**} \in E^{**}).$$

It follows easily that

$$f^{**} \text{ is } w(E^{**}, E^*)\text{–lower semicontinuous,} \tag{38.1}$$

$$x \in E \implies f^{**}(\widehat{x}) \le f(x), \tag{38.2}$$

and

$$t^{**} \in E^{**}, \ f^{**}(t^{**}) \le 0 \text{ and } w^* \in E^* \implies \langle w^*, t^{**}\rangle \le f^*(w^*). \tag{38.3}$$

Furthermore, for all $x^{**} \in E^{**}$, the Fenchel–Young inequality, (8.2), gives

$$\left.\begin{aligned}
f^{**}(x^{**}) &= \sup\nolimits_{x^* \in E^*} \left[\langle x^*, x^{**}\rangle - f^*(x^*)\right] \\
&\le \sup\nolimits_{x^* \in E^*} \left[\|x^{**}\|\|x^*\| - f^*(x^*)\right] \\
&= \sup\nolimits_{x^* \in E^*, \ x \in E, \ \|x\| \le \|x^{**}\|} \left[\langle x, x^*\rangle - f^*(x^*)\right] \\
&\le \sup\nolimits_{x \in E, \ \|x\| \le \|x^{**}\|} f(x).
\end{aligned}\right\} \tag{38.4}$$

Definition 38.1. We write $C\mathcal{L}B(E)$ for the set of all convex functions $f\colon E \to \mathbb{R}$ that are Lipschitz on the bounded subsets of E or equivalently, from Theorem 8.7, bounded above on the bounded subsets of E. Consequently, (38.4) implies that

$$f \in C\mathcal{L}B(E) \implies f^{**} \in C\mathcal{L}B(E^{**}). \tag{38.5}$$

We define the topology $T_{C\mathcal{L}B}(E^{**})$ on E^{**} to be the coarsest topology on E^{**} making all the functions $h^{**}\colon E^{**} \to \mathbb{R}$ $(h \in C\mathcal{L}B(E))$ continuous. Then $T_{C\mathcal{L}B\mathcal{N}}(B^*)$ stands for the topology $T_{C\mathcal{L}B}(E^{**}) \times T_{\|\ \|}(E^*)$ on B^*. If $C \subset B$, we write $C^{C\mathcal{L}B\mathcal{N}}$ for the closure of $\iota(C)$ with respect to $T_{C\mathcal{L}B\mathcal{N}}(B^*)$.

We recall that if θ is a map from a topological space into E^{**} then θ is continuous into $T_{C\mathcal{L}B}(E^{**})$ if, and only if, for all $h \in C\mathcal{L}B(E)$, $h^{**} \circ \theta$ is continuous into $\mathbb{R}$; further, if $\{x_\gamma^{**}\}$ is a net of elements of E^{**} and $x^{**} \in E^{**}$ then $x_\gamma^{**} \to x^{**}$ in $T_{C\mathcal{L}B}(E^{**})$ if, and only if, for all $h \in C\mathcal{L}B(E)$, $h^{**}(x_\gamma^{**}) \to h^{**}(x^{**})$ in $\mathbb{R}$.

In the next lemma, we collect together the basic properties of $T_{C\mathcal{L}B}(E^{**})$. Subtler properties of $T_{C\mathcal{L}B}(E^{**})$ will be considered in Section 45.

Lemma 38.2. *Let E be a nonzero Banach space.*

(a) *If $x^* \in E^*$ then the map from $\left(E^{**}, \mathcal{T}_{CLB}(E^{**})\right)$ into $\mathbb{R}$ defined by $x^{**} \mapsto \widetilde{q}(x^{**}, x^*)$ is continuous.*

(b) *If $x \in E$ then the map from $\left(E^{**}, \mathcal{T}_{CLB}(E^{**})\right)$ into $\mathbb{R}$ defined by $x^{**} \mapsto \|x^{**} - \widehat{x}\|$ is continuous.*

(c) *If $x \in E$ then the map from $\left(E^{**}, \mathcal{T}_{CLB}(E^{**})\right)$ into itself defined by $x^{**} \mapsto x^{**} - \widehat{x}$ is continuous.*

(d) *If $\lambda \in \mathbb{R}$ then the map from $\left(E^{**}, \mathcal{T}_{CLB}(E^{**})\right)$ into itself defined by $x^{**} \mapsto \lambda x^{**}$ is continuous.*

(e) *The map $\widetilde{q}$ is continuous from $\left(B^*, \mathcal{T}_{CLBN}(B^*)\right)$ into $\mathbb{R}$.*

(f) $w(E^{**}, E^*) \subset \mathcal{T}_{CLB}(E^{**}) \subset \mathcal{T}_{\|\ \|}(E^{**})$.

(g) *Let $\{x_\gamma\}$ be a net of elements of E and $x \in E$. Then*

$$\widehat{x_\gamma} \to \widehat{x} \text{ in } \mathcal{T}_{CLB}(E^{**}) \iff x_\gamma \to x \text{ in } \mathcal{T}_{\|\ \|}(E).$$

(h) $\mathcal{T}_{CLB}(E^{**}) = w(E^{**}, E^*) \iff E$ *is finite dimensional.*

Proof. (a) follows since, if $h := x^* \in \mathcal{CLB}(E)$, then $h^{**} = \widetilde{q}(\cdot, x^*)$. Likewise, (b) follows since, if $h := \|\cdot -x\| \in \mathcal{CLB}(E)$, then $h^{**} = \|\cdot -\widehat{x}\|$.

(c) follows since, if $h \in \mathcal{CLB}(E)$ and we define $g \in \mathcal{CLB}(E)$ by $g := h(\cdot -x)$, then $g^{**} = h^{**}(\cdot -\widehat{x})$. Likewise, (d) follows since, if $h \in \mathcal{CLB}(E)$ and we define $g \in \mathcal{CLB}(E)$ by $g := h(\lambda \cdot)$, then $g^{**} = h^{**}(\lambda \cdot)$. (Here, the cases $\lambda > 0$, $\lambda < 0$ and $\lambda = 0$ must be handled separately, and the $\lambda = 0$ case uses the Fenchel–Moreau theorem, Corollary 12.4.)

(e) Let $\{v_\gamma\} = \{(v_{\gamma 1}, v_{\gamma 2})\}$ be a net of elements of B^*, $v \in B^*$ and $v_\gamma \to v$ in $\mathcal{T}_{CLBN}(B^*)$. From (a) with $x^* := v_2$, $\widetilde{q}(v_{\gamma 1}, v_2) \to \widetilde{q}(v)$ and, from (b) with $x := 0$, $\|v_{\gamma 1}\| \to \|v_1\|$, and so $\{v_{\gamma 1}\}$ is eventually bounded. Since

$$|\widetilde{q}(v_\gamma) - \widetilde{q}(v)| = |\widetilde{q}(v_\gamma) - \widetilde{q}(v_{\gamma 1}, v_2) + \widetilde{q}(v_{\gamma 1}, v_2) - \widetilde{q}(v)|$$
$$\le \|v_{\gamma 1}\| \|v_{\gamma 2} - v_2\| + |\widetilde{q}(v_{\gamma 1}, v_2) - \widetilde{q}(v)|,$$

it follows that $\widetilde{q}(v_\gamma) \to \widetilde{q}(v)$ in $\mathbb{R}$.

(f) It is clear from (a) that $w(E^{**}, E^*) \subset \mathcal{T}_{CLB}(E^{**})$, and from (38.5) that $\mathcal{T}_{CLB}(E^{**}) \subset \mathcal{T}_{\|\ \|}(E^{**})$.

(g) ($\impliedby$) is clear from (f) and the fact that $\widehat{\ }$ is an isometry. For ($\implies$), consider the element h of $\mathcal{CLB}(E)$ already used in (b).

(h)($\implies$) If $\mathcal{T}_{CLB}(E^{**}) = w(E^{**}, E^*)$ then, from (g), $\mathcal{T}_{\|\ \|}(E) = w(E, E^*)$. It is well known that this implies that E is finite dimensional. ($\impliedby$) follows from (f). □

We now come to the definition of "type (ED)".

Definition 38.3. We say that $S: E \rightrightarrows E^*$ is *maximally monotone of type (ED)* if S is maximally monotone and $G(\overline{S}) \subset G(S)^{CLBN}$.

It is clear from Lemma 38.2(b,f) that *every maximally monotone multifunction of type (ED) is of type (D)*. On the other hand, in every case where it has been proved that a multifunction is maximally monotone of type (D) then it is also of type (ED) (see Theorem 38.4, Theorem 46.1(c) and Theorem 48.4(b)).

Maximally monotone multifunctions of type (ED) were introduced in [99, Definition 35.1, p. 138] under the name "type (DS)", but their properties were not fully exploited until [100], [101], and [107].

We leave the proof of the following simple result to the reader.

Theorem 38.4. *If E is reflexive then every maximally monotone multifunction S: $E \rightrightarrows E^*$ is of type (ED).*

39 Second application of Theorem 35.8: type (ED) implies type (FPV)

The result of Theorem 39.1 below was first established in [101, Theorem 20, pp. 407–409], using the "free convexification" of a multifunction and a minimax theorem. The proof given here using the Fitzpatrick function and Theorem 35.8 provide an enormous simplification, and avoids the excursion to E^{***} made in [101].

Theorem 39.1. *Let E be a nonzero Banach space and S: $E \rightrightarrows E^*$ be maximally monotone of type (ED). Then S is of type (FPV).*

Proof. Let U be an open convex subset of E, $d \in U \times E^*$, $U \cap D(S) \neq \emptyset$ and

$$s \in G(S) \text{ and } s_1 \in U \implies q(s - d) \geq 0. \tag{36.7}$$

Now let $f := \varphi_S$. Then Theorem 36.3 implies that

$$f \text{ is a } \widetilde{BC}\text{-function} \quad \text{and} \quad \widetilde{\text{pos}} f^* \subset G(\overline{S}). \tag{39.1}$$

By hypothesis, there exists $y \in U \cap D(S)$. Since the segment $[d_1, y]$ is a compact subset of the open set U, we can choose $\varepsilon > 0$ so that

$$[d_1, y] + \{x \in E: \|x\| \leq 2\varepsilon\} \subset U.$$

Let

$$D := [d_1, y] + \{x \in E: \|x\| \leq \varepsilon\}.$$

We now show that the conditions of Corollary 35.10(a) are satisfied. Clearly $d \in \text{int } D \times E^*$. Further, Lemma 23.4 gives us that $\pi_1 \text{dom } f \supset D(S) \ni y$, and so $\pi_1 \text{dom } f \cap \text{int } D \neq \emptyset$. Finally, let $u \in \widetilde{\text{pos}} f^*$ and $u_1 \in \ddot{D}$. Since S is of type (ED), (39.1) and Definition 38.3, give a net $\{s_\gamma\} = \{(s_{\gamma 1}, s_{\gamma 2})\}$ of elements of $G(S)$ such that

$$\iota(s_\gamma) \to u \text{ in } \mathcal{T}_{\mathcal{CLBN}}(E^{**} \times E^*).$$

Since $u_1 \in \ddot{D}$, it follows from the Banach–Alaoglu theorem, Theorem 4.1, that there exists $x \in [d_1, y]$ such that $\|u_1 - \widehat{x}\| \le \varepsilon$. Now Lemma 38.2(b) implies that $\|s_{\gamma 1} - \widehat{x}\| \to \|u_1 - \widehat{x}\|$, and so, eventually, $\|s_{\gamma 1} - \widehat{x}\| \le 2\varepsilon$, from which, eventually, $s_{\gamma 1} \in U$ and so, from (36.7), eventually $\widetilde{q}\big(\iota(s_\gamma) - \iota(d)\big) = q(s_\gamma - d) \ge 0$. Passing to the limit using Lemma 38.2(e), $\widetilde{q}\big(u - \iota(d)\big) \ge 0$, and so (35.12) is satisfied.

Corollary 35.10(a) now implies that $d \in \mathrm{pos}\, f$. From (23.11), $d \in G(S)$. This completes the proof that S is of type (FPV). $\qquad\square$

Remark 39.2. Theorem 39.1 can be strengthened in two ways without a change of proof. Firstly, the assumption that S is maximally monotone of type (ED) can be relaxed to the assumption that S is maximally monotone, $\varphi_S{}^* \ge \widetilde{q}$ on $E^{**} \times E^*$ and $\widetilde{\mathrm{pos}}\, \varphi_S{}^* \subset G(S)^{\mathcal{CLBN}}$. Secondly, in fact, $d \in G(S)$ whenever U is an open convex subset of E such that $U \cap D(S_\varphi) \ne \emptyset$ and $d \in U \times E^*$ satisfies (36.7).

40 Final applications of Theorem 35.8: type (ED) implies strong

The result of Theorem 40.1 below was first established in [101, Theorem 15, pp. 400–402], using the "free convexification" of a multifunction and a minimax theorem. The proof given here using the Fitzpatrick function and Theorem 35.8 provide an enormous simplification, and avoids the excursion to E^{***} made in [101].

Theorem 40.1. Let E be a nonzero Banach space and $S \colon E \rightrightarrows E^*$ be maximally monotone of type (ED). Then S is strongly maximal.

Proof. As usual, let $B := E \times E^*$ and $B^* := E^{**} \times E^*$. Let $h := \varphi_S$. Then Theorem 36.3 implies that h is a $\widetilde{BC}$–function and $\widetilde{\mathrm{pos}}\, h^* \subset G(\overline{S})$.

We suppose first that C is a nonempty $w(E, E^*)$–compact convex subset of E, $y^* \in E^*$ and

$$\text{for all } s \in G(S), \quad \text{there exists } b \in C \times \{y^*\} \text{ such that } \quad q(s - b) \ge 0, \quad (36.8)$$

and we will prove that (36.9) is satisfied. Let $D := -C$. From (36.8), for all $s \in G(S)$, there exists $y \in D$ such that $\langle s_1 + y, s_2 - y^* \rangle \ge 0$, from which

$$s \in G(S) \quad \Longrightarrow \quad \langle s_1, s_2 - y^* \rangle + \sup\langle D, s_2 - y^* \rangle \ge 0. \quad (40.1)$$

We now verify that (35.13) is satisfied. To this end, let $v \in \widetilde{\mathrm{pos}}\, h^* \subset G(\overline{S})$. Since S is of type (D), Definition 36.1 provides a bounded net $\{s_\gamma\} = \{(s_{\gamma 1}, s_{\gamma 2})\}$ of elements of $G(S)$ such that $\iota(s_\gamma) \to v$ in $\mathcal{T}_{\mathcal{WN}}(B^*)$, from which $s_{\gamma 2} \to v_2$ in $\mathcal{T}_{\|\ \|}(E^*)$. Now (40.1) implies that,

$$\text{for all } \gamma, \ \langle s_{\gamma 1}, s_{\gamma 2} - y^* \rangle + \sup \langle D, s_{\gamma 2} - y^* \rangle \ge 0.$$

It is easily seen that the function $\sup\langle D, \cdot\rangle$ is continuous on E^* so, passing to the limit using (35.2), we obtain (35.13).

Corollary 35.10(b) now gives $x \in C$ such that $(x, y^*) \in \operatorname{pos} h$. From (23.11), $(x, y^*) \in G(S) \cap (C \times \{y^*\})$, completing the proof of (36.9).

We now suppose that C is a nonempty $w(E^*, E)$–compact convex subset of E^*, $y \in E$ and

for all $s \in G(S)$, there exists $b \in \{y\} \times C$ such that $q(s - b) \geq 0$, (36.10)

and we will prove that (36.11) is satisfied. Let $D := -C$. From (36.10), for all $s \in G(S)$, there exists $y^* \in D$ such that $\langle s_1 - y, s_2 + y^*\rangle \geq 0$, from which

$$s \in G(S) \quad \Longrightarrow \quad \langle s_1 - y, s_2\rangle + \sup\langle s_1 - y, D\rangle \geq 0. \qquad (40.2)$$

We now verify that (35.7) is satisfied. To this end, let $v \in \widetilde{\operatorname{pos}} h^* \subset G(\overline{S})$. Since S is of type (ED), Definition 38.3 gives a net $\{s_\gamma\} = \{(s_{\gamma 1}, s_{\gamma 2})\}$ of elements of $G(S)$ such that $\iota(s_\gamma) \to v$ in $\mathcal{T}_{\mathcal{CLBN}}(B^*)$, and it follows from Lemma 38.2(c) that $(\widehat{s_{\gamma 1}} - \widehat{y}, s_{\gamma 2}) \to (v_1 - \widehat{y}, v_2)$ in $\mathcal{T}_{\mathcal{CLBN}}(B^*)$. From (40.2),

$$\text{for all } \gamma, \; \langle s_{\gamma 1} - y, s_{\gamma 2}\rangle + \sup\langle s_{\gamma 1} - y, D\rangle \geq 0.$$

Lemma 38.2(e) implies that $\langle s_{\gamma 1} - y, s_{\gamma 2}\rangle \to \langle v_2, v_1 - \widehat{y}\rangle$ and, since the function $\sup\langle \cdot, D\rangle$ is in $\mathcal{CLB}(E)$ and its biconjugate is the function $\sup\langle D, \cdot\rangle$, the definition of $\mathcal{T}_{\mathcal{CLB}}(E^{**})$ implies that $\sup\langle s_{\gamma 1} - y, D\rangle \to \sup\langle D, v_1 - \widehat{y}\rangle$. So, passing to the limit, we obtain (35.7).

Corollary 35.9(a) now gives $x^* \in -D$ such that $(y, x^*) \in \operatorname{pos} h$. From (23.11), $(y, x^*) \in G(S) \cap (\{y\} \times C)$, completing the proof of (36.11). $\square$

Remark 40.2. Theorem 40.1 can be strengthened without a change of proof. The assumption that S is maximally monotone of type (ED) can be relaxed to the assumption that S is maximally monotone, $\varphi_S^* \geq \widetilde{q}$ on $E^{**} \times E^*$ and $\widetilde{\operatorname{pos}} \varphi_S^* \subset G(S)^{\mathcal{CLBN}}$.

Corollary 40.3. *If E is reflexive then every maximally monotone multifunction $S: E \rightrightarrows E^*$ is strongly maximal.*

Proof. This is immediate from Theorem 38.4 and Theorem 40.1.

41 Strong maximality and coercivity

Theorem 41.1. *Let E be a nonzero Banach space and $S: E \rightrightarrows E^*$ be strongly maximally monotone. Suppose that $y \in E$, $K \geq 0$ and*

$$s \in G(S) \text{ and } \|s_2\| > K \quad \Longrightarrow \quad \langle s_1 - y, s_2\rangle \geq 0. \qquad (41.1)$$

Then there exists $x^* \in E^*$ such that $\|x^*\| \leq K$ and $(y, x^*) \in G(S)$ and, in particular, $y \in D(S)$.

Proof. It is easily seen from (41.1) that

$$s \in G(S) \quad \Longrightarrow \quad \langle s_1 - y, s_2 \rangle + K\|s_1 - y\| \geq 0.$$

We write $C := \{x^* \in E^* \colon \|x^*\| \leq K\}$. Now let $s \in G(S)$. From the one–dimensional form of the Hahn–Banach theorem, Corollary 2.4, we can choose $y^* \in C$ such that $\langle s_1 - y, y^* \rangle = -K\|s_1 - y\|$, from which $\langle s_1 - y, s_2 - y^* \rangle \geq 0$. Furthermore, the Banach–Alaoglu theorem, Theorem 4.1, implies that C is $w(E^*, E)$–compact. Taking $b := (y, y^*)$, we see that (36.10) is satisfied. The result now follows from the strong maximal monotonicity of S. $\qquad\square$

Corollary 41.2. *Let E be a nonzero Banach space, $S \colon E \rightrightarrows E^*$ be strongly maximally monotone, and $S^{-1} \colon E^* \rightrightarrows E$ be coercive, that is to say $\inf \langle S^{-1}x^*, x^* \rangle / \|x^*\| \to \infty$ as $\|x^*\| \to \infty$. Then $D(S) = E$.*

Proof. Let y be an arbitrary element of E. Chose $K \geq 0$ so that

$$x^* \in E^* \text{ and } \|x^*\| > K \quad \Longrightarrow \quad \inf \langle S^{-1}x^*, x^* \rangle / \|x^*\| \geq \|y\|.$$

If s is an arbitrary element of $G(S)$ and $\|s_2\| > K$ then $q(s)/\|s_2\| \geq \|y\|$, from which $\langle s_1 - y, s_2 \rangle \geq 0$. The result now follows from Theorem 41.1. $\quad\square$

Theorem 41.3. *Let E be a nonzero reflexive Banach space and $S \colon E \rightrightarrows E^*$ be maximally monotone. Suppose that $y^* \in E^*$, $K \geq 0$ and*

$$s \in G(S) \text{ and } \|s_1\| > K \quad \Longrightarrow \quad \langle s_1, s_2 - y^* \rangle \geq 0.$$

Then there exists $x \in E$ such that $\|x\| \leq K$ and $(x, y^*) \in G(S)$ and, in particular, $y^* \in R(S)$.

Proof. Corollary 40.3 implies that $S^{-1} \colon E^* \rightrightarrows E$ is strongly maximal, and so the result follows from Theorem 41.1. $\qquad\square$

Corollary 41.4. *Let E be a nonzero reflexive Banach space and $S \colon E \rightrightarrows E^*$ be maximally monotone and coercive, that is to say $\inf \langle x, Sx \rangle / \|x\| \to \infty$ as $\|x\| \to \infty$. Then $R(S) = E^*$.*

Proof. Let y^* be an arbitrary element of E^*. Chose $K \geq 0$ so that

$$x \in E \text{ and } \|x\| > K \quad \Longrightarrow \quad \inf \langle x, Sx \rangle / \|x\| \geq \|y^*\|.$$

If s is an arbitrary element of $G(S)$ and $\|s_1\| > K$ then $q(s)/\|s_1\| \geq \|y^*\|$, from which $\langle s_1, s_2 - y^* \rangle \geq 0$. The result now follows from Theorem 41.3. $\quad\square$

Remark 41.5. Corollary 41.4 fails badly if we drop the hypothesis of reflexivity. Gossez proved in [49, Proposition 2, p. 90] that *if $S \colon \ell^1 \to \ell^\infty$ is the skew linear operator defined in Example 35.2 then there exist arbitrarily small $\lambda > 0$ such that $R(S + \lambda J)$ is not even dense in ℓ^∞.* Fitzpatrick–Phelps gave a more explicit example in [45, Example 3.2, pp. 63–64]: they proved that *if the skew linear operator $S \colon L_1[0,1] \to L_\infty[0,1]$ is defined by $Sx(t) := \int_0^t x - \int_t^1 x$ then $x^* \in R(S + J) \Longrightarrow \|x^* - 1\| \geq \frac{1}{3}$.* (The maximal monotonicity of the sums follow from Voisei's theorem, Theorem 51.1 — the coercivity of the sums is obvious.)

42 Type (ED) implies type (ANA) and type (BR)

The first generalization of Torralba's theorem, Theorem 29.9, to the non-reflexive case was established by Revalski–Théra in [74, Theorem 3.7, pp. 512–513]. They proved the following:

Theorem 42.1. *Let* $T\colon E \rightrightarrows E^*$ *be maximally monotone of type (D),* $b \in E \times E^*$, α, $\beta > 0$ *and* $\inf_{t \in G(T)} q(t - b) \geq -\alpha\beta$. *Then there exists* $v \in G(\overline{T})$ *such that* $\|v_1 - \widehat{b_1}\| \leq \alpha$ *and* $\|v_2 - b_2\| \leq \beta$.

The main result of this section is Theorem 42.6, in which we show that if we consider multifunctions of type (ED) rather than of type (D) and change the inequalities from "$\geq$" and "$\leq$" to "$>$" and "$<$", respectively, then there is a result analogous to Theorem 42.1 in which the approximation to w can be taken in $G(T)$ rather than in $G(\overline{T})$. Further, if $w \notin G(T)$ then we can control both the ratios

$$\frac{\|t_1 - w_1\|}{\|t_2 - w_2\|} \quad \text{and} \quad \frac{q(t - w)}{\|t_1 - w_1\|\|t_2 - w_2\|}.$$

This control is best explained using the concept of *negative alignment pair*, which we now describe.

Definition 42.2. *Let* $T\colon E \rightrightarrows E^*$, $b \in E \times E^*$ *and* $\rho, \sigma \geq 0$. *We say that* (ρ, σ) *is a negative alignment pair for* T *with respect to* b *if there exists a sequence* $\{t_m\}_{m \geq 1} = \{(t_{m1}, t_{m2})\}_{m \geq 1}$ *of elements of* $G(T)$ *for which*

$$\lim_{m \to \infty} \|t_{m1} - b_1\| = \rho, \quad \lim_{m \to \infty} \|t_{m2} - b_2\| = \sigma \text{ and } \lim_{m \to \infty} q(t - b) = -\rho\sigma.$$

Theorem 42.3(a) contains an "antimonotone" property of negative alignment pairs, and Theorem 42.3(b) contains a uniqueness theorem for negative alignment pairs — both for the case when T is monotone.

Theorem 42.3. *Let* E *be a nonzero Banach space,* $T\colon E \rightrightarrows E^*$ *be monotone and* $b \in E \times E^*$.
(a) *Let* (ρ, σ) *and* $(\widetilde{\rho}, \widetilde{\sigma})$ *be negative alignment pairs for* T *with respect to* b. *Then*

$$(\rho - \widetilde{\rho})(\sigma - \widetilde{\sigma}) \leq 0.$$

(b) *Suppose now that* α, $\beta > 0$. *Then there exists at most one value of* $\tau \geq 0$ *such that* $(\tau\alpha, \tau\beta)$ *is a negative alignment pair for* T *with respect to* b.

Proof. (a) Let $\{t_m\}_{m\geq 1} = \{(t_{m1}, t_{m2})\}_{m\geq 1}$ and $\{\tilde{t}_n\}_{n\geq 1} = \{(\tilde{t}_{n1}, \tilde{t}_{n2})\}_{n\geq 1}$ be sequences of elements of $G(T)$ such that

$$\lim_{m\to\infty} \|t_{m1} - b_1\| = \rho, \quad \lim_{m\to\infty} \|t_{m2} - b_2\| = \sigma, \quad \lim_{m\to\infty} q(t_m - b) = -\rho\sigma,$$

$$\lim_{n\to\infty} \|\tilde{t}_{n1} - b_1\| = \tilde{\rho}, \quad \lim_{n\to\infty} \|\tilde{t}_{n2} - b_2\| = \tilde{\sigma} \quad \text{and} \quad \lim_{n\to\infty} q(\tilde{t}_n - b) = -\tilde{\rho}\tilde{\sigma}.$$

Then, since T is monotone, for all $m, n \geq 1$,

$$0 \leq q(t_m - \tilde{t}_n)$$
$$= q(t_m - b) - \langle t_{m1} - b_1, \tilde{t}_{n2} - b_2\rangle - \langle\tilde{t}_{n1} - b_1, t_{m2} - b_2\rangle + q(\tilde{t}_n - b)$$
$$\leq q(t_m - b) + \|t_{m1} - b_1\|\|\tilde{t}_{n2} - b_2\| + \|\tilde{t}_{n1} - b_1\|\|t_{m2} - b_2\| + q(\tilde{t}_n - b).$$

Letting $m \to \infty$,

$$0 \leq -\rho\sigma + \rho\|\tilde{t}_{n2} - b_2\| + \|\tilde{t}_{n1} - b_1\|\sigma + q(\tilde{t}_n - b),$$

and then, letting $n \to \infty$,

$$0 \leq -\rho\sigma + \rho\tilde{\sigma} + \tilde{\rho}\sigma - \tilde{\rho}\tilde{\sigma}.$$

This completes the proof of (a).

(b) Suppose that τ, $\tilde{\tau} \geq 0$ and $(\tau\alpha, \tau\beta)$ and $(\tilde{\tau}\alpha, \tilde{\tau}\beta)$ are negative alignment pairs for T with respect to b. We have from (a) that

$$(\tau\alpha - \tilde{\tau}\alpha)(\tau\beta - \tilde{\tau}\beta) \leq 0.$$

It follows easily from this that $\tau = \tilde{\tau}$, which gives (b). $\qquad\square$

We now give an existence theorem for negative alignment pairs for maximally monotone multifunctions of type (ED). Its proof was suggested by that of Revalski–Théra, [74, Proposition 3.3, pp. 510–511]. Lemma 42.4 is a simplified version of our main result, Theorem 42.6.

Lemma 42.4. *Let E be a nonzero Banach space, $B^* = E^{**} \times E^*$ and $S: E \rightrightarrows E^*$ be maximally monotone of type (ED). Then:*
(a) There exists a unique value of $\tau \geq 0$ such that (τ, τ) is a negative alignment pair for S with respect to $(0,0)$.
(b) If $(0,0) \notin G(S)$ then $\tau > 0$.
(c) If $\inf_{s\in G(S)} q(s) > -1$ then $\tau < 1$.

Proof. (a) From Theorem 36.3, there exists $v \in G(\overline{S})$ such that $\|v_1\| = \|v_2\|$ and $\tilde{q}(v) = -\|v_1\|\|v_2\|$. Now let $\tau := \|v_1\| = \|v_2\|$. Then we have $\tau \geq 0$ and $\tilde{q}(v) = -\tau^2$. Since S is of type (ED), there exists a net $\{s_\gamma\} = \{(s_{\gamma 1}, s_{\gamma 2})\}$ of elements of $G(S)$ such that $\iota(s_\gamma) \to v$ in $\mathcal{T}_{\mathcal{CLBN}}(B^*)$. Then $\|s_{\gamma 2}\| \to \|v_2\| = \tau$, from Lemma 38.2(b), $\|s_{\gamma 1}\| = \|\widehat{s_{\gamma 1}}\| \to \|v_1\| = \tau$ and, from Lemma 38.2(e),

$$q(s_\gamma) = \tilde{q}(\iota(s_\gamma)) \to \tilde{q}(v) = -\tau^2. \tag{42.1}$$

It is now easy to see that (τ, τ) is a negative alignment pair for S with respect to $(0,0)$, and the "uniqueness" is immediate from Theorem 42.3(b).

(b) If $(0,0) \notin G(S)$ then, from the maximal monotonicity of S and (36.1), $(0,0) \notin G(\overline{S})$ and so $v \neq (0,0)$. It follows that $\tau = \|v_1\| = \|v_2\| > 0$.

(c) We follow the argument of (a) up to (42.1). The additional hypothesis gives that, $\inf_\gamma q(s_\gamma) > -1$. Passing to the limit, $-\tau^2 > -1$. Hence $\tau < 1$, as required. $\qquad\square$

Lemma 42.5(b) contains a useful stability property of maximally monotone multifunctions of type (ED).

Lemma 42.5. *Let E be a nonzero Banach space, $B := E \times E^*$, $B^* = E^{**} \times E^*$ and $T\colon E \rightrightarrows E^*$ be nontrivial, $b \in B$ and $\alpha, \beta > 0$. Define the (invertible) linear maps $\Delta\colon B \to B$ and $\widetilde{\Delta}\colon B^* \to B^*$ by*

$$\Delta(c_1, c_2) := (c_1/\alpha, c_2/\beta) \quad \text{and} \quad \widetilde{\Delta}(v_1, v_2) := (v_1/\alpha, v_2/\beta)$$

$\big(c = (c_1, c_2) \in B, \ v = (v_1, v_2) \in B^*\big)$. *Let $S\colon E \rightrightarrows E^*$ be defined by $G(S) = \Delta\big(G(T) - b\big)$. Then:*

(a) $G(\overline{S}) = \widetilde{\Delta}\big(G(\overline{T}) - \iota(b)\big)$.

(b) *If T is maximally monotone of type (ED) then so is S.*

Proof. (a) is immediate from the definitions of $\overline{S}$ and $\overline{T}$, and it is also immediate that S is maximally monotone in (b). It remains to prove that S is of type (ED). To this end, let $v \in G(\overline{S})$. Since T is of type (ED), we derive from (a) that there exists a net $\{t_\gamma\}$ of elements of $G(T)$ such that $\iota(t_\gamma) \to \widetilde{\Delta}^{-1}v + \iota(b)$ in $\mathcal{T}_{\mathcal{CLBN}}(B^*)$. It now follows from Lemma 38.2(c,d) and the standard properties of $\mathcal{T}_{\|\ \|}(E^*)$ that $\widetilde{\Delta}\big(\iota(t_\gamma - b)\big) \to v$ in $\mathcal{T}_{\mathcal{CLBN}}(B^*)$. Since, for each γ, $\widetilde{\Delta}\big(\iota(t_\gamma - b)\big) = \iota\big(\Delta(t_\gamma - b)\big) \in \iota(G(S))$, we have proved that S is of type (ED). $\qquad\square$

We now bootstrap Lemma 42.4 to obtain our main result on the existence of negative alignment pairs, and give some simple consequences. With reference to Theorem 42.6(c), if T is *not* of type (ED) then we will show in Remark 42.7 that it may happen that T *is* of type (BR), and we will show in Example 47.9 that it may also happen that T is *not* of type (BR). Both these examples are single–valued, continuous and linear.

Theorem 42.6. *Let E be a nonzero Banach space, $T\colon E \rightrightarrows E^*$ be maximally monotone of type (ED), $b \in E \times E^*$ and $\alpha, \beta > 0$. Then:*

(a) *There exists a unique value of $\tau \geq 0$ such that $(\tau\alpha, \tau\beta)$ is a negative alignment pair for T with respect to b.*

(b) *If $b \in E \times E^* \setminus G(T)$ then $\tau > 0$, and there exists $t \in G(T)$ such that $t_1 \neq b_1$, $t_2 \neq b_2$,*

$$\frac{\|t_1 - b_1\|}{\|t_2 - b_2\|} \ \text{is as near as we please to} \ \frac{\alpha}{\beta}$$

and

$$\frac{q(t-b)}{\|t_1 - b_1\| \|t_2 - b_2\|} \text{ is as near as we please to } -1.$$

In particular, T is of type (ANA) (see Definition 36.11).
(c) If, further, $\inf_{t \in G(T)} q(t-b) > -\alpha\beta$ then $\tau < 1$, and we can take t so that, in addition, $\|t_1 - b_1\| < \alpha$ and $\|t_2 - b_2\| < \beta$. So T is of type (BR) (see Definition 36.13).

Proof. We define S as in Lemma 42.5(a). From Lemma 42.5(b), S is maximally monotone of type (ED). The results now follow from Lemma 42.4. □

Remark 42.7. Let $S \colon \ell^1 \to E^* = \ell^\infty$ be Gossez's skew linear map of Example 35.2 for which $\widetilde{\varphi_S}$ is not a BC-function. Suppose now that $b \in E \times E^*$, α, $\beta > 0$ and and $\inf_{s \in G(S)} q(s - b) > -\alpha\beta$. Arguing as in Example 35.2, $\sup_{x \in E} \langle x, b_2 - Sb_1 \rangle < q(b) + \alpha\beta$, from which $b \in G(S)$. It follows that S is of type (BR). On the other hand, Theorem 36.3 implies that S is not of type (NI), hence not of type (ED).

If $\eta > 0$ then the multifunction $J_\eta \colon E \rightrightarrows E^*$ is defined by declaring that $s \in G(J_\eta)$ when $g_0(s) - q(s) \le \eta$. Thus if $T \colon E \rightrightarrows E^*$ and $\lambda, \eta > 0$, then the statement "$R(T + \lambda J_\eta) = E^*$" means that for all $z^* \in E^*$, there exists $t \in G(T)$ such that

$$\tfrac{1}{2}\|t_1\|^2 - \langle t_1, z^* - t_2 \rangle / \lambda + \tfrac{1}{2}\|z^* - t_2\|^2 / \lambda^2 \le \eta. \tag{42.2}$$

Thus, by virtue of Theorem 48.4(b), Theorem 42.8 below generalizes the result proved by Gossez that if T is a subdifferential then, for all $\lambda, \eta > 0$, $R(T + \lambda J_\eta) = E^*$.

Theorem 42.8. *Let E be a nonzero Banach space, $T \colon E \rightrightarrows E^*$ be maximally monotone of type (ED), and $\lambda, \eta > 0$. Then $R(T + \lambda J_\eta) = E^*$.*

Proof. Let $z^* \in E^*$. Then Theorem 42.6 with $b := (0, z^*)$, $\alpha := 1$ and $\beta := \lambda$ gives $\tau \ge 0$ for which there exists $t \in G(T)$ such that $\|t_1\|$, $\|t_2 - z^*\|$ and $\langle t_1, t_2 - z^* \rangle$ are as near as we please to $\tau, \lambda\tau$ and $-\lambda\tau^2$, respectively. Thus

$$\tfrac{1}{2}\|t_1\|^2 - \langle t_1, z^* - t_2 \rangle / \lambda + \tfrac{1}{2}\|z^* - t_2\|^2 / \lambda^2$$

can be made as near as we please to $\tfrac{1}{2}\tau^2 - \lambda\tau^2/\lambda + \tfrac{1}{2}(\lambda\tau)^2/\lambda^2 = 0$. This gives the required result. □

Definition 42.9. We say that $T \colon E \rightrightarrows E^*$ is *ultramaximally monotone* if T is maximally monotone and $G(\overline{T}) \subset \iota(G(T))$. An ultramaximally monotone multifunction is clearly maximally monotone of type (ED). By appropriately modifying the proofs of Lemma 42.4 and Theorem 42.6, one can prove the following result:

Theorem 42.10. *Let $T: E \rightrightarrows E^*$ be ultramaximally monotone, $b \in E \times E^*$ and $\alpha, \beta > 0$. Then:*
(a) There exists a unique value of $\tau \geq 0$ for which there exists $t \in G(T)$ such that $\|t_1 - b_1\| = \tau\alpha$, $\|t_2 - b_2\| = \tau\beta$ and $q(t - b) = -\tau^2\alpha\beta$.
(b) If $b \in E \times E^* \setminus G(T)$ then $\tau > 0$, and there exists $t \in G(T)$ such that $t \neq b_1, t_2 \neq b_2$,

$$\frac{\|t_1 - b_1\|}{\|t_2 - b_2\|} = \frac{\alpha}{\beta} \quad \text{and} \quad \frac{q(t - b)}{\|t_1 - b_1\|\|t_2 - b_2\|} = -1.$$

(c) If, further, $\inf_{t \in G(T)} q(t - b) \geq -\alpha\beta$ then $\tau \leq 1$, and we can take t so that, in addition, $\|t_1 - b_1\| \leq \alpha$ and $\|t_2 - b_2\| \leq \beta$.

If E is reflexive then every maximally monotone multifunction on E is ultramaximal, and so Theorem 42.10 generalizes Theorem 29.9. The proof of Theorem 42.10 is clearly much simpler than that of Theorem 42.6 (since it does not involve any nets), nevertheless it can be applied in a situation that occurs in the study of certain nonlinear elliptic functional equations — see Browder, [32, Theorem 1, p. 90–91]. Here is a description of the situation: Let $S: E^* \to E$ be single–valued, hemicontinuous and monotone and $T := S^{-1}$. Then, arguing as in Gossez, [48, Proposition 3.1, p. 378], T is ultramaximal and so Theorem 42.10 applies to T. We shall explain in Remark 47.10 why certain nonreflexive Banach spaces cannot support *continuous linear* ultra-maximally monotone operators.

Comments on the "negative alignment set" of T

Now suppose that $T: E \rightrightarrows E^*$ is maximally monotone of type (ED), and $b \in E \times E^* \setminus G(T)$. We close this section by investigating the *negative alignment set* (of T with respect to b), defined by

$$\mathcal{NAS} := \big\{(\rho, \sigma):$$

$$(\rho, \sigma) \text{ is a negative alignment pair for } T \text{ with respect to } b\big\}.$$

We will see in Theorem 42.12 that $\mathcal{NAS}$ is a continuous curve with certain monotonicity and maximality properties. We will need the following elementary lemma:

Lemma 42.11. *Let $\theta \in {]0, \pi/2[}$, $\varphi \in {]0, \pi/2[}$, $\lambda \in \mathbb{R}$ and*

$$(\lambda \cos \theta - \cos \varphi)(\lambda \sin \theta - \sin \varphi) \leq 0.$$

Then

$$\frac{\cos \varphi}{\cos \theta} \wedge \frac{\sin \varphi}{\sin \theta} \leq \lambda \leq \frac{\cos \varphi}{\cos \theta} \vee \frac{\sin \varphi}{\sin \theta}.$$

Proof. This is simply a restatement of the assertion that λ lies between the zeros of the quadratic function $\mathbb{R} \to \mathbb{R}$ defined by

$$\nu \mapsto (\nu \cos \theta - \cos \varphi)(\nu \sin \theta - \sin \varphi),$$

which is true since the leading term $\cos \theta \sin \theta$ of the quadratic is strictly positive.   □

Theorem 42.12.

(a) *There is a continuous function* $g \colon \,]0, \pi/2[\,\to\,]0, \infty[$ *such that*

$$\mathcal{NAS} = \left\{ \big(g(\theta) \cos \theta, g(\theta) \sin \theta \big) \colon 0 < \theta < \pi/2 \right\}. \tag{42.3}$$

(b) *The "x–projection"* $\theta \mapsto g(\theta) \cos \theta$ *is non–increasing on* $(0, \pi/2)$, *and the "y–projection"* $\theta \mapsto g(\theta) \sin \theta$ *is non–decreasing on* $(0, \pi/2)$.

(c) *If* $\gamma > 0$, $\delta > 0$ *and* $(\alpha, \beta) \in \mathcal{NAS} \implies (\gamma - \alpha)(\delta - \beta) \le 0$ *then* $(\gamma, \delta) \in \mathcal{NAS}$. *In other words,* $\mathcal{NAS}$ *is a "maximally antimonotone" subset of the first quadrant.*

Proof. (a) It is clear from Theorem 42.6(a,b) that there exists a unique negative alignment pair for T with respect to b on each open ray from $(0, 0)$ into the interior of the first quadrant. It is immediate from this that there exists a function $g \colon \,]0, \pi/2[\,\to\,]0, \infty[$ satisfying (42.3). It remains to show for (a) that g is continuous on $]0, \pi/2[$. Let $\theta, \varphi \in \,]0, \pi/2[$. It follows from Theorem 42.3(a) that

$$\big(g(\theta) \cos \theta - g(\varphi) \cos \varphi \big) \big(g(\theta) \sin \theta - g(\varphi) \sin \varphi \big) \le 0.$$

Thus, from Lemma 42.11,

$$\frac{\cos \varphi}{\cos \theta} \wedge \frac{\sin \varphi}{\sin \theta} \le \frac{g(\theta)}{g(\varphi)} \le \frac{\cos \varphi}{\cos \theta} \vee \frac{\sin \varphi}{\sin \theta}, \tag{42.4}$$

from which it is clear by letting $\varphi \to \theta$ that g is continuous on $]0, \pi/2[$, which completes the proof of (a).

(b) Suppose now that $0 < \theta \le \varphi < \pi/2$. Since $\cos \theta \ge \cos \varphi$ and $\sin \theta \le \sin \varphi$, it follows from (42.4) that

$$\frac{\cos \varphi}{\cos \theta} \le \frac{g(\theta)}{g(\varphi)} \le \frac{\sin \varphi}{\sin \theta}.$$

Thus $g(\theta) \cos \theta \ge g(\varphi) \cos \varphi$ and $g(\theta) \sin \theta \le g(\varphi) \sin \varphi$, which give the required results.

(c) Let $\gamma > 0$, $\delta > 0$ and $(\alpha, \beta) \in \mathcal{NAS} \implies (\gamma - \alpha)(\delta - \beta) \le 0$. Choose $r > 0$ and $\theta \in \,]0, \pi/2[$ so that $(\gamma, \delta) = (r \cos \theta, r \sin \theta)$. Let $\varphi \in \,]0, \pi/2[$ be arbitrary. Then, since $\big(g(\varphi) \cos \varphi, g(\varphi) \sin \varphi \big) \in \mathcal{NAS}$, we have by hypothesis that

$$\big(r \cos \theta - g(\varphi) \cos \varphi \big) \big(r \sin \theta - g(\varphi) \sin \varphi \big) \le 0.$$

Thus, from Lemma 42.11,

$$\frac{\cos\varphi}{\cos\theta} \wedge \frac{\sin\varphi}{\sin\theta} \le \frac{r}{g(\varphi)} \le \frac{\cos\varphi}{\cos\theta} \vee \frac{\sin\varphi}{\sin\theta}.$$

Letting $\varphi \to \theta$ and using the continuity of g, we obtain $r = g(\theta)$, thus

$$(\gamma, \delta) = (r\cos\theta, r\sin\theta) = \big(g(\theta)\cos\theta, g(\theta)\sin\theta\big) \in \mathcal{NAS},$$

as required.   □

There are various questions that come to mind about $\mathcal{NAS}$ and g:

- Can the set $\mathcal{NAS}$ have horizontal or vertical segments?
- What can be said about the behavior of g near 0 and $\pi/2$?
- At a more general level, what functions g are possible, and what insight does the set $\mathcal{NAS}$ give about T?

43 The closure of the range

Gossez proved in [50, Proposition, p. 360] that *if $S\colon \ell^1 \to \ell^\infty$ is the skew linear operator defined in Example 35.2, $\lambda > 0$ and $\overline{R(S + \lambda J)}$ is convex then $R(S + \lambda J)$ would be dense in ℓ^∞*. Consequently, the result cited in Remark 41.5 shows that *there exist arbitrarily small $\lambda > 0$ such that $\overline{R(S + \lambda J)}$ is not convex*. Similarly, Fitzpatrick–Phelps proved in [45, Example 3.2, pp. 63–64] that *if the skew linear operator $S\colon L_1[0,1] \to L_\infty[0,1]$ is defined by $Sx(t) := \int_0^t x - \int_t^1 x$ then $\overline{R(S + J)}$ is not convex*.

On the other hand, it was essentially proved by Gossez in [48] (see Phelps, [69, Theorem 3.8, p. 22] for an exposition) that *if E is a nonzero Banach space and $S\colon E \rightrightarrows E^*$ is maximally monotone of type (D) then $\overline{R(S)}$ is convex*, and it was proved in Fitzpatrick–Phelps, [44, Theorem 3.5, p. 585] that $\overline{R(S)}$ *is also convex if S is of type (FP)*. (We now know from Theorem 37.1 that the second of these two results subsumes the first one.) Finally, it was proved by Fitzpatrick–Phelps in [45] that $\overline{R(S)}$ *is also convex if S is monotone and there exists $\eta > 0$ such that, for all $\lambda > 0$, $R(S + \lambda J_\eta) = E^*$ (see (42.2))*. In this section, we shall show that, in all of the above situations, the statement "$\overline{R(S)}$ is convex" can be strengthened to "$\overline{R(S)} = \overline{\mathrm{co}\,R(S)} = \overline{R(S_\varphi)}$". Our proofs of all the cases discussed above rely on Theorem 27.6(b).

We first consider the type (FP) case.

Theorem 43.1. *Let E be a nonzero Banach space and $S\colon E \rightrightarrows E^*$ be maximally monotone of type (FP). Then*

$$\overline{R(S)} = \overline{\mathrm{co}\,R(S)} = \overline{R(S_\varphi)}.$$

Proof. Let $z^* \in E^* \setminus \overline{R(S)}$. Let $\alpha := \mathrm{dist}(z^*, R(S)) > 0$, and choose $y^* \in R(S)$ such that $\|z^* - y^*\| \le \frac{5}{4}\alpha$, and $y \in E$ such that

$$\|y\| \le 1 \quad \text{and} \quad \langle y, z^* - y^* \rangle \ge \tfrac{2}{3}\|z^* - y^*\|. \tag{43.1}$$

Let

$$U := [z^*, y^*] + \{x^* \in E: \|x^*\| < \tfrac{1}{4}\alpha\}$$

— U is open and $U \ni z^*$. Furthermore, since $U \ni y^*$, $U \cap R(S) \ne \emptyset$. On the other hand, since $z^* \notin R(S)$, for all $n \ge 1$, $(ny, z^*) \notin G(S)$, and so the fact that S is of type (FP) implies that there exists $s \in G(S)$ (depending on n) such that $s_2 \in U$ and $\langle ny - s_1, z^* - s_2 \rangle < 0$, that is to say

$$\langle s_1, z^* - s_2 \rangle > n\langle y, z^* - s_2 \rangle. \tag{43.2}$$

Now $s_2 \in U$, and so there exists $\lambda \in [0, 1]$ (depending on n) such that

$$\|z^* - s_2 - \lambda(z^* - y^*)\| = \|s_2 - [z^* + \lambda(y^* - z^*)]\| < \tfrac{1}{4}\alpha. \tag{43.3}$$

Consequently, $\|z^* - s_2\| \le \lambda\|z^* - y^*\| + \frac{1}{4}\alpha$ and so, since $s_2 \in R(S)$, the definition of α implies that $\alpha \le \lambda\|z^* - y^*\| + \frac{1}{4}\alpha$, from which

$$\lambda\|z^* - y^*\| \ge \tfrac{3}{4}\alpha. \tag{43.4}$$

(43.1) and (43.3) give

$$\langle y, z^* - s_2 - \lambda(z^* - y^*)\rangle \ge -\|y\|\|z^* - s_2 - \lambda(z^* - y^*)\| \ge -\tfrac{1}{4}\alpha$$

and so, from (43.1) and (43.4),

$$\langle y, z^* - s_2 \rangle \ge \lambda\langle y, z^* - y^* \rangle - \tfrac{1}{4}\alpha \ge \tfrac{2}{3}\lambda\|z^* - y^*\| - \tfrac{1}{4}\alpha \ge \tfrac{2}{3}\tfrac{3}{4}\alpha - \tfrac{1}{4}\alpha = \tfrac{1}{4}\alpha.$$

We now obtain from (43.2) that $\langle s_1, z^* - s_2 \rangle > \frac{1}{4}n\alpha$. Since $s_2 \in U$, we also have $\|z^* - s_2\| \le \|z^* - y^*\| + \frac{1}{4}\alpha \le \frac{5}{4}\alpha + \frac{1}{4}\alpha = \frac{3}{2}\alpha$. Consequently,

$$\frac{\langle s_1, z^* - s_2 \rangle}{\|z^* - s_2\|} \ge \frac{\frac{1}{4}n\alpha}{\frac{3}{2}\alpha} = \frac{n}{6}.$$

The result now follows from Theorem 27.6(b). $\qquad\square$

Theorem 43.2. *Let E be a nonzero Banach space and $S: E \rightrightarrows E^*$ be maximally monotone of type (D). Then*

$$\overline{R(S)} = \overline{\mathrm{co}\,R(S)} = \overline{R(S_\varphi)}.$$

Proof. This is immediate from Theorem 37.1 and Theorem 43.1. $\qquad\square$

Theorem 43.2 suggests the following problem.

Problem 43.3. *If S is maximally monotone of type (NI), is $\overline{R(S)}$ necessarily convex?*

We now give our generalization of Fitzpatrick–Phelps, [45, Theorem 1.2, pp. 54–56]:

Theorem 43.4. *Let E be a nonzero Banach space, $S\colon E \rightrightarrows E^*$ be monotone, and suppose that there exists $\eta > 0$ such that, for all $\lambda > 0$, $R(T+\lambda J_\eta) = E^*$. Then*

$$\overline{R(S)} = \overline{\operatorname{co} R(S)} = \overline{R(S_\varphi)}.$$

Proof. From (42.2), for all $\lambda > 0$ and $z^* \in E^*$, there exists $s \in G(S)$ such that

$$\tfrac{1}{2}\|s_1\|^2 - \langle s_1, z^* - s_2\rangle/\lambda + \tfrac{1}{2}\|z^* - s_2\|^2/\lambda^2 \le \eta.$$

Now suppose that $z^* \in E^* \setminus \overline{R(S)}$ and let $\alpha := \operatorname{dist}\bigl(z^*, R(S)\bigr) > 0$. If now $0 < \lambda \le 1/\eta$ then it follows from the above by dropping the $\|s_1\|^2$ term and using the fact that $s_2 \in R(S)$ that

$$\frac{\langle s_1, z^* - s_2\rangle}{\|z^* - s_2\|} \ge \frac{\alpha}{2\lambda} - \frac{1}{\alpha}.$$

The result now follows from Theorem 27.6(b) by letting $\lambda \to 0$.   □

The chart below sums up what we have proved in Theorem 37.1, Theorem 43.1, Theorem 42.8 and Theorem 43.4.

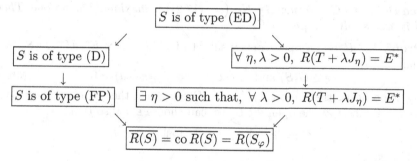

It is interesting to speculate whether there are any relationships between the left hand branch and the right hand branch of the above chart.

Finally, Lemma 31.1 leads us to ask whether $\operatorname{int} R(S) = \operatorname{int} R(S_\varphi)$ in the situations that we have considered in this section. However, Borwein–Fitzpatrick–Vanderwerff proved in [22, Theorem 3.1, p. 68] that if E is not reflexive then there exists a coercive, continuous convex function f on E such that $\operatorname{int} R(\partial f)$ is not convex.

44 The sum problem and the closure of the domain

The problem that has attracted the most interest since maximal monotonicity was introduced more than three decades ago and that has, so far, defied solution, is whether Rockafellar's original sum theorem is true for nonreflexive Banach spaces. Specifically, if E is not reflexive, $S: E \rightrightarrows E^*$ and $T: E \rightrightarrows E^*$ are maximally monotone and

$$D(S) \cap \operatorname{int} D(T) \neq \emptyset.$$

then is $S + T$ maximally monotone?

We have already observed that if $S: E \rightrightarrows E^*$ is maximally monotone and either E is reflexive or S is a subdifferential or S is of type (ED) (in particular, if S has convex graph) then S is of type (FPV). We also made the comment that it could be a very hard problem to find an example of a maximally monotone multifunction that is not of type (FPV). Theorem 44.1 contains the explanation for that comment — its proof is borrowed from that of Fitzpatrick–Phelps, [44, Proposition 3.3, p. 585]. A similar result was proved by Verona–Verona in [114].

Theorem 44.1. *Let E be a nonzero Banach space, $S: E \rightrightarrows E^*$ be maximally monotone and suppose that if C is a nonempty closed convex subset of E and $D(S) \cap \operatorname{int} C \neq \emptyset$ then $S + N_C$ (see (17.2)) is maximally monotone. Then S is necessarily of type (FPV).*

Proof. Let U be an open convex subset of E, $d \in U \times E^*$, $U \cap D(S) \neq \emptyset$ and

$$s \in G(S) \text{ and } s_1 \in U \quad \Longrightarrow \quad q(s - d) \geq 0. \tag{36.7}$$

By hypothesis, there exists $y \in U \cap D(S)$. Since the segment $[d_1, y]$ is a compact subset of the open set U, we can choose $\varepsilon > 0$ so that

$$C := [d_1, y] + \{x \in E: \|x\| \leq \epsilon\} \subset U.$$

From (36.7),

$$s \in G(S) \text{ and } s_1 \in C \quad \Longrightarrow \quad q(s - d) \geq 0. \tag{44.1}$$

Since $d_1 \in C$,

$$(s_1, x^*) \in G(N_C) \quad \Longrightarrow \quad \langle s_1 - d_1, x^* \rangle \geq 0. \tag{44.2}$$

Adding (44.1) and (44.2),

$$s \in G(S) \text{ and } (s_1, x^*) \in G(N_C) \quad \Longrightarrow \quad \langle s_1 - d_1, s_2 + x^* - d_2 \rangle \geq 0,$$

that is to say,

$$t \in G(S + N_C) \quad \Longrightarrow \quad q(t - d) \geq 0. \tag{44.3}$$

Now $D(S) \cap \operatorname{int} D(N_C) = D(S) \cap \operatorname{int} C \ni y$ hence, by assumption, $S + N_C$ is maximally monotone. Thus, from (44.3), $d \in G(S + N_C)$, that is to say, $d_2 \in Sd_1 + N_C(d_1)$. Finally, since $d_1 \in \operatorname{int} C$, $N_C(d_1) = \{0\}$, hence $d_2 \in Sd_1$, that is to say $d \in G(S)$. This completes the proof that S is of type (FPV). $\square$

It was pointed out in Problem 31.3 that it is unknown whether $\overline{D(S)}$ is necessarily convex when S is maximally monotone but E is not reflexive. Now it was proved in Theorem 31.2, that if E is reflexive and S is maximally monotone then $\overline{D(S)} = \overline{D(S_\varphi)}$. We shall prove in Theorem 44.2 that this result remains true even if E is not reflexive, provided that S is of type (FPV). So if it is a hard problem to find an example of a maximally monotone multifunction that is not of type (FPV), it is even harder to find one such that $\overline{D(S)} \neq \overline{D(S_\varphi)}$. It is, course, then harder still to find one such that $\overline{D(S)}$ is not convex. Theorem 44.2, which is "dual" to Theorem 43.1, depends on Theorem 27.5.

Theorem 44.2. *Let E be a nonzero Banach space and $S \colon E \rightrightarrows E^*$ be maximally monotone of type (FPV). Then*

$$\overline{D(S)} = \overline{\operatorname{co} D(S)} = \overline{D(S_\varphi)}.$$

Proof. Let $z \in E \setminus \overline{D(S)}$. Let $\alpha := \operatorname{dist}(z, D(S)) > 0$, and choose $y \in D(S)$ such that $\|z - y\| \leq \frac{5}{4}\alpha$, and $y^* \in E^*$ such that

$$\|y^*\| \leq 1 \quad \text{and} \quad \langle z - y, y^* \rangle \geq \tfrac{2}{3}\|z - y\|. \tag{44.4}$$

Let

$$U := [z, y] + \{x \in E \colon \|x\| < \tfrac{1}{4}\alpha\}$$

— U is open and $U \ni z$. Furthermore, since $U \ni y$, $U \cap D(S) \neq \emptyset$. On the other hand, since $z \notin D(S)$, for all $n \geq 1$, $(z, ny^*) \notin G(S)$ and so the fact that S is of type (FPV) implies that there exists $s \in G(S)$ (depending on n) such that $s_1 \in U$ and $\langle z - s_1, ny^* - s_2 \rangle < 0$, that is to say

$$\langle z - s_1, s_2 \rangle > n\langle z - s_1, y^* \rangle. \tag{44.5}$$

Now $s_1 \in U$, and so there exists $\lambda \in [0, 1]$ (depending on n) such that

$$\left\| z - s_1 - \lambda(z - y) \right\| = \left\| s_1 - [z + \lambda(y - z)] \right\| < \tfrac{1}{4}\alpha. \tag{44.6}$$

Consequently, $\|z - s_1\| \leq \lambda\|z - y\| + \tfrac{1}{4}\alpha$ and so, since $s_1 \in D(S)$, the definition of α implies that $\alpha \leq \lambda\|z - y\| + \tfrac{1}{4}\alpha$, from which

$$\lambda\|z - y\| \geq \tfrac{3}{4}\alpha. \tag{44.7}$$

(44.4) and (44.6) give

$$\langle z - s_1 - \lambda(z - y), y^* \rangle \geq -\left\| z - s_1 - \lambda(z - y) \right\| \|y^*\| \geq -\tfrac{1}{4}\alpha$$

and so, from (44.4) and (44.7),

$$\langle z - s_1, y^* \rangle \geq \lambda\langle z - y, y^* \rangle - \tfrac{1}{4}\alpha \geq \tfrac{2}{3}\lambda\|z - y\| - \tfrac{1}{4}\alpha \geq \tfrac{2}{3}\tfrac{3}{4}\alpha - \tfrac{1}{4}\alpha = \tfrac{1}{4}\alpha.$$

We now obtain from (44.5) that $\langle z - s_1, s_2 \rangle > \tfrac{1}{4}n\alpha$. Since $s_1 \in U$, we also have $\|z - s_1\| \leq \|y - z\| + \tfrac{1}{4}\alpha \leq \tfrac{5}{4}\alpha + \tfrac{1}{4}\alpha = \tfrac{3}{2}\alpha$. Consequently,

$$\frac{\langle z - s_1, s_2 \rangle}{\|z - s_1\|} \geq \frac{\frac{1}{4}n\alpha}{\frac{3}{2}\alpha} = \frac{n}{6}.$$

The result now follows from Theorem 27.5(b). □

45 The biconjugate of a maximum and $\mathcal{T}_{\mathcal{CLB}}(E^{**})$

Let E be a nonzero Banach space, $B := E \times E^*$ and $B^* = E^{**} \times E^*$. In this section, we give some results that we will need in our work on maximally monotone multifunctions with convex graph in Section 46, and in our proof that subdifferentials are maximally monotone of type (ED) in Section 48. These results are based ultimately on the fact that the biconjugate of the pointwise maximum of a finite number of functions is the maximum of their biconjugates. (See Corollary 45.5.) What is curious is that we can establish this fact without having a simple explicit formula for the *conjugate* of the pointwise maximum. As we shall see in Lemma 45.1(a) and Remark 45.20, we have two such formulae, but they are not simple. Corollary 45.5 will be applied in Lemma 45.9 to obtain, among other things, the fundamental property that $\widehat{E}$ is dense in $\left(E^{**}, \mathcal{T}_{\mathcal{CLB}}(E^{**})\right)$. Theorem 45.12 gives an unexpected characterization of the closure of certain convex subsets of B^* with respect to $\mathcal{T}_{\mathcal{CLBN}}(B^*)$ — this will also be used in Section 46. In some senses, Corollary 45.6 leads us to consider a topology different from $\mathcal{T}_{\mathcal{CLB}}(E^{**})$ on E^{**}, namely the topology $\mathcal{T}_{\mathcal{CC}}(E^{**})$, which we will also introduce in this section. We will outline the similarities in the behavior of these topologies, indicating in Example 45.16 why $\mathcal{T}_{\mathcal{CC}}(E^{**})$ is not as good a choice for us as $\mathcal{T}_{\mathcal{CLB}}(E^{**})$.

Lemma 45.1. *Let E be a nonzero Banach space. Suppose that $f, g \in \mathcal{PC}(E)$, $\operatorname{dom} f \cap \operatorname{dom} g \neq \emptyset$, and $w^* \in E^*$.*
(a) $-(f \vee g)^*(w^*) = \max_{\rho \in [0,1]} \inf_{\operatorname{dom}(f \vee g)} \left[\rho f + (1 - \rho)g - w^*\right]$.
(b) *Let g be (finitely) bounded above in some neighborhood of a point of $\operatorname{dom} f$ and $\rho, \sigma > 0$. Then there exist $u^*, v^* \in E^*$ such that $\rho u^* + \sigma v^* = w^*$ and*

$$\rho f^*(u^*) + \sigma g^*(v^*) = \sup_{\operatorname{dom}(f \vee g)} \left[w^* - \rho f - \sigma g\right].$$

Proof. (a) Since $\operatorname{dom}(f \vee g) = \operatorname{dom} f \cap \operatorname{dom} g \neq \emptyset$, this is immediate from Lemma 3.1 with $f_1 := f - w^*$ and $f_2 := g - w$.

(b) Corollary 10.2 gives us $y^*, z^* \in E^*$ such that $y^* + z^* = w^*$ and $(\rho f)^*(y^*) + (\sigma g)^*(z^*) = (\rho f + \sigma g)^*(w^*) = \sup_{\operatorname{dom}(\rho f + \sigma g)} \left[w^* - \rho f - \sigma g\right]$. Since $\operatorname{dom}(\rho f + \sigma g) = \operatorname{dom}(f \vee g)$, the result follows by setting $u^* := y^*/\rho$ and $v^* := z^*/\sigma$. □

Lemma 45.2. *Let E be a nonzero Banach space, $f, g \in \mathcal{PCPC}(E)$, g be (finitely) bounded above in some neighborhood of a point of $\operatorname{dom} f$, $x^{**} \in E^{**}$, $f^{**}(x^{**}) \vee g^{**}(x^{**}) \leq 0$, and $w^* \in E^*$.*
(a) *Let $\rho, \sigma > 0$. Then $\langle w^*, x^{**} \rangle \leq \sup_{\operatorname{dom}(f \vee g)} \left[w^* - \rho f - \sigma g\right]$.*

(b) $\langle w^*, x^{**}\rangle \leq \sup_{\mathrm{dom}(f\vee g)} \left[w^* - g\right]$ and $\langle w^*, x^{**}\rangle \leq \sup_{\mathrm{dom}(f\vee g)} \left[w^* - f\right]$.
(c) Let $\rho \in [0,1]$. Then

$$\langle w^*, x^{**}\rangle \leq \sup_{\mathrm{dom}(f\vee g)} \left[w^* - \rho f - (1-\rho)g\right]. \tag{45.1}$$

(d) $\langle w^*, x^{**}\rangle \leq (f \vee g)^*(w^*)$.
(e) $(f \vee g)^{**}(x^{**}) \leq 0$.

Proof. (a) Choose u^* and v^* as in Lemma 45.1(b). Then, from (38.3),
$\langle w^*, x^{**}\rangle = \langle \rho u^* + \sigma v^*, x^{**}\rangle = \rho\langle u^*, x^{**}\rangle + \sigma\langle v^*, x^{**}\rangle \leq \rho f^*(u^*) + \sigma g^*(v^*)$,
and the result follows from the choice of u^* and v^*.

(b) Since $f \in \mathcal{PCPC}(E)$, we can fix $x^* \in \mathrm{dom}\, f^*$. For all $\rho > 0$, we apply
(a) with w^* replaced by $\rho x^* + w^*$ and $\sigma := 1$ and obtain

$$\langle \rho x^* + w^*, x^{**}\rangle \leq \sup_{\mathrm{dom}(f\vee g)} \left[\rho x^* + w^* - \rho f - g\right]$$
$$\leq \rho \sup_{\mathrm{dom}(f\vee g)} \left[x^* - f\right] + \sup_{\mathrm{dom}(f\vee g)} \left[w^* - g\right].$$

Since $\sup_{\mathrm{dom}(f\vee g)} \left[x^* - f\right] \leq f^*(x^*) < \infty$, it follows by letting $\rho \to 0$ that
$\langle w^*, x^{**}\rangle \leq \sup_{\mathrm{dom}(f\vee g)} \left[w^* - g\right]$. Similarly, if we fix $x^* \in \mathrm{dom}\, g^*$ and, for
all $\sigma > 0$, we apply (a) with w^* replaced by $w^* + \sigma x^*$ and $\rho := 1$, we obtain

$$\langle w^* + \sigma x^*, x^{**}\rangle \leq \sup_{\mathrm{dom}(f\vee g)} \left[w^* - f\right] + \sigma \sup_{\mathrm{dom}(f\vee g)} \left[x^* - g\right],$$

and letting $\sigma \to 0$ gives $\langle w^*, x^{**}\rangle \leq \sup_{\mathrm{dom}(f\vee g)} \left[w^* - f\right]$.

(c) If $\rho = 0$ or $\rho = 1$ then (45.1) follows from (b). If $\rho \in\,]0,1[$ then (45.1)
follows from (a).

(d) follows from (c) and Lemma 45.1(a), and (e) is evident from (d) by
allowing w^* to run over E^*.   □

Theorem 45.3. Let E be a nonzero Banach space, $f, g \in \mathcal{PCPC}(E)$ and g
be (finitely) bounded above in some neighborhood of a point of $\mathrm{dom}\, f$.
(a) Let $x^{**} \in E^{**}$. Then $(f \vee g)^{**}(x^{**}) \leq f^{**}(x^{**}) \vee g^{**}(x^{**})$.
(b) $(f \vee g)^{**} = f^{**} \vee g^{**}$ on E^{**}.

Proof. (a) Let $\alpha := f^{**}(x^{**}) \vee g^{**}(x^{**})$. Since the result is immediate if
$\alpha = \infty$, we can and will suppose that $\alpha \in \mathbb{R}$. We now obtain the result by
applying Lemma 45.2(e) with f and g replaced by $f - \alpha$ and $g - \alpha$.

(b) Since $f \vee g \geq f$ on E, $(f\vee g)^{**} \geq f^{**}$ on E^{**}. Similarly, $(f\vee g)^{**} \geq g^{**}$
on E^{**}. Thus $(f\vee g)^{**} \geq f^{**}\vee g^{**}$ on E^{**}, and the result follows from (a).   □

Definition 45.4. We write $\mathcal{CC}(E)$ for the set of all real convex continuous
functions on E. The standard example of a function $f \in \mathcal{CC}(\ell^2) \setminus \mathcal{CLB}(\ell^2)$
is defined by $f(x) := \sum_{n=1}^{\infty} n x_n^{2n}$ $(x = \{x_n\}_{n\geq 1} \in \ell^2)$. It was proved by
Borwein–Fitzpatrick–Vanderwerff in [22, Theorem 2.2, p. 64] using the deep
Josefson-Nissenzweig theorem that if E is infinite dimensional then $\mathcal{CC}(E) \setminus$
$\mathcal{CLB}(E) \neq \emptyset$.

Corollary 45.5 generalizes a result proved in [99, Theorem 33.3(c), p. 131] and [46, Corollary 7, p. 3558]. This increased generality (i.e., replacing the assumption "$h_0 \in \mathcal{PCLSC}(E)$" by the assumption "$h_0 \in \mathcal{PCPC}(E)$") is necessitated by the application that we will ultimately give in Theorem 45.12. It was pointed out to the author by Radu Ioan Boţ that Corollary 45.5 remains true under the weaker hypothesis that $h_0, \ldots, h_m \in \mathcal{PCPC}(E)$ and $h_1 \vee \cdots \vee h_m$ is (finitely) bounded above in some neighborhood of a point of dom h_0.

Corollary 45.5. *Let E be a nonzero Banach space, $h_0 \in \mathcal{PCPC}(E)$ and $h_1, \ldots, h_m \in \mathcal{CC}(E)$. Then*

$$(h_0 \vee \cdots \vee h_m)^{**} = h_0{}^{**} \vee \cdots \vee h_m{}^{**} \text{ on } E^{**}.$$

Proof. This is immediate from Theorem 45.3(b) and induction. ☐

Corollary 45.6. *Let E be a nonzero Banach space, $f_0 \in \mathcal{PCPC}(E)$ and $f_1, \ldots, f_m \in \mathcal{CC}(E)$. Let $x^{**} \in E^{**}$ and $\varepsilon > 0$. Then there exists $t \in E$ such that,*

$$\text{for all } i = 0, \ldots, m, \quad f_i(t) \leq f_i{}^{**}(x^{**}) + \varepsilon.$$

Proof. Since we can remove those values of i for which $f_i{}^{**}(x^{**}) = \infty$, we can and will suppose that $f_0{}^{**}(x^{**}), \ldots, f_m{}^{**}(x^{**}) \in \mathbb{R}$. For all $i = 0, \ldots, m$, let $h_i := f_i - f_i{}^{**}(x^{**})$. Then $h_i{}^{**}(x^{**}) = 0$, hence $h_0{}^{**}(x^{**}) \vee \cdots \vee h_m{}^{**}(x^{**}) = 0$. From Corollary 45.5, $(h_0 \vee \cdots \vee h_m)^{**}(x^{**}) = 0$ and so, from (38.3) with $f := h_0 \vee \cdots \vee h_m$, and $w^* := 0$, $(h_0 \vee \cdots \vee h_m)^*(0) \geq 0$, that is to say $\inf_E (h_0 \vee \cdots \vee h_m) \leq 0$. The result follows by rewriting this inequality in terms of the functions f_i. ☐

Definition 45.7. We define the topology $\mathcal{T}_{CC}(E^{**})$ on E^{**} to be the coarsest topology on E^{**} such that, for all $f \in \mathcal{CC}(E)$, all of the functions $f^{**} : E^{**} \to]-\infty, \infty]$ are continuous. Clearly, $\mathcal{T}_{CLB}(E^{**}) \subset \mathcal{T}_{CC}(E^{**})$. As usual, $B := E \times E^*$ and $B^* = E^{**} \times E^*$. In line with the usage introduced in Definition 38.1, we write $\mathcal{T}_{CCN}(B^*)$ for the topology $\mathcal{T}_{CC}(E^{**}) \times \mathcal{T}_{\|\ \|}(E^*)$ on B^*. If $C \subset B$, we write C^{CCN} for the closure of $\iota(C)$ with respect to $\mathcal{T}_{CCN}(B^*)$ and C^{WN} for the closure of $\iota(C)$ with respect to $\mathcal{T}_{WN}(B^*)$.

The next two lemmas indicate some of the common features of $\mathcal{T}_{CLB}(E^{**})$ and $\mathcal{T}_{CC}(E^{**})$.

Lemma 45.8. *Let E be a nonzero Banach space, $\{x_\gamma^{**}\}$ be a net of elements of E^{**} and $x^{**} \in E^{**}$. Then $x_\gamma^{**} \to x^{**}$ in $\mathcal{T}_{CLB}(E^{**})$ if, and only if,*

$$h \in \mathcal{CLB}(E) \quad \Longrightarrow \quad \limsup_\gamma h^{**}(x_\gamma^{**}) \leq h^{**}(x^{**}). \qquad (45.2)$$

Similarly, $x_\gamma^{**} \to x^{**}$ in $\mathcal{T}_{CC}(E^{**})$ if, and only if,

$$h \in \mathcal{CC}(E) \quad \Longrightarrow \quad \limsup_\gamma h^{**}(x_\gamma^{**}) \leq h^{**}(x^{**}).$$

Proof. We deal with the "$\mathcal{T}_{\mathcal{CLB}}(E^{**})$" case, the proof of the "$\mathcal{T}_{\mathcal{CC}}(E^{**})$" case is similar. "Only if" is immediate. Suppose, conversely, that (45.2) is true. Let $x^* \in E^*$. Then, since $x^* \in \mathcal{CLB}(E)$ and $-x^* \in \mathcal{CLB}(E)$, we have from (45.2) that

$$\limsup_\gamma \langle x^*, x_\gamma^{**} \rangle \le \langle x^*, x^{**} \rangle \quad \text{and} \quad \liminf_\gamma \langle x^*, x_\gamma^{**} \rangle \ge \langle x^*, x^{**} \rangle,$$

thus $x_\gamma^{**} \to x^{**}$ in $w(E^{**}, E^*)$. From (38.1),

$$h \in \mathcal{CLB}(E) \implies \liminf_\gamma h^{**}(x_\gamma^{**}) \ge h^{**}(x^{**}).$$

Combining this with (45.2),

$$h \in \mathcal{CLB}(E) \implies h^{**}(x_\gamma^{**}) \to h^{**}(x^{**}),$$

that is to say, $x_\gamma^{**} \to x^{**}$ in $\mathcal{T}_{\mathcal{CLB}}(E^{**})$. This completes the proof of "if".   □

Lemma 45.9(a) will be used in Lemma 45.11 and Theorem 48.1, and Lemma 45.9(b) contains a fundamental density property of $\mathcal{T}_{\mathcal{CC}}(E^{**})$ that will be used in Theorem 45.14.

Lemma 45.9. *Let E be a nonzero Banach space.*
(a) *Let $f \in \mathcal{PCPC}(E)$ and $x^{**} \in E^{**}$. Then there exists a net $\{x_\gamma\}$ of elements of E such that $\widehat{x_\gamma} \to x^{**}$ in $\mathcal{T}_{\mathcal{CC}}(E^{**})$, $\widehat{x_\gamma} \to x^{**}$ in $\mathcal{T}_{\mathcal{CLB}}(E^{**})$, $f(x_\gamma) \to f^{**}(x^{**})$ and $f^{**}(\widehat{x_\gamma}) \to f^{**}(x^{**})$.*
(b) *$\widehat{E}$ is a dense subset of $\big(E^{**}, \mathcal{T}_{\mathcal{CC}}(E^{**})\big)$ and $\big(E^{**}, \mathcal{T}_{\mathcal{CLB}}(E^{**})\big)$.*

Proof. (a) From Corollary 45.6, for each nonempty finite subset H of $\mathcal{CC}(E)$ and $\varepsilon > 0$, there exists $x_{H,\varepsilon} \in E$ such that $f(x_{H,\varepsilon}) \le f^{**}(x^{**}) + \varepsilon$ and

$$h \in H \implies h(x_{H,\varepsilon}) \le h^{**}(x^{**}) + \varepsilon.$$

If we direct (H, ε) in the usual (product) way, we can construct a net $\{x_\gamma\}$ of elements of E such that

$$\limsup_\gamma f(x_\gamma) \le f^{**}(x^{**}) \tag{45.3}$$

and $h \in \mathcal{CC}(E) \implies \limsup_\gamma h(x_\gamma) \le h^{**}(x^{**})$. It follows from the Fenchel–Moreau theorem, Corollary 12.4, that, for all γ, $h(x_\gamma) = h^{**}(\widehat{x_\gamma})$, and so Lemma 45.8 implies that $\widehat{x_\gamma} \to x^{**}$ in $\mathcal{T}_{\mathcal{CC}}(E^{**})$, from which $\widehat{x_\gamma} \to x^{**}$ in $\mathcal{T}_{\mathcal{CLB}}(E^{**})$. From Lemma 38.2(f), $\widehat{x_\gamma} \to x^{**}$ in $w(E^{**}, E^*)$ and so (38.1) gives $f^{**}(x^{**}) \le \liminf_\gamma f^{**}(\widehat{x_\gamma})$. (a) follows by combining this with (45.3) and using the fact that $f^{**}(\widehat{x_\gamma}) \le f(x_\gamma)$. (Since we are not assuming that f is lower semicontinuous, we cannot assert that $f^{**}(\widehat{x_\gamma}) = f(x_\gamma)$.)
(b) follows by simply taking $f := 0$ in (a).   □

Corollary 45.10 is a simple application of Lemma 45.9. It is a generalization of Gossez, [48, Corollaire 3.2, p. 379]. Corollary 45.10 is somewhat unexpected since, as we will see in Remark 45.13, $\big(E^{**}, \mathcal{T}_{\mathcal{CLB}}(E^{**})\big)$ and $\big(E^{**}, \mathcal{T}_{\mathcal{CLBN}}(E^{**})\big)$ are not, in general, topological vector spaces.

Corollary 45.10. *Let E be a nonzero Banach space and C be a nonempty convex subset of E. Then the $w(E^{**}, E^*)$–closure of $\widehat{C}$, the $\mathcal{T}_{\mathcal{CLB}}(E^{**})$–closure of $\widehat{C}$ and the $\mathcal{T}_{\mathcal{CC}}(E^{**})$–closure of $\widehat{C}$ are identical.*

Proof. Let $f := \mathbb{I}_C$ and apply Lemma 45.9(a).   □

Lemma 45.11, which is a "higher dimensional" version of Corollary 45.10, is also quite unexpected since $\big(B^*, \mathcal{T}_{\mathcal{CLBN}}(B^*)\big)$ and $\big(B^*, \mathcal{T}_{\mathcal{CCN}}(B^*)\big)$ are not, in general, topological vector spaces. (As usual, $B := E \times E^*$ and $B^* = E^{**} \times E^*$). We recall that the sets $C^{\mathcal{WN}}$, $C^{\mathcal{CLBN}}$ and $C^{\mathcal{CCN}}$ were defined in Definition 38.1 and Definition 45.7. Lemma 45.11 is a stepping stone to Theorem 45.12, which will be used explicitly in Theorem 46.1.

Lemma 45.11. *Let E be a nonzero Banach space, and C be a nonempty convex subset of $E \times E^*$. Then $C^{\mathcal{WN}} = C^{\mathcal{CLBN}} = C^{\mathcal{CCN}}$.*

Proof. By virtue of the obvious relationships, we only have to prove that $C^{\mathcal{WN}} \subset C^{\mathcal{CCN}}$. So let $w \in C^{\mathcal{WN}}$. Then there exists a net $\{c_\gamma\}$ of elements of C such that $\iota(c_\gamma) \to w$ in $\mathcal{T}_{\mathcal{WN}}(E^{**} \times E^*)$. If $x \in \pi_1 C$, let C_x stand for the section $\{x^* \in E^*: (x, x^*) \in C\}$, and define $f: E \to [0, \infty]$ by

$$f(x) := \begin{cases} \operatorname{dist}(w_2, C_x), & \text{if } x \in \pi_1 C; \\ \infty, & \text{if } x \in E \setminus \pi_1 C. \end{cases}$$

It is easily seen that $f \in \mathcal{PC}(E)$. We first show that

$$f^{**}(w_1) = 0. \tag{45.4}$$

Since $f \geq 0$ on E, $f^*(0) = -\inf_E f \leq 0$, and so $f^{**}(w_1) \geq \langle 0, w_1 \rangle - f^*(0) \geq 0$. On the other hand, using (38.2), for all γ,

$$f^{**}\big(\pi_1 \iota(c_\gamma)\big) = f^{**}(\widehat{\pi_1 c_\gamma}) \leq f(\pi_1 c_\gamma) \leq \|\pi_2 c_\gamma - w_2\| \to 0$$

so, from the $w(E^{**}, E^*)$–lower semicontinuity of f^{**} mentioned in (38.1), $f^{**}(w_1) \leq 0$. This completes the proof of (45.4).

Lemma 45.9(a) now gives a net $\{x_\gamma\}$ of elements of E such that $\widehat{x_\gamma} \to w_1$ in $\mathcal{T}_{\mathcal{CC}}(E^{**})$ and $f(x_\gamma) \to f^{**}(w_1)$. (45.4) implies that $f(x_\gamma) \to 0$, so eventually $f(x_\gamma) < \infty$, in which case $x_\gamma \in \pi_1 C$. We can and will suppose that the net $\{x_\gamma\}$ has been truncated to exclude those values of γ for which $x_\gamma \notin \pi_1 C$. Thus, for all γ, $f(x_\gamma) = \operatorname{dist}(w_2, C_{x_\gamma})$, and so, for all $n \geq 1$, there exists $x^*_{\gamma,n} \in C_{x_\gamma}$ such that $\|x^*_{\gamma,n} - w_2\| < f(x_\gamma) + \frac{1}{n}$. We note then that $(x_\gamma, x^*_{\gamma,n}) \in C$. If we direct $\{\gamma, n\}$ in the usual (product) way, then clearly $\|x^*_{\gamma,n} - w_2\| \to 0$, that us to say $x^*_{\gamma,n} \to w_2$ in $\mathcal{T}_{\|\;\|}(E^*)$. Thus $(\widehat{x_\gamma}, x^*_{\gamma,n}) \to w$ in $\mathcal{T}_{\mathcal{CCN}}(B^*)$, and so $w \in C^{\mathcal{CCN}}$. This completes the proof of Lemma 45.11.   □

Theorem 45.12. *Let E be a nonzero Banach space, C be a nonempty convex subset of $E \times E^*$ and $w \in E^{**} \times E^*$. Then $w \in C^{\mathcal{CLBN}}$ if, and only if,*

$$v \in E^{**} \times E^* \quad \Longrightarrow \quad \big\lfloor w, v \big\rceil \leq \sup_{s \in C} \big\lfloor \iota(s), v \big\rceil.$$

Proof. Lemma 45.11 implies that $w \in C^{\mathcal{CLBN}}$ if, and only if, $w \in C^{\mathcal{WN}}$. However, $\mathcal{T}_{\mathcal{WN}}(E^{**} \times E^*)$ is a compatible locally convex topology on the SSD space $E^{**} \times E^*$, and so the result follows from Theorem 8.8. $\square$

Remark 45.13. From Lemma 45.9(b), Lemma 38.2(b) and the remarks in the second paragraph of Section 38, if E is *not reflexive* then $(E^{**}, \mathcal{T}_{\mathcal{CLB}}(E^{**}))$ and $(E^{**}, \mathcal{T}_{\mathcal{CC}}(E^{**}))$ *are not topological vector spaces.*

Theorem 45.14. *Let E be a nonzero Banach space.*
(a) $\mathcal{T}_{\mathcal{CC}}(E^{**}) = \mathcal{T}_{\| \ \|}(E^{**}) \iff E$ *is reflexive.*
(b) $\mathcal{T}_{\mathcal{CLB}}(E^{**}) = \mathcal{T}_{\| \ \|}(E^{**}) \iff E$ *is reflexive*

Proof. In both cases, ($\Longleftarrow$) is obvious, and ($\Longrightarrow$) is immediate from Remark 45.13. $\square$

Lemma 45.15 below will be used in our main result on subdifferentials, Theorem 48.1, and also in our analysis of saddle functions in Lemma 49.3 and Theorem 49.4.

Lemma 45.15. *Let E be a nonzero Banach space, $\{x_\gamma^{**}\}$ be a net of elements of E^{**}, $x^{**} \in E^{**}$ and $x_\gamma^{**} \to x^{**}$ in $\mathcal{T}_{\mathcal{CLB}}(E^{**})$. Let $\{y_\gamma^{**}\}$ be a net of elements of E^{**} and $\|y_\gamma^{**} - x_\gamma^{**}\| \to 0$. Then $y_\gamma^{**} \to x^{**}$ in $\mathcal{T}_{\mathcal{CLB}}(E^{**})$.*

Proof. Let $X := \{z^{**} \in E^{**}: \|z^{**}\| \le \|x^{**}\| + 2\}$. It follows from Lemma 38.2(b) that eventually $\|x_\gamma^{**}\| \le \|x^{**}\| + 1$. Also, eventually $\|y_\gamma^{**} - x_\gamma^{**}\| \le 1$. Thus, eventually both x_γ^{**} and y_γ^{**} are in X. Now let $h \in \mathcal{CLB}(E)$. Since h^{**} is Lipschitz on X and $\|y_\gamma^{**} - x_\gamma^{**}\| \to 0$, $|h^{**}(y_\gamma^{**}) - h^{**}(x_\gamma^{**})| \to 0$. Further, since $x_\gamma^{**} \to x^{**}$ in $\mathcal{T}_{\mathcal{CLB}}(E^{**})$, $h^{**}(x_\gamma^{**}) \to h^{**}(x^{**})$. We now obtain by addition that $h^{**}(y_\gamma^{**}) \to h^{**}(x^{**})$. This gives the required result. $\square$

We have already remarked in Remark 45.13 on the pathology of the two topologies $\mathcal{T}_{\mathcal{CLB}}(E^{**})$ and $\mathcal{T}_{\mathcal{CC}}(E^{**})$. We will see in Example 45.16 that $\mathcal{T}_{\mathcal{CC}}(E^{**})$ can be even more pathological in that the analog of Lemma 45.15 fails with $\mathcal{T}_{\mathcal{CLB}}(E^{**})$ replaced by $\mathcal{T}_{\mathcal{CC}}(E^{**})$.

Example 45.16. We give an example of a function $f \in \mathcal{CC}(c_0)$ such that f^{**} is not continuous on $c_0^{**} = \ell^\infty$. Define $f \in \mathcal{CC}(c_0)$ by

$$f(x) := \sum_{n \ge 1} x_n{}^{2n} \quad (x = \{x_n\}_{n \ge 1} \in c_0).$$

Since $f \in \mathcal{PCLSC}(c_0)$ and $\operatorname{dom} f = c_0$, the dom lemma, Lemma 13.3, and translation imply that f is continuous on c_0. Now let $\xi = \{\xi_n\}_{n \ge 1} \in \ell^\infty$. For $N \ge 1$, let $x_N \in c_0$ be defined by

$$(x_N)_n := \begin{cases} \xi_n, & \text{if } n \le N; \\ 0, & \text{otherwise.} \end{cases}$$

From the Fenchel–Moreau theorem, Corollary 12.4,

$$f^{**}(\widehat{x_N}) = f(x_N) = \sum_{n=1}^{N} x_n{}^{2n}.$$

Now $\widehat{x_N} \to \xi$ in $w(\ell^\infty, \ell^1)$, and so the $w(\ell^\infty, \ell^1)$–lower semicontinuity of f^{**} mentioned in (38.1) implies that

$$f^{**}(\xi) \leq \lim_{N \to \infty} f^{**}(\widehat{x_N}) = \lim_{N \to \infty} f(x_N) = \sum_{n \geq 1} x_n{}^{2n}. \qquad (45.5)$$

For $N \geq 1$, define $g_N \in \mathcal{CC}(\mathbb{R}^N)$ by $g_N(x_1, \ldots, x_N) := \sum_{n=1}^N x_n{}^{2n}$ and $f_N \in \mathcal{CC}(c_0)$ by $f_N(x) := g_N(x_1, \ldots, x_N)$ $(x = \{x_n\}_{n \geq 1} \in c_0)$. It is easily seen that if $y \in \ell^1$ and $f_N{}^*(y) < \infty$ then $y_{N+1} = y_{N+2} = \cdots = 0$, and so, using the Fenchel–Moreau theorem, Corollary 12.4, again,

$$\begin{aligned}
f_N{}^{**}(\xi) &= \sup\left\{\langle y, \xi \rangle - f_N{}^*(y) : y \in \ell^1, y_{N+1} = y_{N+2} = \cdots = 0\right\} \\
&= \sup\left\{\langle z, (\xi_1, \ldots, \xi_N)\rangle - f_N{}^*(z_1, \ldots, z_N, 0, 0, \ldots) : z \in \mathbb{R}^N\right\} \\
&= \sup\left\{\langle z, (\xi_1, \ldots, \xi_N)\rangle - g_N{}^*(z) : z \in \mathbb{R}^N\right\} \\
&= g_N{}^{**}(\xi_1, \ldots, \xi_N) = g_N(\xi_1, \ldots, \xi_N) = \sum_{n=1}^N \xi_n{}^{2n}.
\end{aligned}$$

Now $f \geq f_N$ on c_0, and so $f^{**} \geq f_N{}^{**}$ on ℓ^∞, from which, $f^{**}(\xi) \geq \sum_{n=1}^N \xi_n{}^{2n}$. Letting $N \to \infty$, and combining this with (45.5), we see that

$$f^{**}(\xi) = \sum_{n \geq 1} \xi_n{}^{2n}.$$

Let $x^{**} := \{n^{-1/n}\}_{n \geq 1} \in \ell^\infty$ and, for $N \geq 1$, $y_N^{**} \in \ell^\infty$ be defined by

$$\left(y_N^{**}\right)_n := \begin{cases} n^{-1/n}, & \text{if } n \leq N; \\ 1, & \text{otherwise.} \end{cases}$$

Since $\lim_{n \to \infty} n^{-1/n} = 1$, $y_N^{**} \to x^{**}$ in ℓ^∞ as $N \to \infty$. On the other hand, for all $N \geq 1$,

$$f^{**}(y_N^{**}) \geq \sum_{n > N} 1 = \infty$$

and

$$f^{**}(x^{**}) = \sum_{n \geq 1} \frac{1}{n^2} < \infty.$$

So $f^{**}(y_N^{**}) \not\to f^{**}(x^{**})$ as $N \to \infty$. Consequently, f^{**} is not continuous on ℓ^∞. Let $x_N^{**} := x^{**}$. Then $x_N^{**} \to x^{**}$ in $\mathcal{T}_{CC}(E^{**})$ and $\|y_N^{**} - x_N^{**}\| \to 0$, but $y_N^{**} \not\to x^{**}$ in $\mathcal{T}_{CC}(E^{**})$. In other words, the analog of Lemma 45.15 with $\mathcal{T}_{CLB}(E^{**})$ replaced by $\mathcal{T}_{CC}(E^{**})$ fails.

Another difference between $\mathcal{T}_{CC}(E^{**})$ and $\mathcal{T}_{CLB}(E^{**})$ is exhibited by the following remark (compare Lemma 38.2(f)).

Remark 45.17. Let $f \in \mathcal{CC}(c_0)$ be as in Example 45.16. Then, from the definition of $\mathcal{T}_{CC}(\ell^\infty)$, f^{**} is $\mathcal{T}_{CC}(\ell^\infty)$–continuous. Since f^{**} is not $\mathcal{T}_{\|\ \|}(\ell^\infty)$–continuous, it follows that $\mathcal{T}_{CC}(\ell^\infty) \not\subset \mathcal{T}_{\|\ \|}(\ell^\infty)$.

Problem 45.18. Let E be a general non–reflexive Banach space. Does there always exist $f \in \mathcal{CC}(E)$ such that f^{**} is not continuous?

Remark 45.19. One might be led to suspect by analogy with the inf–convolution formula for the conjugate of a sum referred to in Lemma 45.1(b) that, in the situation of Lemma 45.1(a), $(f \vee g)^*(w^*)$ is given by the formula

$$\min_{\rho \in [0,1],\ u^*,\ v^* \in E^*,\ \rho u^* + (1-\rho)v^* = w^*} \left[\rho f^*(u^*) + (1 - \rho)g^*(v^*) \right], \qquad (45.6)$$

but this is not necessarily true if $f \notin \mathcal{CC}(E)$. The following example where $f \in \mathcal{PCLSC}(\mathbb{R}^2)$, $g \in \mathcal{CC}(\mathbb{R}^2)$ but (45.6) fails is due to S. Fitzpatrick (personal communication). Define f and g by

$$f(x_1, x_2) := \begin{cases} x_2 & \text{if } x_1 \geq 0; \\ \infty & \text{otherwise;} \end{cases}$$

and

$$g(x_1, x_2) := x_1.$$

Then $(f \vee g)^*(0) = -\inf(f \vee g) = 0$. On the other hand, f^* is the indicator function of $(-\infty, 0] \times \{1\}$ and g^* is the indicator function of $\{(1,0)\}$. Consequently, if $\rho \in [0,1]$, $u^* \in \mathbb{R}^2$, $v^* \in \mathbb{R}^2$ and $\rho u^* + (1 - \rho)v^* = 0$ then $\rho f^*(u^*) + (1-\rho)g^*(v^*) = \infty$, and so (45.6) fails. See Remark 45.20 below for more discussion of this question.

Remark 45.20. We refer the reader to [46, Theorem 6, p. 3558 and Theorem 12, pp. 3560–3561] for a proof of the fact that if f, $g \in \mathcal{PCLSC}(E)$ and $\bigcup_{\lambda > 0} \lambda(\operatorname{dom} f - \operatorname{dom} g)$ is a closed subspace of E then $(f \vee g)^{**} = f^{**} \vee g^{**}$ on E^{**}, and $(f \vee g)^*(w^*)$ is given by the explicit formula described below. This formula is defined by a two–stage process as follows. If $w^* \in E^*$ and $\delta > 0$, let $B(w^*, \delta) := \{x^* \in E^* : \|x^* - w^*\| < \delta\}$ and $L(w^*, \delta)$ be the set

$$\{(\rho, \sigma, u^*, v^*) : \rho > 0,\ \sigma > 0,\ u^*,\ v^* \in E^*,\ \rho + \sigma = 1,\ \rho u^* + \sigma v^* \in B(w^*, \delta)\},$$

and

$$(f^* \underset{\delta}{\wedge} g^*)(w^*) := \inf_{(\rho, \sigma, u^*, v^*) \in L(w^*, \delta)} \left[\rho f^*(u^*) + \sigma g^*(v^*) \right].$$

Then the formula is that

$$(f \vee g)^*(w^*) = \sup_{\delta > 0}(f^* \underset{\delta}{\wedge} g^*)(w^*) = \lim_{\delta \to 0}(f^* \underset{\delta}{\wedge} g^*)(w^*).$$

More results on the situation discussed here have been obtained recently by Boţ–Wanka in [28]. However, for the reasons explained in the remarks preceding Corollary 45.5, the restriction that $f \in \mathcal{PCLSC}(E)$ makes this version of the analysis inappropriate for some of the applications that we have in mind.

Problem 45.21. Let $\{x_\gamma^{**}\}$ be a net of elements of E^{**}, $x^{**} \in E^{**}$, $x_\gamma^{**} \to x^{**}$ in $w(E^{**}, E^*)$ and, for all $x \in E$, $\|x_\gamma^{**} - \widehat{x}\| \to \|x^{**} - \widehat{x}\|$. Does it necessarily follow that, for all $f \in \mathcal{CLB}(E)$, $f^{**}(x_\gamma^{**}) \to f^{**}(x^{**})$?

46 Maximally monotone multifunctions with convex graph

In this section, we consider the maximally monotone multifunctions with convex graph. This is a very small subclass of the maximally monotone multifunctions. However, it is an important subclass since it includes all affine maximally monotone operators, and all maximally monotone multifunctions whose inverse is an affine function. In Theorem 46.1, we consider some of the subclasses of the maximally monotone multifunctions introduced in Section 36 and Section 38 with reference to those with convex graph. In particular, we prove in Theorem 46.1(a,b) that there is no point in looking among the maximally monotone multifunctions with convex graph for an example of a maximally monotone multifunction that is not strongly maximally monotone or not of type (FPV). In Theorem 46.3, we give a sum theorem for maximally monotone multifunctions with convex graph.

Theorem 46.1. *Let E be a nonzero Banach space, $S: E \rightrightarrows E^*$ be maximally monotone and $G(S)$ be convex. Then:*
(a) *S is strongly maximal.*
(b) *S is of type (FPV).*
(c) *If S is of type (NI) then S is of type (ED) (and hence of type (D) and type (FP)).*

Proof. (a) Let C be a nonempty $w(E, E^*)$–compact convex subset of E, $y^* \in E^*$ and

$$\text{for all } s \in G(S), \quad \text{there exists } b \in C \times \{y^*\} \text{ such that} \quad q(s-b) \geq 0, \quad (36.8)$$

from which

$$\inf_{s \in G(S)} \max_{x \in C} q\big(s - (x, y^*)\big) \geq 0.$$

Now the function $q\big(s - (\cdot, y^*)\big)$ is affine and continuous on C and, from Lemma 19.7, the function $q\big(\cdot - (x, y^*)\big)$ is convex on $G(S)$. From the minimax theorem, Theorem 3.2, with $X := G(S)$, $Y := C$ and the function

$$(s, x) \mapsto q\big(s - (x, y^*)\big),$$

there exists $x \in C$ such that, for all $s \in G(S)$, $q\big(s - (x, y^*)\big) \geq 0$. Since S is maximally monotone, it follows from this that $(x, y^*) \in G(S)$ (see Definition 20.1). This gives (36.9). Similarly, if C is a nonempty $w(E^*, E)$–compact subset of E^*, $y \in E$ and

$$\text{for all } s \in G(S), \quad \text{there exists } b \in \{y\} \times C \text{ such that} \quad q(s-b) \geq 0, \quad (36.10)$$

then, using the function $(s, x^*) \mapsto q\big(s - (y, x^*)\big)$, (36.11) is satisfied. This completes the proof of (a).

(b) Let U be an open convex subset of E, $d \in U \times E^*$, $U \cap D(S) \neq \emptyset$ and

$$s \in G(S) \text{ and } s_1 \in U \quad \Longrightarrow \quad q(s - d) \geq 0. \qquad (36.7)$$

Our aim is to prove that

$$d \in G(S). \qquad (46.1)$$

Since $U \cap D(S) \neq \emptyset$, we can fix $t \in G(S)$ such that $t_1 \in U$. Let $M := q(t - d)$. From (36.7), $M \geq 0$. Let $0 < \delta < \text{dist}(t_1, E \setminus U)$. We first prove that

$$s \in G(S) \text{ and } x \in U \quad \Longrightarrow \quad q(s - d) + \frac{M}{\delta}\|s_1 - x\| \geq 0. \qquad (46.2)$$

Let $s \in G(S)$ and $x \in U$. If $s_1 = x$ then this is immediate from (36.7), so we can and will suppose that $s_1 \neq x$. Let $\eta := \|s_1 - x\| > 0$. Write

$$c := \frac{\delta s + \eta t}{\delta + \eta} \in G(S).$$

Then

$$\frac{(\delta + \eta)c_1 - \delta x}{\eta} = \frac{\delta s_1 + \eta t_1 - \delta x}{\eta} = \delta\frac{s_1 - x}{\eta} + t_1 \in U,$$

consequently,

$$c_1 \in \frac{\delta x + \eta U}{\delta + \eta} \subset U.$$

Thus, from (36.7), $q(c - d) \geq 0$. From Lemma 19.7, the function $q(\cdot - d)$ is convex on $G(S)$, hence

$$\frac{\delta q(s - d) + \eta q(t - d)}{\delta + \eta} \geq 0,$$

which gives (46.2). From the Hahn–Banach–Lagrange theorem, Theorem 1.11, with $C := G(S) \times U$, $k(s, x) := q(s - d)$ and $j(s, x) := s_1 - x$, there exists $x^* \in E^*$ such that

$$(s, x) \in G(S) \times U \quad \Longrightarrow \quad q(s - d) + \langle s_1 - x, x^* \rangle \geq 0$$
$$\Longrightarrow \quad \langle d_1, s_2 \rangle + \langle s_1, d_2 - x^* \rangle - q(s) \leq q(d) - \langle x, x^* \rangle,$$

or equivalently, using the definition of φ_S and (23.11), for all $x \in U$,

$$\langle x, x^* \rangle \leq q(d) - \varphi_S(d_1, d_2 - x^*) \leq q(d) - q(d_1, d_2 - x^*) = \langle d_1, x^* \rangle. \qquad (46.3)$$

Now $d_1 \in U$ and U is open, so it follows that $x^* = 0$. Substituting this back in (46.3), we obtain that $\varphi_S(d) \leq q(d)$, and (23.10) implies that $d \in G(S)$, i.e., (46.1) is satisfied. This completes the proof of (b).

(c) Let $w \in G(\overline{S})$. From (36.1) and (36.3),

$$v \in B^* \quad \Longrightarrow \quad \inf_{s \in G(S)} \widetilde{q}(\iota(s) - w) \geq 0 \geq \inf_{s \in G(S)} \widetilde{q}(\iota(s) - w - v).$$

Now Lemma 19.4, with B and q replaced by B^* and $\widetilde{q}$, respectively and $A := \iota(G(S)) - w$ gives

$$v \in B^* \quad \Longrightarrow \quad \big\lfloor w, v \big\rceil \leq \sup_{s \in G(S)} \big\lceil \iota(s), v \big\rceil,$$

Thus, from Theorem 45.12, $w \in G(S)^{\mathcal{CLBN}}$, which completes the proof that S is of type (ED). $\square$

Remark 46.2. It is also possible to prove Theorem 46.1(a) using the Hahn–Banach–Lagrange theorem, Theorem 1.11, rather than a minimax theorem. As a hint, we point out that, in (40.1), the map $s \mapsto \langle s_1, s_2 - y^* \rangle$ is convex on $G(S)$, and the map $\sup \langle D, \cdot \rangle$ is sublinear on E^* and, in (40.2), the map $s \mapsto \langle s_1 - y, s_2 \rangle$ is convex on $G(S)$, and the map $\sup \langle \cdot, D \rangle$ is sublinear on E. The problem is to show that a linear map on E^* dominated by $\sup \langle D, \cdot \rangle$ on E^* is necessarily of the form $\langle x, \cdot \rangle$ for some $x \in D$. This is true, but more technical. It is for this reason that we have opted to use a minimax proof.

Theorem 46.3. *Let E be a nonzero Banach space, $S, T \colon E \rightrightarrows E^*$ be maximally monotone, $G(S)$ and $G(T)$ be convex and $\bigcup_{\lambda > 0} \lambda [D(S) - D(T)]$ be a closed subspace of E. Then $S + T$ is maximally monotone.*

Proof. We will prove this result using (23.10). So let $b \in E \times E^*$ and $\varphi_{S+T}(b) \leq q(b)$, that is to say b is monotonically related to $G(S+T)$. Now let $(x, y^*) \in G(S) - b$ and $(x, z^*) \in G(T) - b$. Then

$$(x + b_1, y^* + z^* + 2b_2) \in G(S + T),$$

from which

$$q(x, y^*) + q(x, z^*) + b_2 \circ \pi_1(x, z^*) = \langle x, y^* + z^* + b_2 \rangle$$
$$= \langle (x + b_1) - b_1, (y^* + z^* + 2b_2) - b_2 \rangle \geq 0.$$

Thus we have proved that

$$\left. \begin{aligned} (x, y^*) \in G(S) - b \text{ and } (x, z^*) \in G(T) - b \quad \Longrightarrow \\ q(x, y^*) + q(x, z^*) + b_2 \circ \pi_1(x, z^*) \geq 0. \end{aligned} \right\} \tag{46.4}$$

Now let $p := q + \mathbb{I}_{G(S)-b}$ and $r := q + \mathbb{I}_{G(T)-b} + b_2 \circ \pi_1$. Then (46.4) gives

$$(x, y^*, z^*) \in E \times E^* \times E^* \quad \Longrightarrow \quad p(x, y^*) + r(x, z^*) \geq 0.$$

Now give $E \times E^*$ the topology $\mathcal{T}_{\| \ \|}(E \times E^*)$. It is clear from Lemma 19.7, the convexity and the closedness of $G(S)$ and $G(T)$, and the continuity of q that $p, r \in \mathcal{PCLSC}(E \times E^*)$. Furthermore,

$$\pi_1 \mathrm{dom}\, p - \pi_1 \mathrm{dom}\, r = (D(S) - b_1) - (D(T) - b_1) = D(S) - D(T).$$

Thus Lemma 16.2 gives $x^* \in E^*$ such that

$$p^*(0, -x^*) + r^*(0, x^*) \leq 0. \tag{46.5}$$

However, by direct computation and (23.11),

$$p^*(0, -x^*) = \varphi_S(b_1, b_2 - x^*) - q(b_1, b_2 - x^*) \geq 0$$

and

$$r^*(0, x^*) = \varphi_T(b_1, x^*) - q(b_1, x^*) \geq 0,$$

and so substituting into (46.5) yields

$$\varphi_S(b_1, b_2 - x^*) = q(b_1, b_2 - x^*) \quad \text{and} \quad \varphi_T(b_1, x^*) = q(b_1, x^*).$$

Thus, from (23.11) again, $(b_1, b_2 - x^*) \in G(S)$ and $(b_1, x^*) \in G(T)$. Consequently, $b = (b_1, (b_2 - x^*) + x^*) \in G(S + T)$. This completes the proof of (23.10), and establishes the maximal monotonicity of $S + T$. $\qquad\square$

Problem 46.4. Are there any maximally monotone multifunctions with convex graph which are not either affine or with an affine inverse?

47 Possibly discontinuous positive linear operators

Let $D(S)$ be a linear subspace of E and $S\colon D(S) \to E^*$ be a (possibly discontinuous) linear operator such that

$$x \in D(S) \quad \Longrightarrow \quad \langle x, Sx \rangle \geq 0.$$

In Section 17, we observed that if $D(S) = E$ then S is maximally monotone. Furthermore, S is automatically continuous (exercise!). In this section, we consider the situation when $D(S)$ is a *proper* subspace of E. It is still true and easy to see that S is monotone.

We first give a characterization of the maximal monotonicity of S in Theorem 47.1, and give in Theorem 47.3 a sufficient condition for the sum of maximally monotone linear operators to be maximally monotone. We then turn to some of the subclasses of the maximally monotone multifunctions introduced in Section 36 and Section 38, with reference to these linear operators. It was proved in Bauschke–Borwein, [9, Theorem 4.1, pp. 10–12], that if $S\colon E \to E^*$ is continuous, linear and positive then

$$S \text{ is of type (NI)} \iff S \text{ is of type (FP)}. \tag{47.1}$$

In fact, there are many other equivalent conditions in [9, Theorem 4.1], for instance that S^* be positive. We show (among other things) in Theorem 47.5(c) that the implication ($\Longrightarrow$) in (47.1) remains true in the discontinuous case. Finally, we prove in Theorem 47.7 that there is no point in looking among the continuous linear operators for an example of a maximally monotone multifunction that is not of type (ANA).

Some of the results in this section appear in Phelps–Simons, [70], with proofs based on the Eidelheit separation theorem, and in [99] with proofs based on a minimax theorem.

We start off by giving a characterization of the maximal monotonicity of S in terms of subsets of E (rather than of $E \times E^*$).

Theorem 47.1. *Let* $S \colon D(S) \to E^*$ *be monotone and linear. Then* S *is maximally monotone if, and only if,* $D(S)$ *is dense and*

$$\pi_1\{b \in E \times E^* \colon \varphi_S(b) \leq q(b)\} \subset D(S). \tag{47.2}$$

Proof. ($\Longrightarrow$) Suppose that S is maximally monotone. Let $z^* \in E^*$ and $\langle D(S), z^* \rangle = \{0\}$. Then

$$\varphi_S(0, z^*) = \sup_{x \in D(S)} \left[0 + 0 - \langle x, Sx \rangle \right] = 0 = q(0, z^*),$$

and so, from (23.11), $(0, z^*) \in G(S)$, from which $z^* = S0 = 0$. Thus we have proved that if $z^* \in E^*$ then $\langle D(S), z^* \rangle = \{0\} \Longrightarrow z^* = 0$. Corollary 4.6 now implies that $D(S)$ is dense. Furthermore, we have from (23.10) that

$$\{b \in E \times E^* \colon \varphi_S(b) \leq q(b)\} \subset G(S),$$

from which (47.2) is immediate.

($\Longleftarrow$) Let $D(S)$ be dense and (47.2) be satisfied. Let $b \in E \times E^*$ and $\varphi_S(b) \leq q(b)$. From (47.2), $b_1 \in D(S)$. Now let $x \in D(S)$, and λ be an arbitrary real number. Then $(b_1 + \lambda x, Sb_1 + \lambda Sx) \in G(S)$, from which

$$\langle b_1, Sb_1 + \lambda Sx \rangle + \langle b_1 + \lambda x, b_2 \rangle - \langle b_1 + \lambda x, Sb_1 + \lambda Sx \rangle \leq \varphi_S(b) \leq q(b) = \langle b_1, b_2 \rangle.$$

This implies that $\lambda^2 \langle x, Sx \rangle + \lambda \langle x, Sb_1 - b_2 \rangle \geq 0$. Since this holds for all $\lambda \in \mathbb{R}$, we deduce that $\langle x, Sb_1 - b_2 \rangle = 0$ (exercise!). The density of $D(S)$ now implies that $Sb_1 = b_2$, i.e., $b \in G(S)$, and the maximal monotonicity of S follows from (23.10). $\qquad \square$

Now let

$$H(S) := \pi_1\{b \in E \times E^* \colon \varphi_S(b) \leq q(b)\}.$$

Lemma 47.2 below gives a characterization of $H(S)$, which implies, in particular, that $H(S)$ is closed under multiplication by scalars, and also that $H(S)$ is an F_σ.

Lemma 47.2. *Let* $S \colon D(S) \to E^*$ *be monotone and linear and* $z \in E$. *Then* $z \in H(S)$ *if, and only if, there exists* $M \geq 0$ *such that*

$$x \in D(S) \Longrightarrow \langle z - x, Sx \rangle \leq M \|z - x\|. \tag{47.3}$$

Proof. If $z \in H(S)$ then there exists $b \in E \times E^*$ such that $b_1 = z$ and $\sup_{x \in D(S)} \left[\langle b_1, Sx \rangle + \langle x, b_2 \rangle - \langle x, Sx \rangle \right] = \varphi_S(b) \le q(b) = \langle b_1, b_2 \rangle$. Thus

$$x \in D(S) \quad \Longrightarrow \quad \langle z - x, Sx \rangle \le \langle z - x, b_2 \rangle \le \|z - x\| \|b_2\|,$$

and (47.3) follows with $M = \|b_2\|$. Suppose, conversely, that $M \ge 0$ and (47.3) is satisfied. Then $\inf_{x \in D(S)} \left[M \|z - x\| + \langle x - z, Sx \rangle \right] \ge 0$. Now the function $x \mapsto \langle x - z, Sx \rangle$ is convex and so, from the Hahn–Banach–Lagrange theorem, Theorem 1.11, there exists $z^* \in E^*$ such that $\|z^*\| \le M$ and $\inf_{x \in D(S)} \left[\langle z - x, z^* \rangle + \langle x - z, Sx \rangle \right] \ge 0$. Since this can be rewritten $\varphi_S(z, z^*) \le q(z, z^*)$, we have $z \in H(S)$. $\qquad\square$

Our next result is a substantial generalization of Phelps–Simons, [70, Theorem 7.2, p. 325].

Theorem 47.3. *Let E be a nonzero Banach space, $S \colon D(S) \to E^*$ and $T \colon D(T) \to E^*$ be maximally monotone and linear and $\bigcup_{\lambda > 0} \lambda [D(S) - D(T)]$ be a closed subspace of E. Then $S + T$ is maximally monotone.*

Proof. This is immediate from Theorem 46.3. $\qquad\square$

Examples 47.4. The following examples of discontinuous maximally monotone linear operators are taken from Phelps–Simons, [70, Example 4.3, p. 311, Example 6.4, p. 318, and Example 7.4, p. 326], to which we refer the reader for a more comprehensive analysis.

• Let $E := L^1[0, 1]$,

$$D(S) = \{x \in L^1 \colon x \text{ is Lipschitz and } x(0) = 0\}.$$

Define $S \colon D(S) \to L^\infty$ by $Sx := x'$. Then S is maximally monotone.

• Let $E := L^1[0, 1]$,

$$D(S) = \{x \in L^1 \colon x \text{ is Lipschitz and } x(0) = x(1)\}.$$

Define $S \colon D(S) \to L^\infty$ by $Sx := x'$. Then S is maximally monotone.

• Let $E := \ell^2$, and define $V, W \colon \ell^2 \to \ell^2$ by

$$Vx := (x_1, x_2 - x_1, x_3 - x_2, \dots) \quad \text{and} \quad Wx := (x_1 - x_2, x_2 - x_3, x_3 - x_4, \dots)$$

for $x = \{x_n\}_{n \ge 1} \in \ell^2$. Both V and W are injective, so we can define S and T by $S := V^{-1}$ (with $D(S) = R(V)$) and $T := W^{-1}$ (with $D(T) = R(W)$). S and T are maximally monotone. Even though $D(S) - D(T)$ is dense in ℓ^2, $S + T$ is not maximally monotone (exercise!). Compare this example with Theorem 24.1. (In fact, it is even true that $D(S) \cap D(T)$ is dense in ℓ^2 — see [70].)

Theorem 47.5(a) first appeared in Bauschke–Simons, [11, Theorem 1.1, pp. 166–167], Theorem 47.5(b) first appeared in [99, Theorem 38.2, pp. 146–147] and Theorem 47.5(c) first appeared in [70, Theorem 6.7, p. 320–323]. (We point out that the "decomposition technique" used in [9] does not seem to be applicable to the discontinuous case.)

Theorem 47.5. *Let* $S\colon D(S) \to E^*$ *be linear and maximally monotone. Then:*
(a) *S is strongly maximally monotone.*
(b) *S is of type (FPV).*
(c) *If S is of type (NI) then S is of type (ED) (and hence of type (D) and type (FP)).*

Proof. These results are all immediate from the corresponding parts of Theorem 46.1. □

Problem 47.6. Is the converse of Theorem 47.5(c) true? That is to say, if $S\colon D(S) \to E^*$ is linear and of type (FP) then is S necessarily maximally monotone of type (NI)?

Theorem 47.7 appears in Bauschke–Simons, [11, Theorem 2.1, pp. 167–168].

Theorem 47.7. *Let $S\colon E \to E^*$ be positive and linear. Then S is maximally monotone of type (ANA).*

Proof. Suppose that $(x, x^*) \in E \times E^* \setminus G(S)$. Then $Sx \neq x^*$. For all $n \geq 1$, we can find $z_n \in E$ such that $\|z_n\| = 1$ and

$$\langle z_n, Sx - x^* \rangle \to -\|Sx - x^*\| \quad \text{as } n \to \infty. \tag{47.4}$$

For all $n \geq 1$, let $y_n := x + z_n/n$. Then $\|Sy_n - Sx\| = \|Sz_n\|/n \leq \|S\|/n$ hence

$$\|Sy_n - Sx\| \to 0 \quad \text{and} \quad \|Sy_n - x^*\| \to \|Sx - x^*\| \neq 0 \quad \text{as } n \to \infty. \tag{47.5}$$

Now, for all sufficiently large $n \geq 1$, we have the inequality

$$\frac{|\langle y_n - x, Sy_n - Sx \rangle|}{\|y_n - x\| \|Sy_n - x^*\|} \leq \frac{\|Sy_n - Sx\|}{\|Sy_n - x^*\|}.$$

Combining this with (47.5), we obtain that

$$\frac{\langle y_n - x, Sy_n - Sx \rangle}{\|y_n - x\| \|Sy_n - x^*\|} \to 0 \quad \text{as } n \to \infty. \tag{47.6}$$

On the other hand, from (47.4) and (47.5),

$$\frac{\langle y_n - x, Sx - x^* \rangle}{\|y_n - x\| \|Sy_n - x^*\|} = \frac{\langle z_n, Sx - x^* \rangle}{\|Sy_n - x^*\|} \to \frac{-\|Sx - x^*\|}{\|Sx - x^*\|} = -1 \quad \text{as } n \to \infty.$$

Adding this to (47.6), we obtain that

$$\frac{\langle y_n - x, Sy_n - x^* \rangle}{\|y_n - x\| \|Sy_n - x^*\|} \to -1 \quad \text{as } n \to \infty.$$

This completes the proof that S is of type (ANA).   □

Remark 47.8. As we have already observed in Problem 36.12, we do not know if S is necessarily of type (ANA) if $D(S)$ is a subspace of E and $S \colon D(S) \to E^*$ is linear and maximally monotone.

In our next example, we show that the "tail" operator serves to distinguish type (ANA) and type (BR).

Example 47.9. Let $E = \ell^1$, and $S \colon \ell^1 \to \ell^\infty = E^*$ be the positive linear operator defined in Remark 36.6. Theorem 47.7 clearly implies that S is of type (ANA). Now, by direct computation, $Se^{(1)} = e^{(1)}$ and $S^*\widehat{e^{(1)}} = e$. Let $x^* := Se^{(1)} + S^*\widehat{e^{(1)}} \in \ell^\infty$. Then, for all $x \in \ell^1$,

$$\begin{aligned}
\langle x, Sx - x^* \rangle &= \langle x, Sx \rangle - \langle x, Se^{(1)} \rangle - \langle x, S^*\widehat{e^{(1)}} \rangle \\
&= \langle x, Sx \rangle - \langle x, Se^{(1)} \rangle - \langle e^{(1)}, Sx \rangle \\
&= \langle x - e^{(1)}, Sx - Se^{(1)} \rangle - \langle e^{(1)}, Se^{(1)} \rangle \\
&\geq 0 - 1 = -1 > -4 \cdot \tfrac{1}{2}.
\end{aligned}$$

So if S were of type (BR) then there would exist $x \in \ell^1$ such that

$$\|x - 0\| \leq 4 \quad \text{and} \quad \|Sx - x^*\| \leq \tfrac{1}{2}.$$

This is clearly impossible since, for all $x \in E$, $Sx \in c_0$ and so $\|Sx - x^*\| \geq 1$. So S is not of type (BR).

Remark 47.10. The author is grateful to Heinz Bauschke for pointing out to him that certain Banach spaces cannot support continuous linear ultra-maximally monotone operators. If $S \colon E \to E^*$ is continuous, linear and ultramaximal monotone then it can be seen that

$$x^{**} \in E^{**} \text{ and } S^{**}x^{**} \in \widehat{E^*} \quad \Longrightarrow \quad x^{**} \in \widehat{E},$$

that is, S is *Tauberian* — see Wilansky, [117, p. 175]. It follows from [117, Theorem 11–4–2, pp. 174–175] that if E is not reflexive then the closure in E^* of the image under S of the unit ball of E is not weakly compact in E^* hence, in the notation of Saab–Saab, [86, Definition 6, p. 378], E does not have "property (w)". Spaces with this property are discussed in [86, pp. 378–380 and Proposition 47, p. 386]. In particular, E cannot be of the form $c_0(\Gamma)$ or $C(\Omega)$ (Ω compact Hausdorff). See also the discussion in Bauschke, [6, pp. 167–169].

48 Subtler properties of subdifferentials

The main result of this section is Theorem 48.1, in which we show that if $f \in \mathcal{PCLSC}(E)$ then $\iota(G(\partial f))$ is dense in $G^{-1}(\partial f^*)$ in $\mathcal{T}_{\mathcal{CLBN}}(E^{**} \times E^*)$. Theorem 48.1, Theorem 48.4(a) and Lemma 48.9 represent sharpenings of results proved by Gossez and Rockafellar, which used some rather delicate functional analysis. (See the comments preceding Lemma 48.9 for a more detailed discussion.) Our proof uses Lemma 45.9(a), which depends ultimately on the formula for the biconjugate of a maximum that we established in Theorem 45.3. It is then but a short step to Theorem 48.4(b), in which we prove that subdifferentials are maximally monotone of type (ED). Theorem 48.4(c–g) contain a number of consequences of this. Further, every maximally monotone multifunction of type (ED) is automatically of "dense type" in the sense introduced by Gossez in [48, p. 375]. Thus Theorem 48.4(b) extends the result proved in [48, Théorème 3.1, pp. 376–378] that subdifferentials are maximally monotone of dense type. Corollary 48.8 contains a result that is approximately a considerable generalization of the Brøndsted–Rockafellar theorem.

Theorem 48.1. *Let E be a nonzero Banach space and $f \in \mathcal{PCLSC}(E)$. Then $G^{-1}(\partial f^*) \subset (G(\partial f))^{\mathcal{CLBN}}$.*

Proof. Let v be an arbitrary element of $G^{-1}(\partial f^*)$. Lemma 45.9(a), provides a net $\{x_\gamma\}$ of elements of E such that

$$\widehat{x_\gamma} \to v_1 \text{ in } \mathcal{T}_{\mathcal{CLB}}(E^{**}) \quad \text{and} \quad f(x_\gamma) \to f^{**}(v_1). \qquad (48.1)$$

Since $v \in G^{-1}(\partial f^*)$, we have $f^{**}(v_1) + f^*(v_2) = \widetilde{q}(v)$, from which $f^{**}(v_1) \in \mathbb{R}$, and so we can and will suppose that, for all γ, $f(x_\gamma) \in \mathbb{R}$. Let $\eta_\gamma := f(x_\gamma) + f^*(v_2) - q(x_\gamma, v_2) = f(x_\gamma) + f^*(v_2) - \widetilde{q}(\widehat{x_\gamma}, v_2)$. From (48.1) and Lemma 38.2(a), $\eta_\gamma \to f^{**}(v_1) + f^*(v_2) - \widetilde{q}(v) = 0$. The Fenchel–Young inequality, (8.2), implies that $\eta_\gamma \geq 0$. Now if $\eta_\gamma > 0$ then it follows from Theorem 18.6 that there exists $s_\gamma = (s_{\gamma 1}, s_{\gamma 2}) \in G(\partial f)$ such that $\|s_{\gamma 1} - x_\gamma\| \leq \sqrt{\eta_\gamma}$ and $\|s_{\gamma 2} - v_2\| \leq \sqrt{\eta_\gamma}$; if, on the other hand, $\eta_\gamma = 0$, this remains true with $s_\gamma := (x_\gamma, v_2)$. Since $\|\widehat{s_{\gamma 1}} - \widehat{x_\gamma}\| = \|s_{\gamma 1} - x_\gamma\| \to 0$, the first part of (48.1) and Lemma 45.15 imply that, $\widehat{s_{\gamma 1}} \to v_1$ in $\mathcal{T}_{\mathcal{CLB}}(E^{**})$. The result now follows since $s_{\gamma 2} \to v_2$ in $\mathcal{T}_{\|\ \|}(E^*)$. $\qquad \square$

We now give a simple corollary of Theorem 48.1, which should be compared with Gossez, [48, Corollaire 3.1, pp. 378–379].

Corollary 48.2. *Let E be a nonzero Banach space, C be a nonempty closed convex subset of E and $\ddot{C}$ be the $w(E^{**}, E^*)$–closure of $\widehat{C}$ in E^{**}. If $x^{**} \in \ddot{C}$, $x^* \in E^*$ and $\langle x^*, x^{**} \rangle = \sup \langle C, x^* \rangle$ then there exists a net $\{s_\gamma\}$ of elements of $G(N_C)$ such that $\iota(s_\gamma) \to (x^{**}, x^*)$ in $\mathcal{T}_{\mathcal{CLBN}}(E^{**} \times E^*)$.*

Proof. Let $f := \mathbb{I}_C$ and apply Theorem 48.1. $\qquad \square$

We do not know if the analogs of Theorem 48.1 and Corollary 48.2 with $\mathcal{T}_{\mathcal{CLB}}(E^{**})$ replaced by $\mathcal{T}_{\mathcal{CC}}(E^{**})$ are true. Specifically:

Problem 48.3. Let E be a nonzero Banach space. Let $f \in \mathcal{PCLSC}(E)$. Then is it always true that $G^{-1}(\partial f^*) \subset (G(\partial f))^{\mathcal{CCN}}$? Let C be a nonempty proper closed convex subset of E, $\ddot{C}$ be the $w(E^{**}, E^*)$–closure of $\widehat{C}$ in E^{**}, $x^* \in E^*$ and $\langle x^*, x^{**} \rangle = \sup\langle C, x^* \rangle$. Then does there necessarily exist a net $\{s_\gamma\}$ of elements of $G(N_C)$ such that $\iota(s_\gamma) \to (x^{**}, x^*)$ in $\mathcal{T}_{\mathcal{CCN}}(E^{**} \times E^*)$?

Theorem 48.4. Let E be a nonzero Banach space and $f \in \mathcal{PCLSC}(E)$.
(a) Let $w \in E^{**} \times E^*$. Then the conditions (48.2)–(48.4) are equivalent:

$$w \in G(\overline{\partial f}) \tag{48.2}$$

$$\inf_{v \in G^{-1}(\partial f^*)} \widetilde{q}(v - w) \geq 0 \tag{48.3}$$

$$w \in G^{-1}(\partial f^*). \tag{48.4}$$

(b) ∂f is maximally monotone of type (ED).
(c) ∂f is maximally monotone of type (FP).
(d) ∂f is maximally monotone of type (FPV).
(e) ∂f is strongly maximally monotone.
(f) ∂f is maximally monotone of type (ANA).
(g) ∂f is maximally monotone of type (BR).

Proof. (a) ((48.2)$\Longrightarrow$(48.3)) Let v be an arbitrary element of $G^{-1}(\partial f^*)$. From Theorem 48.1 and Lemma 38.2(f), there exists a bounded net $\{s_\gamma\}$ of elements of $G(\partial f)$ such that $\iota(s_\gamma) \to v$ in $\mathcal{T}_{\mathcal{WN}}(E^{**} \times E^*)$. From (36.1) and (48.2), for all γ, $\widetilde{q}(\iota(s_\gamma) - w) \geq 0$ hence, by passing to the limit and using (35.2), $\widetilde{q}(v - w) \geq 0$. Thus we have established (48.3).

((48.3)$\Longrightarrow$(48.2)) This is immediate from (36.1) since, from the Fenchel–Moreau theorem, Corollary 12.4, $s \in G(\partial f) \iff \iota(s) \in G^{-1}(\partial f^*)$.

((48.3)$\Longleftrightarrow$(48.4)) This equivalence follows since, from Rockafellar's maximal monotonicity theorem, Theorem 18.7, $\partial f^*: E^* \rightrightarrows E^{**}$ is maximally monotone.

(b) is immediate from (a) and Theorem 48.1, (c) from (b) and Theorem 37.1, (d) from (b) and Theorem 39.1, (e) from (b) and Theorem 40.1, (f) from (b) and Theorem 42.6(b), and (g) from (b) and Theorem 42.6(c). $\square$

Remark 48.5. Theorem 48.4(c) was first proved in [92]. Theorem 48.4(d) was first proved by Fitzpatrick–Phelps in [45, Corollary 3.4, p. 66] and Verona–Verona in [113, Theorem 3, p. 269]. Theorem 48.4(e) was first proved in [94, Theorem 6.1 and Theorem 6.2, p. 1386]. Theorem 48.4(f) was first proved in [97, Theorem 13, p. 229 and Theorem 26, p. 237].

Our next result is immediate from Theorem 48.4(b) and Theorem 42.6. Of course, Theorem 48.6 simultaneously extends both Theorem 48.4(f) and Theorem 48.4(g).

Theorem 48.6. *Let E be a nonzero Banach space, $f \in \mathcal{PCLSC}(E)$, $b \in E \times E^* \setminus G(\partial f)$ and α, $\beta > 0$. Then:*
(a) *There exists $s \in G(\partial f)$ such that $s_1 \neq b_1$, $s_2 \neq b_2$,*

$$\frac{\|s_1 - b_1\|}{\|s_2 - b_2\|} \text{ as near as we please to } \frac{\alpha}{\beta}$$

and

$$\frac{q(s - b)}{\|s_1 - b_1\|\|s_2 - b_2\|} \text{ as near as we please to } -1.$$

(b) *If, further,* $\inf_{s \in G(\partial f)} q(s - b) > -\alpha\beta$, *then we can take s so that, in addition,* $\|s_1 - b_1\| < \alpha$ *and* $\|s_2 - b_2\| < \beta$.

Theorem 48.6(b) leads to a new version of the Brøndsted–Rockafellar theorem for subdifferentials, which we shall state as Corollary 48.8. The bridge between these two results is provided by Lemma 48.7 below, which was essentially proved by Martínez-Legaz–Théra in [62]. It was also shown in [62] that the inequality (48.5) can easily be strict. For instance, we can take $E := \mathbb{R}$, and $f(t) := t^2$. Then, for all $b \in \mathbb{R}^2$,

$$\inf_{s \in G(\partial f)} q(s - b) = -(2b_1 - b_2)^2/8 = -\big(f'(b_1) - b_2\big)^2/8$$

and

$$f(b_1) + f^*(b_2) - q(b) = (2b_1 - b_2)^2/4 = \big(f'(b_1) - b_2\big)^2/4.$$

So the inequality (48.5) is always strict in this case when $b \notin G(\partial f)$.

Lemma 48.7. *Let E be a nonzero Banach space, $f \in \mathcal{PCLSC}(E)$ and $b \in E \times E^*$. Then*

$$\inf_{s \in G(\partial f)} q(s - b) \geq -\big[f(b_1) + f^*(b_2) - q(b)\big]. \tag{48.5}$$

Proof. This follows from the observation that if $s \in G(\partial f)$ then, using the Fenchel–Young inequality, (8.2),

$$q(s - b) + f(b_1) + f^*(b_2) - q(b)$$
$$\geq q(s - b) + \langle b_1 - s_1, s_2 \rangle + f(s_1) + f^*(b_2) - q(b)$$
$$\geq q(s - b) + \langle b_1 - s_1, s_2 \rangle + \langle s_1, b_2 \rangle - q(b) = 0. \qquad \square$$

Corollary 48.8 tells us that, provided that we replace "$\leq$" by "$<$" in the appropriate places, in addition to the other conclusions that we had in the classical Theorem 18.6, we can exert considerable control over the values of $\|s_1 - b_1\|$, $\|s_2 - b_2\|$ and $q(s - b)$. Of course, in any case when the inequality in (48.5) is *strict*, we can obtain a generalization of Theorem 18.6 by applying Theorem 48.6 and Lemma 48.7 with slightly smaller values of α and β. We leave details of this to the reader.

Corollary 48.8. *Let E be a nonzero Banach space, $f \in \mathcal{PCLSC}(E)$, $b \in E \times E^* \setminus G(\partial f)$, $\alpha, \beta > 0$ and $f(b_1) + f^*(b_2) < q(b) + \alpha\beta$. Then there exists $s \in G(\partial f)$ such that $\|s_1 - b_1\| \in \,]0, \alpha[\,$, $\|s_2 - b_2\| \in \,]0, \beta[\,$,*

$$\frac{\|s_1 - b_1\|}{\|s_2 - b_2\|} \text{ as near as we please to } \frac{\alpha}{\beta}$$

and

$$\frac{q(s - b)}{\|s_1 - b_1\|\|s_2 - b_2\|} \text{ as near as we please to } -1.$$

Proof. This is immediate from Theorem 42.6 and Lemma 48.7. □

Theorem 48.4(e) lends some credence to the following

Conjecture: If $f \in \mathcal{PCLSC}(E)$, C is a nonempty $w(E, E^*)$–compact convex subset of E, C^* is a nonempty $w(E^*, E)$–compact convex subset of E^* and,

for all $s \in G(\partial f)$, there exists $b \in C \times C^*$ such that $q(s - b) \geq 0$

then

$$G(\partial f) \cap (C \times C^*) \neq \emptyset.$$

It was proved by Kum in [57, Theorem 2, pp. 374–375], Luc in [58, Theorem 2.2, p. 368] and Zagrodny in [118, Theorem 3.1, p. 305] that if $E = \mathbb{R}$ then this conjecture is true.

There are also examples in [58, pp. 368–370] and [118, Example 3.3, pp. 306–307] that if $E = \mathbb{R}^2$ then the conjecture fails (even with f a C^1 function).

Finally, it was proved in [118, Theorem 4.1, pp. 307–308] that if C and C^* satisfy the further condition that there exists $b_0 \in C \times C^*$ satisfying

$$b \in C \times C^* \implies q(b - b_0) = 0 \qquad (48.6)$$

then the conjecture is true and further, by an extremely intricate and ingenious argument, in [118, Theorem 5.2, pp. 309–314] that if $E = \mathbb{R}^2$ and the conjecture is true then there exists $b_0 \in C \times C^*$ satisfying (48.6).

Lemma 48.9 below is the final step needed to obtain sharpenings of results proved by Gossez in [48, Théorème 3.1 and Lemme 3.1, pp. 376–378] which were, in turn, sharpenings of results proved by Rockafellar in [81, Proposition 1, pp. 211–212]. We note that Lemma 48.9 is *not* needed to obtain Theorem 48.4(b) above.

Lemma 48.9. *Let E be a nonzero Banach space, $f \in \mathcal{PCLSC}(E)$, $v \in G^{-1}(\partial f^*)$ and $\{s_\gamma\} = \{(s_{\gamma 1}, s_{\gamma 2})\}$ be a bounded net of elements of $G(\partial f)$ such that $\iota(s_\gamma) \to v$ in $\mathcal{T}_{WN}(E^{**} \times E^*)$. Then $f(s_{\gamma 1}) \to f^{**}(v_1)$.*

Proof. (35.2) implies that $q(s_\gamma) = \widetilde{q}(\iota(s_\gamma)) \to \widetilde{q}(v)$. For each γ, $s_\gamma \in G(\partial f)$ and so

$$f(s_{\gamma 1}) + f^*(s_{\gamma 2}) = q(s_\gamma).$$

Hence, passing to the limit,

$$\limsup_\gamma f(s_{\gamma 1}) + \liminf_\gamma f^*(s_{\gamma 2}) \leq \widetilde{q}(v).$$

Since $s_{\gamma 2} \to v_2$ in $\mathcal{T}_{\| \ \|}(E^*)$ and $f^* \in \mathcal{PCLSC}(E^*)$, $\liminf_\gamma f^*(s_{\gamma 2}) \geq f^*(v_2)$, thus, using the assumption that $v \in G^{-1}(\partial f^*)$,

$$\limsup_\gamma f(s_{\gamma 1}) \leq \widetilde{q}(v) - f^*(v_2) = f^{**}(v_1). \tag{48.7}$$

On the other hand, since $\widehat{s_{\gamma 1}} \to v_1$ in $w(E^{**}, E^*)$, it follows from the $w(E^{**}, E^*)$–lower semicontinuity of f^{**} mentioned in (38.1) that

$$\liminf_\gamma f(s_{\gamma 1}) \geq \liminf_\gamma f^{**}(\widehat{s_{\gamma 1}}) \geq f^{**}(v_1).$$

Combining this with (48.7), we derive that $\lim_\gamma f(s_{\gamma 1}) = f^{**}(v_1)$. $\quad\square$

49 Saddle functions and type (ED)

The main result of this section is Theorem 49.6, in which we prove that if E and F are nonzero Banach spaces and F is reflexive then the "sub-differential" of a closed saddle–function on $E \times F$ is maximally mono-tone of type (ED). Our main preliminary result is Theorem 49.4. Let E, F and H be nonzero Banach spaces. As explained in Notation 16.1, we identify $(E \times H)^*$ with $H^* \times E^*$, and consequently we shall identify $(E \times H)^{**}$ with $E^{**} \times H^{**}$. We shall use without justification the easily verifiable result that, for all $(x, z) \in E \times H$, $(\widehat{x}, \widehat{z}) = \widehat{(x, z)}$. We shall also write $\mathcal{T}_{\mathcal{CLBN}}(E^{**} \times H)$ for the topology $\mathcal{T}_{\mathcal{CLB}}(E^{**}) \times \mathcal{T}_{\| \ \|}(H)$ on $E^{**} \times H$ — this usage is consistent with that introduced in Definition 38.1.

Lemma 49.1. Let E and H be nonzero Banach spaces and $(x^{**}, z^{**}, z) \in E^{**} \times H^{**} \times H$.

(a) Suppose that $\{(x^{**}_\gamma, z^{**}_\gamma)\}$ is a net of elements of $E^{**} \times H^{**}$ such that

$$(x^{**}_\gamma, z^{**}_\gamma) \to (x^{**}, z^{**}) \text{ in } \mathcal{T}_{\mathcal{CLB}}(E^{**} \times H^{**}).$$

Then

$$(x^{**}_\gamma, z^{**}_\gamma) \to (x^{**}, z^{**}) \text{ in } \mathcal{T}_{\mathcal{CLB}}(E^{**}) \times \mathcal{T}_{\mathcal{CLB}}(H^{**}).$$

(b) Suppose that $\{(x^{**}_\gamma, z_\gamma)\}$ is a net of elements of $E^{**} \times H$ such that

$$(x^{**}_\gamma, \widehat{z_\gamma}) \to (x^{**}, \widehat{z}) \text{ in } \mathcal{T}_{\mathcal{CLB}}(E^{**} \times H^{**}).$$

Then

$$(x^{**}_\gamma, z_\gamma) \to (x^{**}, z) \text{ in } \mathcal{T}_{\mathcal{CLBN}}(E^{**} \times H).$$

Proof. (a) Let h be an arbitrary element of $CLB(E)$, and define $g \in CLB(E \times H)$ by $g := h \circ \pi_1$. Then, by direct computation,

$$g^{**} = h^{**} \circ \pi_1 \in CLB(E^{**} \times H^{**}).$$

The definition of $T_{CLB}(E^{**} \times H^{**})$ now gives

$$g^{**}(x_\gamma^{**}, z_\gamma^{**}) \to g^{**}(x^{**}, z^{**}),$$

that is to say $h^{**}(x_\gamma^{**}) \to h^{**}(x^{**})$. It then follows from the definition of $T_{CLB}(E^{**})$ that $x_\gamma^{**} \to x^{**}$ in $T_{CLB}(E^{**})$. The proof that $z_\gamma^{**} \to z^{**}$ in $T_{CLB}(H^{**})$ is similar. This completes the proof of (a). (b) is immediate from (a) and Lemma 38.2(g). $\qquad \square$

Remark 49.2. If E is not reflexive then the converse of Lemma 49.1(a) is false. To see this, let $x^{**} \in E^{**} \setminus \widehat{E}$. From Lemma 45.9(b), there exists a net $\{x_\gamma\}$ of elements of E such that $\widehat{x_\gamma} \to x^{**}$ in $T_{CLB}(E^{**})$. Then

$$(\widehat{x_\gamma}, x^{**}) \to (x^{**}, x^{**}) \text{ in } T_{CLB}(E^{**}) \times T_{CLB}(E^{**}).$$

Define $h \in CLB(E \times E)$ by $h(y, z) := \|y - z\|$. Then, by direct computation, for all $(y^{**}, z^{**}) \in E^{**} \times E^{**}$, $h^{**}(y^{**}, z^{**}) = \|y^{**} - z^{**}\|$. Thus we have

$$h^{**}(\widehat{x_\gamma}, x^{**}) = \|\widehat{x_\gamma} - x^{**}\| \nrightarrow 0 = \|x^{**} - x^{**}\| = h^{**}(x^{**}, x^{**}),$$

and so $(\widehat{x_\gamma}, x^{**}) \nrightarrow (x^{**}, x^{**})$ in $T_{CLB}(E^{**} \times E^{**})$. The converse of Lemma 49.1(b) is true, and will be established in Theorem 49.4.

It is worth pointing out that we give an indirect proof of Lemma 49.3 below since we do not have a simple formula for g^* in terms of h^*.

Lemma 49.3. Let $z \in H$ and $h \in CLB(E \times H)$. Define $g \in CLB(E)$ by: $g(x) := h(x, z)$ for $x \in E$. Let $x^{**} \in E^{**}$. Then $g^{**}(x^{**}) = h^{**}(x^{**}, \widehat{z})$.

Proof. It follows from Lemma 45.9(b) that there exists a net $\{(x_\gamma, z_\gamma)\}$ of elements of $E \times H$ such that

$$(\widehat{x_\gamma}, \widehat{z_\gamma}) \to (x^{**}, \widehat{z}) \text{ in } T_{CLB}(E^{**} \times H^{**}), \tag{49.1}$$

and then Lemma 49.1(b) implies that

$$(\widehat{x_\gamma}, z_\gamma) \to (x^{**}, z) \text{ in } T_{CLBN}(E^{**} \times H). \tag{49.2}$$

Since

$$\|(\widehat{x_\gamma}, \widehat{z_\gamma}) - (\widehat{x_\gamma}, \widehat{z})\| = \|(0, \widehat{z_\gamma} - \widehat{z})\| = \|\widehat{z_\gamma} - \widehat{z}\| = \|z_\gamma - z\| \to 0,$$

(49.1) and Lemma 45.15 give $(\widehat{x_\gamma}, \widehat{z}) \to (x^{**}, \widehat{z})$ in $T_{CLB}(E^{**} \times H^{**})$, from which

$$h^{**}(\widehat{x_\gamma}, \widehat{z}) \to h^{**}(x^{**}, \widehat{z}).$$

For all γ, two applications of the Fenchel–Moreau theorem, Corollary 12.4, and (49.2) imply that

$$h^{**}(\widehat{x_\gamma}, \widehat{z}) = h(x_\gamma, z) = g(x_\gamma) = g^{**}(\widehat{x_\gamma}) \to g^{**}(x^{**}).$$

The result follows by comparing these two expressions for $\lim_\gamma h^{**}(\widehat{x_\gamma}, \widehat{z})$. $\qquad \square$

Theorem 49.4. *Let E and H be nonzero Banach spaces, $(x^{**}, z) \in E^{**} \times H$ and $\{(x_\gamma^{**}, z_\gamma)\}$ be a net of elements of $E^{**} \times H$. Then*

$$(x_\gamma^{**}, \widehat{z_\gamma}) \to (x^{**}, \widehat{z}) \text{ in } \mathcal{T}_{\mathcal{CLB}}(E^{**} \times H^{**}) \Longleftrightarrow$$

$$(x_\gamma^{**}, z_\gamma) \to (x^{**}, z) \text{ in } \mathcal{T}_{\mathcal{CLBN}}(E^{**} \times H).$$

Proof. ($\Longrightarrow$) was proved in Lemma 49.1(b). As for ($\Longleftarrow$), let h be an arbitrary element of $\mathcal{CLB}(E \times H)$, and define $g \in \mathcal{CLB}(E)$ as in Lemma 49.3. The definition of $\mathcal{T}_{\mathcal{CLB}}(E^{**})$ and Lemma 49.3 imply that $g^{**}(x_\gamma^{**}) \to g^{**}(x^{**})$, and so $h^{**}(x_\gamma^{**}, \widehat{z}) \to h^{**}(x^{**}, \widehat{z})$. The definition of $\mathcal{T}_{\mathcal{CLB}}(E^{**} \times H^{**})$ now gives

$$(x_\gamma^{**}, \widehat{z}) \to (x^{**}, \widehat{z}) \text{ in } \mathcal{T}_{\mathcal{CLB}}(E^{**} \times H^{**}).$$

Since

$$\|(x_\gamma^{**}, \widehat{z_\gamma}) - (x_\gamma^{**}, \widehat{z})\| = \|(0, \widehat{z_\gamma} - \widehat{z})\| = \|\widehat{z_\gamma} - \widehat{z}\| = \|z_\gamma - z\| \to 0,$$

Lemma 45.15 implies that $(x_\gamma^{**}, \widehat{z_\gamma}) \to (x^{**}, \widehat{z})$ in $\mathcal{T}_{\mathcal{CLB}}(E^{**} \times H^{**})$. $\square$

We now prove a lemma on "partially inverting" a multifunction.

Lemma 49.5. *Let E and F be nonzero Banach spaces, F be reflexive and $S: E \times F^* \rightrightarrows F^{**} \times E^*$ be maximally monotone of type (ED). Define $P: E \times F \rightrightarrows F^* \times E^*$ by declaring that $((x, y), (y^*, x^*)) \in G(P)$ exactly when $((x, y^*), (\widehat{y}, x^*)) \in G(S)$. Then P is maximally monotone of type (ED).*

Proof. We leave to the reader the proof (using the reflexivity of F) that P is maximally monotone. Now suppose that $((x^{**}, \widehat{y}), (y^*, x^*)) \in G(\overline{P})$. Then, by direct computation, $((x^{**}, \widehat{y^*}), (\widehat{y}, x^*)) \in G(\overline{S})$. Since S is of type (ED), there exists a net $\{((x_\gamma, y_\gamma^*), (\widehat{y_\gamma}, x_\gamma^*))\}$ of elements of $G(S)$ such that

$$((\widehat{x_\gamma}, \widehat{y_\gamma^*}), (\widehat{y_\gamma}, x_\gamma^*)) \to ((x^{**}, \widehat{y^*}), (\widehat{y}, x^*)) \text{ in } \mathcal{T}_{\mathcal{CLBN}}((E^{**} \times F^{***}) \times (F^{**} \times E^*)).$$

Using Theorem 49.4 with $H := F^*$, this equivalent to:

$$((\widehat{x_\gamma}, y_\gamma^*), (\widehat{y_\gamma}, x_\gamma^*)) \to ((x^{**}, y^*), (\widehat{y}, x^*)) \text{ in } \mathcal{T}_{\mathcal{CLBN}}(E^{**} \times F^*) \times \mathcal{T}_{\| \, \|}(F^{**} \times E^*):$$

this is, in turn, equivalent to:

$$(\widehat{x_\gamma}, y_\gamma^*, \widehat{y_\gamma}, x_\gamma^*) \to (x^{**}, y^*, \widehat{y}, x^*) \text{ in } \mathcal{T}_{\mathcal{CLB}}(E^{**}) \times \mathcal{T}_{\| \, \|}(F^* \times F^{**} \times E^*).$$

Permuting the components and using the fact that $\,\widehat{}\,$ is an isometry, this is equivalent to

$$(\widehat{x_\gamma}, y_\gamma, y_\gamma^*, x_\gamma^*) \to (x^{**}, y, y^*, x^*)) \text{ in } \mathcal{T}_{\mathcal{CLB}}(E^{**}) \times \mathcal{T}_{\| \, \|}(F \times F^* \times E^*),$$

which is, in turn, equivalent to:

$$((\widehat{x_\gamma}, y_\gamma), (y_\gamma^*, x_\gamma^*)) \to ((x^{**}, y), (y^*, x^*)) \text{ in } \mathcal{T}_{\mathcal{CLBN}}(E^{**} \times F) \times \mathcal{T}_{\| \, \|}(F^* \times E^*).$$

Using Theorem 49.4 with $H := F$, this equivalent to

$$((\widehat{x_\gamma}, \widehat{y_\gamma}), (y_\gamma^*, x_\gamma^*)) \to ((x^{**}, \widehat{y}), (y^*, x^*)) \text{ in } \mathcal{T}_{\mathcal{CLBN}}((E^{**} \times F^{**}) \times (F^* \times E^*)).$$

Now, for all γ, $((x_\gamma, y_\gamma^*), (\widehat{y_\gamma}, x_\gamma^*)) \in G(S)$, from which $((x_\gamma, y_\gamma), (y_\gamma^*, x_\gamma^*)) \in G(P)$. Thus we have proved that P is of type (ED), as required. $\square$

A function $k\colon E \times F \to [-\infty, \infty]$ is said to be a *saddle–function* if, for all $x \in E$, the function $k_x := k(x, \cdot)$ is convex on F and, for all $y \in F$, the function $-k^y := -k(\cdot, y)$ is convex on E. If k is a saddle–function, we write

$$\operatorname{dom} k := \big\{(x,y) \in E \times F\colon k_x(F) \subset\,]-\infty, \infty]\ \text{and}\ -k^y(E) \subset\,]-\infty, \infty]\big\}.$$

We suppose that $\operatorname{dom} k \neq \emptyset$. We define the multifunction $\sigma k\colon E \times F \rightrightarrows F^* \times E^*$ by declaring that $\big((x,y), (y^*, x^*)\big) \in G(\sigma k)$ exactly when

$$(x, y) \in \operatorname{dom} k, \quad x^* \in \partial(-k^y)(x) \quad \text{and} \quad y^* \in \partial(k_x)(y).$$

Rockafellar proved in [82, Theorem 3, p. 248] that if F is reflexive and k is "closed" in a sense made specific there then σk is maximally monotone. It is also noted in [82, p. 249] that if all the functions k_x for $x \in E$ and all the functions $-k^y$ for $y \in F$ are lower semicontinuous then k is closed. We will show in Theorem 49.6 below that, in the situation of [82, Theorem 3], σk is in fact maximally monotone of type (ED), so all the desirable properties of such multifunctions outlined in Sections 39–42 are valid for σk.

Theorem 49.6. *Let E and F be nonzero Banach spaces, F be reflexive and k be a closed saddle–function on $E \times F$ such that $\operatorname{dom} k \neq \emptyset$. Then σk is maximally monotone of type (ED).*

Proof. It is shown by Rockafellar on [82, p. 248] that there exists $f \in \mathcal{PCLSC}(E \times F^*)$ such that

$$(y^*, x^*) \in \sigma k(x, y) \iff (\widehat{y}, x^*) \in \partial f(x, y^*).$$

The result now follows from Theorem 48.4(b) and Lemma 49.5.   □

It is not so clear what happens if F is not assumed to be reflexive. Rockafellar proved in [82, Theorem 2, p. 245–247] that if k is finite–valued and separately continuous then σk is maximally monotone. However, we do not know the answer to the following problem:

Problem 49.7. Let E and F be nonzero Banach spaces and k be a finite–valued and separately continuous saddle–function on $E \times F$. Is σk necessarily of type (D)?

VII The sum problem for general Banach spaces

50 Introductory comments

Andrew Eberhard and Jonathan Borwein have announced the following result: if E is a nonzero Banach space, $S, T: E \rightrightarrows E^*$ are maximally monotone and $D(S) \cap \operatorname{int} D(T) \neq \emptyset$ then $S + T$ is maximally monotone. (This is Rockafellar's constraint qualification.) This result would have far–reaching implications. It would follow from Theorems 44.1 and 44.2 that every maximally monotone multifunction, S, would be of type (FPV) and $\overline{D(S)} = \operatorname{co} D(S) = D(S_\varphi)$. Clearly, Theorems 46.1(b), 47.5(b) and 48.4(d) would be unnecessary, as would Corollary 51.2, Corollary 52.3 and all subsequent results in this Chapter. See [20] for a preprint of this result.

51 Voisei's theorem

In this section we give a slight generalization of a recent beautiful result of Voisei. See [116]. For further results in this direction, see Borwein, [18].

Theorem 51.1. *Let E be a nonzero Banach space and $S, T: E \rightrightarrows E^*$ be maximally monotone. Let $D(S)$ and $D(T)$ be closed and convex and $\bigcup_{\lambda > 0} \lambda[D(S) - D(T)]$ be a closed subspace of E. Then $S + T$ is maximally monotone.*

Proof. We first establish (24.2). To this end, suppose that $b \in E \times E^*$ and $\varphi_{S+T}(b) \leq q(b)$. Lemma 28.5 and Corollary 18.3 give

$$\left. \begin{aligned} S + T &= S + N_{D(S)} + T + N_{D(T)} = S + T + N_{D(S)} + N_{D(T)} \\ &= S + T + N_{D(S) \cap D(T)} = S + T + N_{D(S+T)}, \end{aligned} \right\} \quad (51.1)$$

and Lemma 28.4 with $U := S + T$ and $C := D(S + T)$ gives

$$\pi_1 \operatorname{dom} \varphi_{S+T+N_{D(S+T)}} = D(S + T). \quad (51.2)$$

Since $b_1 \in \pi_1 \operatorname{dom} \varphi_{S+T}$, (51.1) and (51.2) imply that $b_1 \in D(S + T)$. This completes the proof of (24.2). From Theorem 28.6, $D(S) = D(S_\varphi)$ and $D(T) = D(T_\varphi)$, and so Theorem 24.1(c) implies that $S + T$ is maximally monotone. $\qquad \Box$

Corollary 51.2. *Let E be a nonzero Banach space, $S: E \rightrightarrows E^*$ be maximally monotone and $D(S)$ be closed and convex. Then S is maximally monotone of type (FPV).*

Proof. This is immediate from Theorem 51.1 and Theorem 44.1.   □

52 Sums with normality maps

Theorem 52.1 is a multifunction version of the following result proved by Rockafellar in [80, Theorem 3, pp. 77 and 84]: *Let C be a nonempty closed convex subset of E, $S: E \to E^*$ be single–valued and monotone, $D(S) \supset C$, and S be continuous on all line segments in C with respect to the topology $w(E^*, E)$. Then $S + N_C$ is maximally monotone.* Theorem 52.1 first appeared in [99, Theorem 41.1, p. 156–157].

Theorem 52.1. *Let C be a nonempty closed convex subset of a Banach space E and $S: E \rightrightarrows E^*$ be monotone. Suppose that $D(S) \supset C$ and*

$$\left. \begin{array}{c} \textit{for all } x \in C, \textit{ there exists a sublinear functional } P^x \textit{ on } E \\ \textit{such that } Sx \textit{ is the set of linear functionals } L \textit{ on } E \\ \textit{such that } L \leq P^x \end{array} \right\} \quad (52.1)$$

and

$$\left. \begin{array}{c} \textit{for all } y \in C - C, \textit{ the map } x \mapsto P^x(y) \textit{ is} \\ \textit{upper semicontinuous on all line–segments in } C. \end{array} \right\} \quad (52.2)$$

Then $S + N_C$ is maximally monotone.

Proof. Let $b \in E \times E^*$ and $\varphi_{S+N_C}(b) \leq q(b)$. We shall show that $b \in G(S + N_C)$, and the result will then follow from (23.10). From Lemma 28.4 with $U := S$, $\pi_1 \mathrm{dom}\, \varphi_{S+N_C} = C$, and so $b_1 \in C$.

Now let x be an arbitrary element of C and $n \geq 1$. Write $u_n := \frac{1}{n}x + (1 - \frac{1}{n})b_1 \in C$. Since $C \subset D(S)$, there exists $u_n^* \in Su_n$. Now $(u_n, u_n^*) = (u_n, u_n^* + 0) \in G(S + N_C)$ thus

$$\frac{1}{n}\langle b_1, u_n^* \rangle + \frac{1}{n}\langle x, b_2 - u_n^* \rangle + (1 - \frac{1}{n})\langle b_1, b_2 \rangle$$
$$= \langle b_1, u_n^* \rangle + \frac{1}{n}\langle x, b_2 - u_n^* \rangle + (1 - \frac{1}{n})\langle b_1, b_2 - u_n^* \rangle$$
$$= \langle b_1, u_n^* \rangle + \langle u_n, b_2 \rangle - \langle u_n, u_n^* \rangle \leq \varphi_{S+N_C}(b) \leq q(b) = \langle b_1, b_2 \rangle,$$

from which we derive that $\langle x - b_1, u_n^* \rangle \geq \langle x - b_1, b_2 \rangle$. (52.1) now implies that $P^{u_n}(x - b_1) + \langle b_1 - x, b_2 \rangle \geq 0$. Letting $n \to \infty$ and using (52.2),

$$P^{b_1}(x - b_1) + \langle b_1 - x, b_2 \rangle \geq 0.$$

(52.1) and the Hahn–Banach–Lagrange theorem, Theorem 1.11, with $j := \cdot - b_1$ and $k := \langle b_1 - \cdot, b_2 \rangle$ now provide us with an element x^* of

Sb_1 such that, for all $x \in C$, $\langle x - b_1, x^* \rangle + \langle b_1 - x, b_2 \rangle \geq 0$, that is to say, $\langle b_1, b_2 - x^* \rangle \geq \langle x, b_2 - x^* \rangle$. Now this means that $b_2 - x^* \in N_C(b_1)$, from which $b_2 = x^* + (b_2 - x^*) \subset (S + N_C)(b_1)$ that is to say, $b \in G(S + N_C)$. This completes the proof of Theorem 52.1. □

Theorem 52.2. *Let C be a nonempty closed convex subset of a Banach space E and $S \colon E \rightrightarrows E^*$ be maximally monotone. Suppose also that $D(S) \supset C$ and $\bigcup_{\lambda > 0} \lambda [D(S_\varphi) - C]$ is a closed subspace of E. Then $S + N_C$ is maximally monotone.*

Proof. Rockafellar's maximal monotonicity theorem, Theorem 18.7, and Theorem 28.6 imply that N_C is maximally monotone and $D\big((N_C)_\varphi\big) = D(N_C) = C$. Thus $\bigcup_{\lambda > 0} \lambda [D(S_\varphi) - D((N_C)_\varphi)]$ is a closed subspace of E. From Lemma 28.4 with $U := S$, $\pi_1 \mathrm{dom}\, \varphi_{S + N_C} = C$, thus if $b \in E \times E^*$ and $\varphi_{S + N_C}(b) \leq q(b)$ then $b_1 \in C = D(S) \cap C = D(S) \cap D(N_C) = D(S + N_C)$. Theorem 24.1(c) now implies that $S + N_C$ is maximally monotone. □

Corollary 52.3 implies a result that first appeared in [99, Theorem 41.2, p. 158]. We mention parenthetically that the proof of [99, Lemma 41.3] is incorrect, and we do not know whether it, [99, Theorem 41.5] and [99, Theorem 41.6] are true.

Corollary 52.3. *Let C be a nonempty closed convex subset of a Banach space E, $S \colon E \rightrightarrows E^*$ be maximally monotone and $D(S) \supset C$. Suppose that either $\mathrm{int}\, D(S) \cap C \neq \emptyset$ or $D(S) \cap \mathrm{int}\, C \neq \emptyset$. Then $S + N_C$ is maximally monotone.*

Proof. This is immediate from Theorem 52.2 since, in either case, $\bigcup_{\lambda > 0} \lambda [D(S_\varphi) - C] = E$. □

53 A theorem of Verona–Verona

In this section, we give a proof of a result of Verona–Verona, which was established in [115, Corollary 2.9(a), pp. 124–125] using the theory of "regular" maximally monotone multifunctions.

Theorem 53.1. *Let E be a nonzero Banach space, $f \in \mathcal{PCLSC}(E)$, $T \colon E \rightrightarrows E^*$ be maximally monotone and $D(T) = E$. Then the multifunction $\partial f + T$ is maximally monotone.*

Proof. Let $b \in E \times E^*$ and $\varphi_{\partial f + T}(b) \leq q(b)$. We will prove that

$$b_1 \in D(\partial f + T). \tag{53.1}$$

It will then follow from (24.2) that $\partial f + T$ is maximally monotone. Let y be an arbitrary element of $\mathrm{dom}\, f$. Since the segment $[b_1, y]$ is compact and $D(T) = E$, the local boundedness theorem, Theorem 26.1, provides $\eta > 0$

and $M \geq 0$ such that if C is the closed convex set $[b_1, y] + \{x \in E: \|x\| \leq \eta\}$ then

$$t \in G(T) \text{ and } t_1 \in C \implies \|t_2\| \leq M. \tag{53.2}$$

Let $h := f + \mathbb{I}_C \in \mathcal{PCLSC}(E)$. Since $\text{dom} f \cap \text{int} \text{dom} \mathbb{I}_C \ni y$, Theorem 18.1 implies that $\partial h = \partial f + \partial \mathbb{I}_C = \partial f + N_C$, from which $D(\partial h) = D(\partial f) \cap C$. Let $g := h(\cdot + b_1) - b_2 \in \mathcal{PCLSC}(E)$. By direct computation, $G(\partial g) = G(\partial h) - b$. We will prove that

$$0 \in D(\partial g). \tag{53.3}$$

Once (53.3) is known, then

$$b_1 \in D(\partial h) = D(\partial f) \cap C \subset D(\partial f) = D(\partial f) \cap D(T) = D(\partial f + T).$$

which gives (53.1), as required.

We now establish (53.3). If $\inf_E g = g(0)$ then $(0,0) \in G(\partial g)$, and (53.3) is obviously satisfied, so we can and will suppose that $\inf_E g < g(0)$. Let $\inf_E g < \lambda < g(0)$, and write $K_\lambda := \sup_{g(x) < \lambda} (\lambda - g(x)) / \|x\|$. We will prove that

$$K_\lambda \leq M. \tag{53.4}$$

From [97, Theorem 4, pp. 221–221], $K_\lambda \in]0, \infty[$ and, for all $\varepsilon \in]0, 1[$, there exists $c \in G(\partial g)$ such that $\langle c_1, c_2 \rangle \leq -(1 - \varepsilon) K_\lambda \|c_1\| < 0$. Since $G(\partial g) = G(\partial h) - b$, there exists $a \in G(\partial h)$ such that

$$\langle a_1 - b_1, a_2 - b_2 \rangle \leq -(1 - \varepsilon) K_\lambda \|a_1 - b_1\| < 0. \tag{53.5}$$

As we have already observed, $\partial h = \partial f + N_C$, thus there exist $s \in G(\partial f)$ and $n \in G(N_C)$ such that $s_1 = n_1 = a_1$ and $s_2 + n_2 = a_2$. In particular, since $b_1 \in C$, the definition of N_C gives $\langle a_1 - b_1, n_2 \rangle = \langle n_1 - b_1, n_2 \rangle \geq 0$. Consequently, it follows from (53.5) and the equality $s_2 + n_2 = a_2$ that

$$\langle a_1 - b_1, s_2 - b_2 \rangle \leq \langle a_1 - b_1, s_2 + n_2 - b_2 \rangle \leq -(1 - \varepsilon) K_\lambda \|a_1 - b_1\| < 0. \tag{53.6}$$

Since $a_1 = n_1 \in C$, there exists $t \in G(T)$ such that $t_1 = a_1$ and, using (53.2), $\|t_2\| \leq M$. Now $\varphi_{\partial f + T}(b) \leq q(b)$ and $(a_1, s_2 + t_2) \in G(\partial f + T)$, from which $\langle a_1 - b_1, s_2 + t_2 - b_2 \rangle \geq 0$. This implies in turn that

$$\langle a_1 - b_1, s_2 - b_2 \rangle \geq -\langle a_1 - b_1, t_2 \rangle \geq -M \|a_1 - b_1\|.$$

Combining this with (53.6), we derive that

$$-M \|a_1 - b_1\| \leq -(1 - \varepsilon) K_\lambda \|a_1 - b_1\| < 0,$$

and so $(1 - \varepsilon) K_\lambda \leq M$. (53.4) now follows by letting $\varepsilon \downarrow 0$. Now let $g(x) < g(0)$ and λ be an arbitrary element of $]g(x), g(0)[$. Then the above analysis shows that $(\lambda - g(x)) / \|x\| \leq M$, and so $\lambda \leq g(x) + M \|x\|$. Letting $\lambda \uparrow g(0)$ gives us that $(0 \in \text{dom} g$ and) $g(0) \leq g(x) + M \|x\|$. Of course, this is trivially satisfied if $g(x) \geq g(0)$, and so we have proved that

$$x \in E \implies g(x) + M \|x\| \geq g(0).$$

As we saw in Example 7.1, this implies that $\partial g(0) \neq \emptyset$, and so we have established (53.3). This completes the proof of Theorem 53.1. $\qquad \square$

Remark 53.2. One interesting feature of the above proof is that we first establish that $K_\lambda < \infty$ and then that $K_\lambda \leq M$. This seems, at the very least, uneconomical. The proof that $K_\lambda < \infty$ does not actually appear explicitly in [97, Theorem 4]. A proof of this using a separation theorem in $E \times \mathbb{R}$ can be found in [91, Lemma 2.2(a), pp. 130–131], and a proof that does not use a separation theorem can be found in [93, Main Theorem(a), pp. 329–330]. This issue is very close to those discussed in Theorem 12.2.

Remark 6.

VIII Open problems

Problem 11.5. Let C be a nonempty bounded closed convex subset of a Banach space E, x_0 be an extreme point of E, $y^* \in E^*$ and $\varepsilon > 0$. Then does there always exist $M \geq 0$ such that, for all $u, v \in C$, $M\|u + v - 2x_0\| \geq \langle v - x_0, y^* \rangle - \varepsilon$? (This is true if E is reflexive.)

Problem 11.6. Do there exist a nonzero finite dimensional Banach space E and $f, g \in \mathcal{PC}(E)$ such that the pair f, g is totally Fenchel unstable?

Problem 14.7. Let F be a Banach space, h, $k \in \mathcal{PC}(F)$ be Borel functions and $\operatorname{dom} h - \operatorname{dom} k$ surround 0. Is $h \ominus k$ necessarily (finitely) bounded above in some neighborhood of 0 in F? In particular: Let C and D be convex Borel sets in F and $C - D$ be absorbing. Is $C - D$ necessarily a neighborhood of 0 in F?

Problem 22.12. Let E be a nonzero Banach space, $B := E \times E^*$, $f, g \in \mathcal{PCLSC}(B)$ be BC–functions and

$$\bigcup_{\lambda > 0} \lambda[\pi_2 \operatorname{dom} f - \pi_2 \operatorname{dom} g] \text{ be a closed subspace of } E^*.$$

Then is $f \oplus_1 g$ a BC–function?

Problem 28.3. Let E be a nonzero Banach space and $S \colon E \rightrightarrows E^*$ be maximally monotone. Then is it necessarily true that

$$\overline{D(S_\varphi)} = \overline{\operatorname{co} D(S)}?$$

Problem 31.3. Is $\overline{D(S)}$ necessarily convex when E is not reflexive, S is maximally monotone and $\operatorname{sur} D(S_\varphi) = \emptyset$?

Problem 34.7. Let E be a nonzero reflexive Banach space, f, $g \in \mathcal{PCLSC}(E \times E^*)$ be BC–functions and $\pi_1 \operatorname{dom} f \cap \pi_1 \operatorname{dom} g \neq \emptyset$. Suppose that there exists an increasing function $j \colon [0, \infty[\to [0, \infty[$ such that

$$\left. \begin{aligned} s \in \operatorname{pos} f \times \operatorname{pos} g, \ s_1 \neq s_3 \text{ and } \langle s_1 - s_3, s_4 \rangle = \|s_1 - s_3\|\|s_4\| \\ \implies \quad \|s_4\| \leq j(\|s_1\| + \|s_2 + s_4\| + \|s_1 - s_3\|\|s_4\|). \end{aligned} \right\}$$

Then is it true that, for all $b \in E \times E^*$, there exist $a, c \in b \in E \times E^*$ such that $a_1 = c_1 = b_1$, $a_2 + c_2 = b_2$ and $f^@(a) + g^@(c) \leq (f \oplus_2 g)^@(a)$? If the answer to this question is "yes", then the results of Section 34 can be more completely integrated into the theory of BC–functions.

Problem 36.4. If S is maximally monotone of type (NI) then does it necessarily follow that S is maximally monotone of type (D)?

Problem 36.8. Is every maximally monotone multifunction of type (FPV)?

Problem 36.10. Is every maximally monotone multifunction strongly maximally monotone?

Problem 36.12. Is every maximally monotone multifunction of type (ANA)? (We do not even know what the situation is for discontinuous positive linear operators.)

Problem 43.3. If S is maximally monotone of type (NI), is $\overline{R(S)}$ necessarily convex?

Problem 45.18. Let E be a general non–reflexive Banach space. Does there always exist $f \in \mathcal{CC}(E)$ such that f^{**} is not continuous? (This is true for $E = c_0$.)

Problem 45.21. Let $\{x_\gamma^{**}\}$ be a net of elements of E^{**}, $x^{**} \in E^{**}$, $x_\gamma^{**} \to x^{**}$ in $w(E^{**}, E^*)$ and, for all $x \in E$, $\|x_\gamma^{**} - \widehat{x}\| \to \|x^{**} - \widehat{x}\|$. Does it necessarily follow that, for all $f \in \mathcal{CLB}(E)$, $f^{**}(x_\gamma^{**}) \to f^{**}(x^{**})$?

Problem 46.4. Are there any maximally monotone multifunctions with convex graph which are not either affine or wih an affine inverse?

Problem 47.6. If $T\colon D(T) \to E^*$ is linear and of type (FP) then is T necessarily maximally monotone of type (NI)?

Problem 48.3. Let E be a nonzero Banach space. Let $f \in \mathcal{PCLSC}(E)$. Then is it always true that $G^{-1}(\partial f^*) \subset (G(\partial f))^{\mathcal{CCN}}$? Let C be a nonempty proper closed convex subset of E, $\breve{C}$ be the $w(E^{**}, E^*)$–closure of $\widehat{C}$ in E^{**}, $x^* \in E^*$ and $\langle x^*, x^{**} \rangle = \sup\langle C, x^* \rangle$. Then does there necessarily exist a net $\{s_\gamma\}$ of elements of $G(N_C)$ such that $\iota(s_\gamma) \to (x^{**}, x^*)$ in $\mathcal{T}_{\mathcal{CCN}}(E^{**} \times E^*)$?

Problem 49.7. Let E and F be nonzero Banach spaces and k be a finite–valued and separately continuous saddle–function on $E \times F$. Is σk necessarily of type (D)?

IX Glossary of classes of multifunctions

For the convenience of the reader, we collect together the definitions of the different classes of multifunctions introduced in the text. In all of these definitions, E is a nonzero Banach space and $S \colon E \rightrightarrows E^*$.

Definition 36.1. We say that S is *maximally monotone of type (D)* if S is maximally monotone and, for all $v \in G(\overline{S})$, there exists a bounded net $\{s_\gamma\}$ of elements of $G(S)$ such that $\iota(s_\gamma) \to v$ in $\mathcal{T}_{\mathcal{WN}}(E^{**} \times E^*)$.

Definition 36.2. Let S be maximally monotone. S is said to be *of type (NI)* if $\quad (v_1, v_2) \in E^{**} \times E^* \implies \inf_{s \in G(S)} \langle s_2 - v_2, \widehat{s_1} - v_1 \rangle \le 0.$ This can be rewritten $\quad v \in E^{**} \times E^* \implies \inf_{s \in G(S)} \widetilde{q}(\iota(s) - v) \le 0.$

Definition 36.5. A monotone multifunction S is said to be *of type (FP)* if, for any open convex subset U of E^* and $d \in E \times U$ such that $U \cap R(S) \ne \emptyset$ and $\quad s \in G(S)$ and $s_2 \in U \implies q(s - d) \ge 0 \quad$ then $d \in G(S)$.

Definition 36.7. A monotone multifunction S is said to be *of type (FPV)* if, for any open convex subset U of E and $d \in U \times E^*$ such that $U \cap D(S) \ne \emptyset$ and $\quad s \in G(S)$ and $s_1 \in U \implies q(s - d) \ge 0 \quad$ then $d \in G(S)$.

Definition 36.9. We say that S is *strongly maximally monotone* if S is monotone and whenever C is a nonempty $w(E, E^*)$-compact convex subset of E, $y^* \in E^*$ and, for all $s \in G(S)$, there exists $b \in C \times \{y^*\}$ such that $q(s - b) \ge 0$ then $\quad G(S) \cap (C \times \{y^*\}) \ne \emptyset$, and, further, whenever C is a nonempty $w(E^*, E)$-compact convex subset of E^*, $y \in E$ and, for all $s \in G(S)$, there exists $b \in \{y\} \times C$ such that $\quad q(s - b) \ge 0 \quad$ then $\quad G(S) \cap (\{y\} \times C) \ne \emptyset$.

Definition 36.11. Let S be maximally monotone. We say that S is *maximally monotone of type (ANA)* if, whenever $b \in E \times E^* \setminus G(S)$ then, for all $n \ge 1$, there exists $s_n = (s_{n1}, s_{n2}) \in G(S)$ such that $s_{n1} \ne b_1$, $s_{n2} \ne b_2$ and

$$\frac{q(s_n - b)}{\|s_{n1} - b_1\| \|s_{n2} - b_2\|} \to -1 \quad \text{as } n \to \infty.$$

Definition 36.13. Let S be maximally monotone. We say that S is *maximally monotone of type (BR)* if, whenever $b \in E \times E^*$, $\alpha, \beta > 0$ and $\inf_{s \in G(S)} q(s - b) > -\alpha\beta$ then there exists $s \in G(S)$ such that $\|s_1 - b_1\| < \alpha$ and $\|s_2 - b_2\| < \beta$.

Definition 38.3. We say that S is *maximally monotone of type (ED)* if S is maximally monotone and $G(\overline{S}) \subset G(S)^{\mathcal{CLBN}}$.

Definition 42.9. We say that S is *ultramaximally monotone* if S is maximally monotone and $G(\overline{S}) \subset \iota(G(S))$.

X A selection of results

Functional analysis

Lemma 1.2: Sublinear form of the Hahn–Banach theorem. *Let E be a nonzero vector space and $P\colon E \to \mathbb{R}$ be sublinear. Then there exists a linear functional L on E such that $L \leq P$ on E.*

Remark 1.4: Extended sublinear functionals. An *extended sublinear functional on E* is a subadditive and positively homogeneous map $P\colon E \to\,]{-\infty}, \infty]$ such that $P(0) = 0$. We give an example of an extended sublinear functional for which the analog of Lemma 1.2 fails.

Lemma 1.6: Mazur–Orlicz theorem. *Let E be a nonzero vector space, $P\colon E \to \mathbb{R}$ be sublinear and D be a nonempty convex subset of E. Then there exists a linear functional L on E such that $L \leq P$ on E and*

$$\inf{}_D L = \inf{}_D P.$$

Definition 1.8. Let C be a nonempty convex subset of a vector space. Then $\mathcal{PC}(C)$ stands for the set of all proper convex functions $k\colon C \to\,]{-\infty}, \infty]$.

Definition 1.9. Let E be a nonzero vector space and $P\colon E \to \mathbb{R}$ be sublinear. Define the vector ordering "$\leq_P$" on E by declaring that $y \leq_P z$ if $P(y - z) \leq 0$. Let C be a nonempty convex subset of a vector space and $j\colon C \to E$. We say that j is *P–convex* if

$$x_1, x_2 \in C,\ \mu_1, \mu_2 > 0 \text{ and } \mu_1 + \mu_2 = 1$$
$$\implies\ j(\mu_1 x_1 + \mu_2 x_2) \leq_P \mu_1 j(x_1) + \mu_2 j(x_2).$$

Remark 1.10. "P–convex" can mean different things under different circumstances. Consider the special case when $E = \mathbb{R}$. If $P(y) := |y|$, $P(y) := y$, $P(y) := -y$ or $P(y) := 0$, respectively, then "P–convex" means "affine", "convex", "concave" or "arbitrary", respectively.

Theorem 1.11: Hahn–Banach–Lagrange theorem. *Let E be a nonzero vector space and $P\colon E \to \mathbb{R}$ be sublinear. Let C be a nonempty convex subset of a vector space, $k \in \mathcal{PC}(C)$ and $j\colon C \to E$ be P–convex. Then there exists a linear functional L on E such that $L \leq P$ on E and*

$$\inf{}_C \left[L \circ j + k \right] = \inf{}_C \left[P \circ j + k \right].$$

Corollary 2.1: Sandwich theorem. *Let E be a nonzero vector space, $P: E \to \mathbb{R}$ be sublinear, $k \in \mathcal{PC}(E)$ and $-k \leq P$ on E. Then there exists a linear functional L on E such that $-k \leq L \leq P$ on E.*

Corollary 2.2: Extension form of the Hahn–Banach theorem. *Let E be a nonzero vector space, F be a linear subspace of E, $P: E \to \mathbb{R}$ be sublinear, $M: F \to \mathbb{R}$ be linear and $M \leq P$ on F. Then there exists a linear functional L on E such that $L \leq P$ on E and $L|_F = M$.*

Remark 2.3. The analog of Corollary 2.2 for extended sublinear functionals fails, even for $E = \mathbb{R}^2$.

Corollary 2.4: One–dimensional form of the Hahn–Banach theorem. *Let P be a sublinear functional on E and $x \in E$. Then there exists a linear functional L on E such that $L \leq P$ on E and $L(x) = P(x)$.*

Notation. If $\lambda, \mu \in \mathbb{R}$, we write $\lambda \vee \mu$ for the maximum value of λ and μ, and $\lambda \wedge \mu$ for the minimum value of λ and μ.

Lemma 3.1: Fan–Glicksberg–Hoffman theorem. *Let C be a nonempty convex subset of a vector space and $f_1, \ldots, f_m$ be convex real functions on C. Then there exist $\lambda_1, \ldots, \lambda_m \geq 0$ such that $\lambda_1 + \cdots + \lambda_m = 1$ and*

$$\inf_C[f_1 \vee \cdots \vee f_m] = \inf_C[\lambda_1 f_1 + \cdots + \lambda_m f_m].$$

Theorem 3.2: Fan's minimax theorem. *Let X be a nonempty convex subset of a vector space, Y be a nonempty convex subset of a vector space and Y also be a compact Hausdorff topological space. Let $h: X \times Y \to \mathbb{R}$ be convex on X, and concave and upper semicontinuous on Y. Then*

$$\inf_X \max_Y h = \max_Y \inf_X h.$$

Theorem 4.4: Separation theorem. *Let C be a nonempty convex subset of a normed space E and $x \in E \setminus \overline{C}$. Then there exists $z^* \in E^*$ such that $\sup_C z^* < \langle x, z^* \rangle$.*

Corollary 4.5. *If F is a nonempty closed convex subset of a normed space E then F is $w(E, E^*)$–closed.*

Corollary 4.6. *If D is a subspace of a normed space E and*

$$z^* \in E^* \text{ and } \langle y, z^* \rangle = 0 \text{ for all } y \in D \quad \Longrightarrow \quad z^* = 0,$$

then D is dense in E.

Theorem 5.1(c). *Let F be a nonzero normed space with dual F^*. Let Y be a nonempty bounded convex subset of F^*. Finally, let $\mathcal{CS}$ stand for the set of all nonempty convex subsets of the unit ball of F, and $\mathcal{CCS}$ stand for the set of all nonempty closed convex subsets of the unit ball of F. Then the three conditions below are equivalent.*

$\overline{Y}$ is $w(F^*, F)$–compact.

For all $X \in \mathcal{CS}$, $\inf_{x \in X} \sup_{y^* \in Y} \langle x, y^* \rangle = \sup_{y^* \in Y} \inf_{x \in X} \langle x, y^* \rangle$.

For all $X \in \mathcal{CCS}$, $\inf_{x \in X} \sup_{y^* \in Y} \langle x, y^* \rangle = \sup_{y^* \in Y} \inf_{x \in X} \langle x, y^* \rangle$.

Notation If E is a normed space, the *bidual*, E^{**}, of E, is defined to be the dual of the normed space E^*.

Corollary 5.3. *Let F be a nonzero normed space, X be a nonempty bounded convex subset of F and Y be a nonempty convex subset of F^* such that the $w(F^*, F^{**})$–closure of Y is $w(F^*, F)$–compact. Then*

$$\inf_{x \in X} \sup_{y^* \in Y} \langle x, y^* \rangle = \sup_{y^* \in Y} \inf_{x \in X} \langle x, y^* \rangle.$$

Corollary 5.4. *Let F be a nonzero reflexive Banach space, X be a nonempty bounded convex subset of F and Y be a nonempty bounded convex subset of F^*. Then*

$$\inf_{x \in X} \sup_{y^* \in Y} \langle x, y^* \rangle = \sup_{y^* \in Y} \inf_{x \in X} \langle x, y^* \rangle.$$

Theorem 5.6: Minimax criterion for weak compactness. *Let E be a nonzero normed space with dual E^*, $\mathcal{CS}$ be the set of all nonempty convex subsets of the unit ball of E^* and $\mathcal{CCS}$ be the set of all nonempty norm–closed convex subsets of the unit ball of E^*. Let Z be a nonempty bounded norm–complete convex subset of E. Then the three conditions below are equivalent.*

$$Z \text{ is } w(E, E^*)\text{–compact.}$$

For all $X \in \mathcal{CS}$, $\inf_{x^* \in X} \sup_{z \in Z} \langle z, x^* \rangle = \sup_{z \in Z} \inf_{x^* \in X} \langle z, x^* \rangle$.

For all $X \in \mathcal{CCS}$, $\inf_{x^* \in X} \sup_{z \in Z} \langle z, x^* \rangle = \sup_{z \in Z} \inf_{x^* \in X} \langle z, x^* \rangle$.

Lagrange multipliers and KKT functionals

Notation. Let $(E, \| \cdot \|)$ be a nonzero normed space and $\preceq$ be a partial ordering on E compatible with its vector space structure. Let N be the negative cone $\{z \in E : z \preceq 0\}$, and $D_N, D_{E \setminus N} \colon E \to [0, \infty[$ be defined by $D_N := \operatorname{dist}(\cdot, N)$ and $D_{E \setminus N} := \operatorname{dist}(\cdot, E \setminus N)$. Let C be a nonempty convex subset of a vector space and $j \colon C \to E$ be convex with respect to $\preceq$, that is to say

$$w, x \in C, \text{ and } \lambda \in]0, 1[\implies j(\lambda w + (1 - \lambda)x) \preceq \lambda j(w) + (1 - \lambda)j(x).$$

Now let $k \in \mathcal{PC}(C)$, and $\mu \in \mathbb{R}$ be the *constrained infimum*

$$\mu = \inf_{j^{-1}N} k = \inf \{k(x) : x \in C, \ j(x) \preceq 0\}.$$

In order to exclude trivial cases, we shall suppose that $\inf_C k < \mu$. Let $W := \{w \in C : k(w) < \mu\} \neq \emptyset$.

Definition 6.3. A *Lagrange multiplier* for the infimization problem above is an element z^* of E^* such that z^* is $\preceq$-*positive*, that is to say, $z^* \le 0$ on N, and $\inf_{x \in C} \left[\langle j(x), z^* \rangle + k(x) \right] = \mu$.

Theorem 6.4: A sharp Lagrange multiplier theorem. (a) *There exists a Lagrange multiplier if, and only if*

$$\text{there exists } M \ge 0 \text{ such that} \qquad M D_N \circ j + k \ge \mu \text{ on } C.$$

In this case, there exists a Lagrange multiplier z^* such that $\|z^*\| \le M$.

(b) *If z^* is a Lagrange multiplier then $j(W) \subset E \setminus \overline{N}$ and*

$$0 < \sup_W \frac{\mu - k}{D_N \circ j} \le \|z^*\| < \infty.$$

(c) *If $j(W) \subset E \setminus \overline{N}$ and $0 < M := \sup_W \dfrac{\mu - k}{D_N \circ j} < \infty$ then*

$$\min \left\{ \|z^*\| : \ z^* \text{ is a Lagrange multiplier} \right\} = M.$$

Theorem 6.6: Slater condition result with a bound. *Suppose that* $V = \left\{ v \in \operatorname{dom} k : j(v) \in \operatorname{int} N \right\} \ne \emptyset.$ *Then:*
(a) $\mu = \inf_V k.$
(b) *There exists a Lagrange multiplier z^* such that* $\|z^*\| \le \inf_V \dfrac{k - \mu}{D_{E \setminus N} \circ j}.$

Notation. Let C be a vector space, and $x_0 \in C$. Let $G \colon C \to E$ and $f \colon C \to \mathbb{R}$, and suppose that $G(x_0) \preceq 0$,

$$\text{for all } x \in C, \quad d^+ G(x) := \lim_{\alpha \to 0+} \frac{G(x_0 + \alpha x) - G(x_0)}{\alpha} \qquad \text{exists in } E,$$

$$\text{for all } x \in C, \quad d^+ f(x) := \lim_{\alpha \to 0+} \frac{f(x_0 + \alpha x) - f(x_0)}{\alpha} \qquad \text{exists in } \mathbb{R},$$

$d^+ G \colon C \to E$ is $\preceq$-convex and $d^+ f \colon C \to \mathbb{R}$ is convex. Let

$$V := \left\{ v \in C : \ G(x_0) + d^+ G(v) \in \operatorname{int} N \right\} \ne \emptyset,$$

and

$$W := \left\{ w \in C : \ d^+ f(w) < 0 \right\} \ne \emptyset.$$

Suppose, finally, that

$$\min \left\{ f(x) : \ x \in C, \ G(x) \preceq 0 \right\} = f(x_0).$$

Definition 6.8. A *KKT functional* for the minimization problem described above is a $\preceq$-*positive* element z^* of E^* such that

$$\text{for all } x \in C, \quad \langle G(x_0) + d^+ G(x), z^* \rangle + d^+ f(x) \ge 0.$$

Theorem 6.9: A sharp version of the KKT theorem. *Suppose that* $G(x_0) + \mathrm{d}^+G(W) \subset E \setminus \overline{N}$, *and*

$$\sup_W \frac{-\mathrm{d}^+f}{D_N \circ (G(x_0) + \mathrm{d}^+G)} < \infty.$$

Then there exists a KKT functional for the minimization problem above, and

$$\min\left\{\|z^*\|\colon z^* \text{ is a KKT functional}\right\} = \sup_W \frac{-\mathrm{d}^+f}{D_N \circ (G(x_0) + \mathrm{d}^+G)}$$

$$\leq \inf_V \frac{\mathrm{d}^+f}{D_{E\setminus N} \circ (G(x_0) + \mathrm{d}^+G)}.$$

Convex analysis

Notation. Let E be a nonzero normed space, $k \in \mathcal{PC}(E)$ and $x \in E$. Then the *subdifferential* of k at x is defined by

$$\partial k(x) := \{z^* \in E^*\colon \quad y \in E \implies k(x) + \langle y - x, z^* \rangle \leq k(y)\}.$$

The *Fenchel conjugate*, $k^*\colon E^* \to]-\infty, \infty]$, of k is defined by

$$k^*(x^*) := \sup_E(x^* - k).$$

Example 7.1. *Let E be a nonzero normed space, $k \in \mathcal{PC}(E)$ and $x \in E$. Then $\partial k(x) \neq \emptyset$ if, and only if, $x \in \mathrm{dom}\, k$ and there exists $M \geq 0$ such that*

$$y \in E \quad \implies \quad k(x) - M\|y - x\| \leq k(y).$$

Remark 7.3: Fenchel functionals. Let E be a nonzero normed space with dual E^*, and $f, g \in \mathcal{PC}(E)$. We will say that $z^* \in E^*$ is a *Fenchel functional* for f and g if $f^*(-z^*) + g^*(z^*) \leq 0$.

Theorem 7.4: A sharp version of the Fenchel duality theorem. *Let E be a nonzero normed space and $f, g \in \mathcal{PC}(E)$. Then:*
(a) *f and g have a Fenchel functional if, and only if,*

$$\text{there exists } M \geq 0 \text{ such that,}$$
$$x, y \in E \quad \implies \quad f(x) + g(y) + M\|x - y\| \geq 0.$$

(b) *If $z^* \in E^*$ is a Fenchel functional for f and g then*

$$\sup_{x, y \in E, \ x \neq y} \frac{-f(x) - g(y)}{\|x - y\|} \leq \|z^*\| < \infty.$$

(c) *If $f + g \geq 0$ on E and* $\displaystyle \sup_{x, y \in E, \ x \neq y} \frac{-f(x) - g(y)}{\|x - y\|} < \infty$ *then*

$$\min \left\{ \|z^*\| \colon \ z^* \text{ is a Fenchel functional for } f \text{ and } g \right\}$$
$$= \sup_{x, y \in E, \ x \neq y} \frac{-f(x) - g(y)}{\|x - y\|} \vee 0.$$

Notation: The Fenchel conjugate with respect to a bilinear form. Let E and E^* be nonzero real vector spaces, and $\langle \cdot, \cdot \rangle \colon E \times E^* \to \mathbb{R}$ be a bilinear form that separates the points of E and also separates the points of E^*. If $f \in \mathcal{PC}(E)$, the *Fenchel conjugate f^* with respect to* $\langle \cdot, \cdot \rangle$ is defined by

$$f^*(x^*) := \sup_{x \in E} \left[\langle x, x^* \rangle - f(x) \right].$$

If $k \colon E^* \to \,]-\infty, \infty]$ is convex, the function ${}^*k \colon E \to [-\infty, \infty]$ is defined by

$${}^*k(x) := \sup_{x^* \in E^*} \left[\langle x, x^* \rangle - k(x^*) \right].$$

If $f, g \in \mathcal{PC}(E)$, a *Fenchel functional for f and g* is an element z^* of E^* such that $f^*(-z^*) + g^*(z^*) \leq 0$. The definitions of f^* and "Fenchel functional" are compatible with those introduced above for normed spaces if we take $\langle \cdot, \cdot \rangle$ to be the canonical bilinear form on $E \times E^*$. If $E^* = E$, we will write $f^{@}$ instead of f^*. We say that a locally convex topology $\mathcal{T}$ on E is *E^*–compatible* if the $\mathcal{T}$–dual of E is exactly $\left\{ \langle \cdot, x^* \rangle \colon \ x^* \in E^* \right\}$.

Notation 8.3. Let E be a nonzero vector space and $f, g \in \mathcal{PC}(E)$. If $w \in E$, we write $(f \ominus g)(w) := \inf_{z \in E} \left[f(z) + g(z - w) \right]$.

Theorem 8.4(b): The $\ominus$ version of the Fenchel duality theorem. Let $f, \ g \in \mathcal{PC}(E)$, $f + g \geq 0$ on E and

$$F := \bigcup_{\lambda > 0} \lambda (\mathrm{dom}\, g - \mathrm{dom}\, f) \ni 0.$$

Suppose that $\mathcal{T}$ is an E^*–compatible topology on E and $f \ominus g$ is (finitely) bounded above in some $\mathcal{T}$–neighborhood of 0 relative to F. Then there exists a Fenchel functional for f and g.

Corollary 8.5: Rockafellar's version of Fenchel duality. Let $f, g \in \mathcal{PC}(E)$, $f + g \geq 0$ on E, $\mathcal{T}$ be an E^*–compatible topology on E, and g be (finitely) bounded above in some $\mathcal{T}$–neighborhood of a point of $\mathrm{dom}\, f$. Then there exists a Fenchel functional for f and g.

Theorem 8.7. Let E be a nonzero vector space, $f \in \mathcal{PC}(E)$, $z_0 \in E$, $K \in \mathbb{R}$, and $P \colon E \to \mathbb{R}$ be a seminorm such that

$$z \in E \text{ and } P(z - z_0) \leq 1 \quad \Longrightarrow \quad f(z) \leq K.$$

Then

$$x, y \in E, P(x - z_0) \leq \tfrac{1}{2} \text{ and } P(y - z_0) \leq \tfrac{1}{2} \quad \Longrightarrow$$
$$|f(x) - f(y)| \leq 4(K - f(z_0)) P(x - y).$$

Theorem 8.8: Bipolar theorem. *Let C be a nonempty convex subset of E, $\mathcal{T}$ be an E^*–compatible topology on E, $C^{\mathcal{T}}$ be the closure of C with respect to $\mathcal{T}$ and $x \in E$. Then $x \in C^{\mathcal{T}}$ if, and only if,*

$$x^* \in E^* \implies \langle x, x^* \rangle \leq \sup \langle C, x^* \rangle.$$

Theorem 10.1. *Let $f, g \in \mathcal{PC}(E)$ and $F := \bigcup_{\lambda > 0} \lambda [\operatorname{dom} g - \operatorname{dom} f] \ni 0$. Let $\mathcal{T}$ be a E^*–compatible topology on E, $x^* \in E^*$ and $(f - x^*) \ominus g$ be (finitely) bounded above in some $\mathcal{T}$–neighborhood of 0 relative to F. Then*

$$(f + g)^*(x^*) = \min_{z^* \in E^*} \left[f^*(x^* - z^*) + g^*(z^*) \right].$$

Corollary 10.2: Rockafellar's formula for the conjugate of a sum. *Let $f, g \in \mathcal{PC}(E)$, $\mathcal{T}$ be a E^*–compatible topology on E, and g be (finitely) bounded above in some $\mathcal{T}$–neighborhood of a point of $\operatorname{dom} f$ and $x^* \in E^*$. Then*

$$(f + g)^*(x^*) = \min_{z^* \in E^*} \left[f^*(x^* - z^*) + g^*(z^*) \right].$$

Notation. Let E be a nonzero Banach space and $f, g \in \mathcal{PC}(E)$. We say that f and g satisfy *Fenchel duality* if there exists $z^* \in E^*$ such that

$$f^*(-z^*) + g^*(z^*) = (f + g)^*(0).$$

We shall say that the pair f, g is *totally Fenchel unstable* if f and g satisfy Fenchel duality but

$$y^*, z^* \in E^* \text{ and } f^*(y^*) + g^*(z^*) = (f + g)^*(y^* + z^*) \implies y^* + z^* = 0.$$

Example 11.1. We give an example of proper, convex lower semicontinuous functions f and g on $\mathbb{R}^2$ that satisfy Fenchel duality but, for most $r \in \left(\mathbb{R}^2 \right)^* = \mathbb{R}^2$, it is not true that there exist $p, q \in \mathbb{R}^2$ such that $p + q = r$ and $f^*(p) + g^*(q) = (f + g)^*(r)$.

Example 11.3. We give an example of a totally Fenchel unstable pair $f, g \in \mathcal{PC}(\ell^2)$.

Notation: Conjugates and biconjugates with respect to a bilinear form. Let E and E^* be nonzero real vector spaces, and $\langle \cdot, \cdot \rangle \colon E \times E^* \to \mathbb{R}$ be a bilinear form that separates the point of E and also separates the points of E^*. We define *the restricted biconjugate of f* to be $^*(f^*) \colon E \to [-\infty, \infty]$. To simplify notation, we shall abbreviate this to $^*f^*$. It follows easily from the definitions that $f \geq {}^*f^*$ on E. We say that $x \in E$ is a *Fenchel–Moreau point* of f if $f(x) = {}^*f^*(x)$.

Example 12.1: Non Fenchel–Moreau points. Let E be an infinite–dimensional normed space. Then there always exists $f \in \mathcal{PC}(E)$ such that f is lower semicontinuous at 0 but that 0 is *not* a Fenchel–Moreau point of f.

Theorem 12.2: Existence of Fenchel–Moreau points. *Let $\mathcal{T}$ be an E^*–compatible topology on E, $f \in PC(E)$ be (finitely) bounded below in a $\mathcal{T}$–neighborhood of an element of dom f, and f be $\mathcal{T}$–lower semicontinuous at an element y of E. Then y is a Fenchel–Moreau point of f, and $f^* \in PC(E^*)$.*

Definition 12.3. If E is a nonzero Hausdorff locally convex space, we write $PCLSC(E)$ for the set

$$\{f \in PC(E)\colon f \text{ is lower semicontinuous on } E\}.$$

Corollary 12.4: Fenchel–Moreau theorem. *Let $\mathcal{T}$ be an E^*–compatible topology on E and $f \in PCLSC(E, \mathcal{T})$. Then $f^* \in PC(E^*)$ and $^*f^* = f$ on E.*

Results that use Baire's theorem

Notation. Let E be a nonzero vector space and $A \subset E$. We write "$x \in$ sur A" and say that "A surrounds x" if, for each $w \in E \setminus \{0\}$, there exists $\delta > 0$ such that $x + \delta w \in A$.

Corollary 13.5. *Let E be a nonzero Banach space and $f \in PCLSC(E)$. Then   sur $(\text{dom } f) = \text{int} (\text{dom } f)$.*

Theorem 14.2: $\ominus$–theorem. *Let F be a nonzero Banach space, $h, k \in PCLSC(F)$, and dom h − dom k surround 0. Then $h \ominus k$ is (finitely) bounded above in a neighborhood of 0 in F.*

Corollary 14.3. *Let F be a nonzero Banach space and f, $k \in PCLSC(F)$. Then sur$(\text{dom } f - \text{dom } k) = \text{int}(\text{dom } f - \text{dom } k)$, and so sur$(\text{dom } f - \text{dom } k)$ is open.*

Theorem 15.1: Attouch–Brezis theorem. *Let E be a nonzero Banach space, f, $g \in PCLSC(E)$, $\bigcup_{\lambda>0} \lambda\big[\text{dom } f - \text{dom } g\big]$ be a closed subspace of E and $f + g \geq 0$ on E. Then there exists a Fenchel functional for f and g.*

Notation 16.1. If E and F are nonzero Banach spaces, we norm $E \times F$ by $\|b\| := \sqrt{\|b_1\|^2 + \|b_2\|^2}$ $(b = (b_1, b_2) \in E \times F)$. The dual of $E \times F$ is $F^* \times E^*$ under the pairing

$$\langle b, v \rangle := \langle b_1, v_2 \rangle + \langle b_2, v_1 \rangle \quad (b = (b_1, b_2) \in E \times F, \ v = (v_1, v_2) \in F^* \times E^*),$$

and the dual norm of $F^* \times E^*$ is given by $\|(v_1, v_2)\| = \sqrt{\|v_1\|^2 + \|v_2\|^2}$. We define the *projection maps* π_1, π_2 by $\pi_1(x, y) := x$ and $\pi_2(x, y) := y$.

Definition 16.3. Let E and F be nonzero Banach spaces, $B := E \times F$ and $f, g \in PC(B)$. For all $b \in B$, let

$$(f \oplus_2 g)(b) := \inf \big\{ f(a) + g(c)\colon a, c \in B, \ a_1 = c_1 = b_1, \ a_2 + c_2 = b_2 \big\}.$$

So $(f \oplus_2 g)(x, \cdot)$ is the *inf–convolution* of $f(x, \cdot)$ and $g(x, \cdot)$.

Theorem 16.4(a): A bivariate version of the Attouch–Brezis theorem. Let E and F be nonzero Banach spaces, $B := E \times F$ and $f, g \in \mathcal{PCLSC}(B)$. Write $B^* = F^* \times E^* = (E \times F)^*$. Let

$$\bigcup_{\lambda > 0} \lambda \big[\pi_1 \mathrm{dom}\, f - \pi_1 \mathrm{dom}\, g \big] \text{ be a closed subspace of } E$$

and, for all $b \in B$, $(f \oplus_2 g)(b) > -\infty$. Then, for all $v \in B^* = (E \times F)^*$,

$$(f \oplus_2 g)^*(v) = \min \big\{ f^*(u) + g^*(w)\colon u, w \in B^*,\ u_1 = w_1 = v_1,\ u_2 + w_2 = v_2 \big\}.$$

In particular, $(f \oplus_2 g)^* = f^* \oplus_2 g^*$ on B^*.

Multifunctions

Notation. We now introduce some general notation for "multifunctions" or "set–valued maps". If X and Y are nonempty sets, we write $S\colon X \rightrightarrows Y$ if, for all $x \in X$, Sx is a (possibly nonempty) subset of Y. We define

$$G(S) := \{(x, y)\colon x \in X,\ y \in Sx\} \quad \text{and} \quad G^{-1}(S) := \{(y, x)\colon x \in X,\ y \in Sx\}.$$

$G(S)$ is the *graph* of S and $G^{-1}(S)$ is the *inverse graph* of S. We shall always suppose that $G(S) \neq \emptyset$ — we shall emphasize this by saying that S is *nontrivial*. We write $D(S) := \{x \in X\colon Sx \neq \emptyset\}$ and $R(S) := \bigcup_{x \in X} Sx$. $R(S)$ is the *range* of S. Finally, if $S\colon X \rightrightarrows Y$, we define $S^{-1}\colon Y \rightrightarrows X$ by $S^{-1}y := \{x \in X\colon Sx \ni y\}$.

Remark 17.1. Let E be a nonzero Banach space and $B := E \times E^*$. For all $b = (b_1, b_2)$ and $c = (c_1, c_2) \in B$, we set $\lfloor b, c \rfloor := \langle b_1, c_2 \rangle + \langle c_1, b_2 \rangle$. Then $\lfloor \cdot, \cdot \rfloor\colon B \times B \to \mathbb{R}$ is a symmetric bilinear form that separates the points of B. We define the quadratic form q on B by $q(b) := \frac{1}{2} \lfloor b, b \rfloor$. Then

$$q(b_1, b_2) = \tfrac{1}{2} \big[\langle b_1, b_2 \rangle + \langle b_1, b_2 \rangle \big] = \langle b_1, b_2 \rangle.$$

Consequently, if $b = (b_1, b_2)$ and $c = (c_1, c_2) \in B$ then

$$\langle b_1 - c_1, b_2 - c_2 \rangle = q(b_1 - c_1, b_2 - c_2) = q\big((b_1, b_2) - (c_1, c_2)\big) = q(b - c).$$

Let $S\colon E \rightrightarrows E^*$. We say that S is *monotone* if

$$b, c \in G(S) \quad \Longrightarrow \quad q(b - c) \geq 0.$$

If S is monotone then we say that S is *maximally monotone* when

$$\text{if } b \in B \quad \text{and} \quad \big(a \in G(S) \Longrightarrow q(b - a) \geq 0\big) \quad \text{then} \quad b \in G(S).$$

Theorem 18.1: Rockafellar's formula for the subdifferential of a sum. Let E be a nonzero normed space, $f, g \in \mathcal{PC}(E)$, and g be (finitely) bounded above in some neighborhood of a point of $\mathrm{dom}\, f$. Then $\partial(f + g) = \partial f + \partial g$.

Theorem 18.2: Attouch–Brezis's formula for the subdifferential of a sum. *Let E be a nonzero Banach space, f, $g \in \mathcal{PCLSC}(E)$ and $\bigcup_{\lambda>0} \lambda[\mathrm{dom}\, f - \mathrm{dom}\, g]$ be a closed subspace of E. Then $\partial(f+g) = \partial f + \partial g$.*

Theorem 18.6: Brøndsted–Rockafellar theorem. *Let E be a nonzero Banach space, $f \in \mathcal{PCLSC}(E)$, $\alpha, \beta > 0$, $b \in E \times E^*$ and $f(b_1) + f^*(b_2) \leq q(b) + \alpha\beta$. Then there exists $s \in G(\partial f)$ such that $\|s_1 - b_1\| \leq \alpha$ and $\|s_2 - b_2\| \leq \beta$.*

Theorem 18.7: Rockafellar's maximal monotonicity theorem. *Let E be a nonzero Banach space and $f \in \mathcal{PCLSC}(E)$. Then $\partial f\colon E \rightrightarrows E^*$ is maximally monotone.*

SSD spaces

Definition 19.1. We will say that B $\bigl($more precisely, $(B, \lfloor \cdot, \cdot \rfloor)\bigr)$ is a *symmetrically self–dual space (SSD space)* if B is a nonzero real vector space and $\lfloor \cdot, \cdot \rfloor\colon B \times B \to \mathbb{R}$ is a symmetric bilinear form that separates the points of B. We define the quadratic form q on B by $q(b) := \frac{1}{2}\lfloor b, b \rfloor$. If $f \in \mathcal{PC}(B)$, we write $f^{@}$ for the Fenchel conjugate of f with respect to the pairing $\lfloor \cdot, \cdot \rfloor$. We will say that a locally convex topology $\mathcal{T}$ on B is *B–compatible* if the $\mathcal{T}$–dual of B is exactly $\bigl\{ \lfloor \cdot, c \rfloor\colon c \in B \bigr\}$.

Notation. If E is a normed space then $\mathcal{T}_{\|\ \|}(E)$ stands for the norm topology of E.

Examples 19.2(a). Let E be a nonzero Banach space and B and $\lfloor \cdot, \cdot \rfloor$ be defined as in Remark 17.1. Then B is a SSD space. Let $\mathcal{T}_{\mathcal{NW}}(B)$ be the topology $\mathcal{T}_{\|\ \|}(\dot{E}) \times w(E^*, E)$ on B. Then $\mathcal{T}_{\mathcal{NW}}(B)$ is B–compatible. We introduce a norm on B by $\|b\| := \sqrt{\|b_1\|^2 + \|b_2\|^2}$. If E is not reflexive then the topology $\mathcal{T}_{\|\ \|}(B)$ is not B–compatible.

Definition 19.5: q–positive sets. Let B be a SSD space and $\emptyset \neq A \subset B$. We say that A is *q–positive* if

$$b, c \in A \Longrightarrow q(b-c) \geq 0.$$

We say that A is *q–negative* if

$$b, c \in A \Longrightarrow q(b-c) \leq 0.$$

Examples 19.6(a). In Example 19.2(a), the q–positive sets are exactly the sets $G(S)$, where $S\colon E \rightrightarrows E^*$ is nontrivial and monotone. (See the discussion in Remark 17.1.)

Lemma 19.8. Let B be a SSD space, $f \in \mathcal{PC}(B)$ and $f \geq q$ on B. Let $\mathrm{pos}\, f := \bigl\{ b \in B\colon f(b) = q(b) \bigr\}$. If $\mathrm{pos}\, f \neq \emptyset$ then $\mathrm{pos}\, f$ is a q–positive subset of B. Let $g \in \mathcal{PC}(B)$ and $g \geq -q$ on B. Let $\mathrm{neg}\, g := \bigl\{ b \in B\colon g(b) = -q(b) \bigr\}$. If $\mathrm{neg}\, g \neq \emptyset$ then $\mathrm{neg}\, g$ is a q-negative subset of B.

Definition 19.11. Let B be a SSD space. We say that $f \in \mathcal{PC}(B)$ is a *BC–function* if
$$b \in B \quad \Longrightarrow \quad f^{@}(b) \geq f(b) \geq q(b).$$

"BC" stands for "bigger conjugate".

Lemma 19.12. *Let B be a SSD space and $f \in \mathcal{PC}(B)$ be a BC–function. Then*
$$\operatorname{pos} f^{@} = \operatorname{pos} f \subset \operatorname{dom} f.$$

Definition 19.14. Let B be a SSD space. We say that $g \in \mathcal{PC}(B)$ is a *TBC–function* if
$$b \in B \quad \Longrightarrow \quad g^{@}(-b) \geq g(b) \geq -q(b).$$

"TBC" stands for "twisted bigger conjugate".

Theorem 19.16: Transversality theorem. *Let B be a SSD space, $f \in \mathcal{PC}(B)$ be a BC–function and $g \colon B \to \mathbb{R}$ be a TBC–function that is continuous with respect to a B–compatible topology. Then* $\operatorname{pos} f - \operatorname{neg} g = B$.

Definition 19.17. Let B be a SSD space and A be a nonempty q–positive subset of B. We define the function $\Phi_A \in \mathcal{PC}(B)$ associated with A by
$$\Phi_A(b) := \sup_{a \in A} \left[\lfloor b, a \rfloor - q(a) \right].$$

Then $\Phi_A = q$ on A, $\Phi_A{}^{@} \leq q$ on A, $\Phi_A{}^{@} \geq \Phi_A \vee q$ on B and $A \subset \operatorname{pos} \Phi_A{}^{@}$. In fact, $\operatorname{pos} \Phi_A{}^{@}$ is the largest q–positive subset C of B such that $\Phi_C = \Phi_A$ on B. Writing "co" for "convex hull",
$$A \subset \operatorname{co} A \subset \operatorname{dom} \Phi_A{}^{@} \subset \operatorname{dom} \Phi_A.$$

Example 19.20. We give an example of a BC–function on $\mathbb{R}^2$ which is not of the form Φ_A for any nonempty q–positive subset A of $\mathbb{R}^2$.

Definition 20.1. Let B be a SSD space and A be a nonempty q–positive subset of B. We say that A is *maximally q–positive* if A is not properly contained in any other q–positive set. If A is a maximally q–positive subset of B then Φ_A is a BC–function and $\operatorname{pos} \Phi_A{}^{@} = \operatorname{pos} \Phi_A = A$.

SSDB spaces

Definition 21.1. We will say that B (more precisely, $(B, \lfloor \cdot, \cdot \rfloor)$) is a *symmetrically self–dual Banach space (SSDB space)* if B is a SSD space and a Banach space, $\mathcal{T}_{\| \ \|}(B)$ is B–compatible, and the norm of $\lfloor \cdot, c \rfloor$ as a functional on B is identical with $\|c\|$. In this case, the quadratic form q is continuous and, for all $b \in B$,
$$|q(b)| = \tfrac{1}{2} \big| \lfloor b, b \rfloor \big| \leq \tfrac{1}{2} \|b\|^2.$$

Let $g_0 := \tfrac{1}{2} \| \cdot \|^2$ on B. Then g_0 is a BC–function and a TBC–function.

Examples 21.2(a). If E is a nonzero reflexive Banach space, we can define B and its associated norm as in Remark 17.1 and Example 19.2(a). Then B is a SSDB space.

Theorem 21.4(b). *Let B be a SSDB space and $f \in PC(B)$ be a BC–function. Then pos f is maximally q–positive.*

Theorem 21.7: Criterion for maximal q–positivity. *Suppose that B is a SSDB space and A is a nonempty q–positive subset of B. Then*

$$A \text{ is maximally } q\text{–positive} \iff A - \text{neg } g_0 = B.$$

Definition 21.8. Let B be a SSDB space and $h \in PC(B)$. We say that h is *autoconjugate* if $h^@ = h$ on B.

Lemma 21.9 *Let B be a SSDB space and $h \in PC(B)$ be autoconjugate. Then h is a BC–function and pos $h^@$ is maximally q–positive.*

Theorem 21.10: Existence of autoconjugates. *Let B be a SSDB space, $f \in PCLSC(B)$ and $f^@ \geq f$ on B. For all $b \in B$, let*

$$h(b) := \inf_{c \in B} \left[\tfrac{1}{2} f(b+c) + \tfrac{1}{2} f^@(b-c) + g_0(c) \right].$$

Then:
(a) h is autoconjugate.
(b) $f \vee q \leq h \leq f^@$ on B and pos h is a maximally q–positive superset of pos $f^@$.
(c) $b \in \text{pos } h$ if, and only if,

$$\text{there exists } d \in \text{pos } g_0 \text{ such that } (b-d, b+d) \in G(\partial f).$$

Theorem 21.11. *Let B be a SSDB space, A be a nonempty q–positive subset of B and, for all $b \in B$*

$$h(b) := \inf_{c \in B} \left[\tfrac{1}{2} \Phi_A(b+c) + \tfrac{1}{2} \Phi_A{}^@(b-c) + g_0(c) \right].$$

Then:
(a) h is autoconjugate.
(b) $\Phi_A \vee q \leq h \leq \Phi_A{}^@$ on B and pos h is a maximally q–positive superset of A. If A is maximally q–positive then pos $h = A$.
(c) $b \in \text{pos } h$ if, and only if,

$$\text{there exists } d \in \text{pos } g_0 \text{ such that } (b-d, b+d) \in G(\partial \Phi_A).$$

Theorem 21.12: Local transversality theorem. *Let B be a SSDB space, $f \in PCLSC(B)$ be a BC–function and $g \in PCLSC(B)$ be a TBC–function. Then*

$$\text{int} \left(\text{pos } f - \text{neg } g \right) = \text{sur} \left(\text{dom } f - \text{dom } g \right).$$

Consequently, $\text{int} \left(\text{pos } f - \text{neg } g \right)$ is convex and $\text{sur} \left(\text{dom } f - \text{dom } g \right)$ is open.

The SSD space $E \times E^*$

Definition 22.3. Let E be a nonzero Banach space and $f \in \mathcal{PC}(E \times E^*)$. We define the convex lower semicontinuous function $f_{\div} \colon E \to]-\infty, \infty]$ by

$$f_{\div}(x) := \sup_{b=(b_1, b_2) \in E \times E^*} \frac{\langle x, b_2 \rangle - f(b)}{1 + \|b_1\|}.$$

Lemma 22.4. Let E be a nonzero Banach space, $f \in \mathcal{PC}(E \times E^*)$, and $x \in E$. Then

$$f_{\div}(x) \vee 0 = \min_{a \in (\pi_1)^{-1} x} \left(f^{@}(a) \vee \|a_2\| \right).$$

Theorem 22.5. Let E be a nonzero Banach space, and $f \in \mathcal{PC}(E \times E^*)$. Then $\pi_1 \mathrm{dom}\, f^{@} = \mathrm{dom}\, f_{\div}$. Consequently, $f_{\div}$ is proper if, and only if, $f^{@}$ is proper.

Corollary 22.6. Let E be a nonzero Banach space and $f, g \in \mathcal{PC}(E \times E^*)$. Then

$$\mathrm{sur}\left[\pi_1 \mathrm{dom}\, f^{@} - \pi_1 \mathrm{dom}\, g^{@}\right] = \mathrm{int}\left[\pi_1 \mathrm{dom}\, f^{@} - \pi_1 \mathrm{dom}\, g^{@}\right].$$

Consequently, $\mathrm{sur}\left[\pi_1 \mathrm{dom}\, f^{@} - \pi_1 \mathrm{dom}\, g^{@}\right]$ is open.

Theorem 22.8(a). Let E be a nonzero Banach space, $f \in \mathcal{PC}(E \times E^*)$, $f \geq q$ and $f^{@} \geq q$ on $E \times E^*$. Then:

$$\mathrm{int}\, \pi_1 \mathrm{pos}\, f^{@} = \mathrm{int}\, \pi_1 \mathrm{dom}\, f^{@} = \mathrm{int}\, \mathrm{dom}\, f_{\div}$$
$$= \mathrm{sur}\, \pi_1 \mathrm{pos}\, f^{@} = \mathrm{sur}\, \pi_1 \mathrm{dom}\, f^{@} = \mathrm{sur}\, \mathrm{dom}\, f_{\div}.$$

Consequently, $\mathrm{int}\, \pi_1 \mathrm{pos}\, f^{@}$ is convex and $\mathrm{sur}\, \pi_1 \mathrm{pos}\, f^{@}$ is open.

Reminder. We recall that BC–functions were defined in Definition 19.11, and the binary operator $\oplus_2$ was defined in Definition 16.3.

Lemma 22.9. Let E be a nonzero Banach space, $B := E \times E^*$, $f, g \in \mathcal{PCLSC}(B)$ be BC–functions and

$$\bigcup_{\lambda > 0} \lambda\left[\pi_1 \mathrm{dom}\, f - \pi_1 \mathrm{dom}\, g\right] \text{ be a closed subspace of } E.$$

Then $f \oplus_2 g$ is a BC–function. Furthermore, $b \in \mathrm{pos}\left(f \oplus_2 g\right)^{@} = \mathrm{pos}\left(f \oplus_2 g\right)$ if, and only if, there exist $a \in \mathrm{pos}\, f^{@} = \mathrm{pos}\, f$ and $c \in \mathrm{pos}\, g^{@} = \mathrm{pos}\, g$ such that $a_1 = c_1 = b_1$ and $a_2 + c_2 = b_2$.

Monotone multifunctions on general Banach spaces

Lemma 23.1. Let E be a nonzero Banach space, $f \colon E \times E^* \to]-\infty, \infty]$ be proper and convex and, for all $(x, x^*) \in E \times E^*$, $f(x, x^*) \geq \langle x, x^* \rangle$. Let

$$\mathrm{pos}\, f = \left\{ (x, x^*) \in E \times E^* \colon f(x, x^*) = \langle x, x^* \rangle \right\} \neq \emptyset.$$

Then the multifunction defined by $G(S) := \mathrm{pos}\, f$ is monotone.

Definition 23.2: Fitzpatrick function. Let E be a nonzero Banach space and $S\colon E \rightrightarrows E^*$ be nontrivial and monotone. We define the *Fitzpatrick function* $\varphi_S\colon E \times E^* \to\,]-\infty, \infty]$ associated with S by

$$\varphi_S(x, x^*) := \sup_{(s, s^*) \in G(S)} \left[\langle s, x^* \rangle + \langle x, s^* \rangle - \langle s, s^* \rangle \right].$$

Then $\varphi_S = \Phi_{G(S)}$, consequently, from the properties in Definition 19.17,

$$s \in G(S) \quad \Longrightarrow \quad \varphi_S(s) = \langle s_1, s_2 \rangle,$$

$$b \in E \times E^* \quad \Longrightarrow \quad \varphi_S{}^@(b) \geq \varphi_S(b) \vee \langle b_1, b_2 \rangle.$$

Finally, $\mathrm{pos}\,\varphi_S{}^@$ is the graph of the largest monotone multifunction $T\colon E \rightrightarrows E^*$ such that $\varphi_T = \varphi_S$ on $E \times E^*$.

If S is maximally monotone then, from the properties in Definition 20.1,

$$\varphi_S \text{ is a BC--function} \quad \text{and} \quad \mathrm{pos}\,\varphi_S{}^@ = \mathrm{pos}\,\varphi_S = G(S).$$

Example 19.20 provides an example of a BC--function that is not the Fitzpatrick function of any maximally monotone multifunction.

Definition 23.3. Let E be a nonzero Banach space and $S\colon E \rightrightarrows E^*$ be nontrivial and monotone. We define the multifunction $S_\varphi\colon E \rightrightarrows E^*$ by $G(S_\varphi) = \mathrm{dom}\,\varphi_S$. In general, S_φ is not monotone but it does, of course, have a convex graph. We note then that

$$D(S_\varphi) = \pi_1 \mathrm{dom}\,\varphi_S \quad \text{and} \quad R(S_\varphi) = \pi_2 \mathrm{dom}\,\varphi_S.$$

Lemma 23.4. Let E be a nonzero Banach space and $S\colon E \rightrightarrows E^*$ be nontrivial and monotone. Then

$$G(S) \subset \mathrm{co}\,G(S) \subset \mathrm{dom}\,\varphi_S{}^@ \subset \mathrm{dom}\,\varphi_S = G(S_\varphi),$$

$$D(S) \subset \mathrm{co}\,D(S) \subset \pi_1\big(\mathrm{dom}\,\varphi_S{}^@\big) \subset D(S_\varphi)$$

and

$$R(S) \subset \mathrm{co}\,R(S) \subset \pi_2\big(\mathrm{dom}\,\varphi_S{}^@\big) \subset R(S_\varphi).$$

Definition 23.5: Duality map. The *duality map* $J\colon E \rightrightarrows E^*$ is defined by:

$$x^* \in Jx \iff \tfrac{1}{2}\|x\|^2 + \tfrac{1}{2}\|x^*\|^2 = \langle x, x^* \rangle.$$

J is maximally monotone and $G(J_\varphi) = \mathrm{dom}\,\varphi_J = E \times E^*$. So, even if S is maximally monotone, $G(S_\varphi)$ can be much larger than $G(S)$.

Theorem 24.1: Sum theorem. *Let E be a nonzero Banach space Suppose that S, T: $E \rightrightarrows E^*$ are maximally monotone and $\bigcup_{\lambda > 0} \lambda[D(S_\varphi) - D(T_\varphi)]$ be a closed subspace of E.*
(a) If E is reflexive then $S + T$ is maximally monotone.
(b) Even if E is not reflexive, if $\varphi_{S+T} \geq q$ on $E \times E^*$ then $S+T$ is maximally monotone.
(c) Even if E is not reflexive, if

$$b \in E \times E^* \text{ and } \varphi_{S+T}(b) \leq q(b) \quad \Longrightarrow \quad b_1 \in D(S + T)$$

then $S + T$ is maximally monotone.

Theorem 25.1: Bounded range. *Let E be a nonzero Banach space, S: $E \rightrightarrows E^*$ be maximally monotone and $R(S)$ be bounded. Then $D(S) = E$.*

Notation. *Let E be a nonzero Banach space, S: $E \rightrightarrows E^*$ be nontrivial and $z \in E$. We say that S is locally bounded at z if there exist η, $K > 0$ such that*

$$s \in G(S) \text{ and } \|s_1 - z\| < \eta \quad \Longrightarrow \quad \|s_2\| \leq K.$$

Theorem 26.1: Local boundedness theorem. *Let E be a nonzero Banach space, S: $E \rightrightarrows E^*$ be nontrivial and monotone and $z \in \text{sur} D(S_\varphi)$. Then S is locally bounded at z.*

Remark 26.2. *Let S: $\mathbb{R}^2 \rightrightarrows \mathbb{R}^2$ be monotone and the four points $(\pm 1, \pm 1)$ lie in $D(S)$. Then S is locally bounded at 0 (even if $0 \notin D(S)$).*

Theorem 27.1: Six set theorem. *Let E be a nonzero Banach space and S: $E \rightrightarrows E^*$ be maximally monotone. Then*

$$\text{int} D(S) = \text{int} (\text{co} D(S)) = \text{int} D(S_\varphi)$$
$$= \text{sur} D(S) = \text{sur} (\text{co} D(S)) = \text{sur} D(S_\varphi).$$

Consequently, $\text{int} D(S)$ is convex and $\text{sur} D(S_\varphi)$ is open.

Remark 27.2. *If S: $\mathbb{R}^2 \rightrightarrows \mathbb{R}^2$ is maximally monotone and the four points $(\pm 1, \pm 1)$ are in $D(S)$ then Theorem 27.1 implies that $]-1, 1[\times]-1, 1[\subset D(S)$ (even if we do not assume that $0 \in D(S)$).*

Theorem 27.3: Nine set theorem. *Let E be a nonzero Banach space, S: $E \rightrightarrows E^*$ be maximally monotone and $\text{sur} D(S_\varphi) \neq \emptyset$. Then*

$$\overline{D(S)} = \overline{\text{co} D(S)} = \overline{D(S_\varphi)}$$
$$= \overline{\text{int} D(S)} = \overline{\text{int} (\text{co} D(S))} = \overline{\text{int} D(S_\varphi)}$$
$$= \overline{\text{sur} D(S)} = \overline{\text{sur} (\text{co} D(S))} = \overline{\text{sur} D(S_\varphi)}.$$

Theorem 27.5(b): The closure of the domain. *Let E be a nonzero Banach space, $S\colon E \rightrightarrows E^*$ be monotone, and $z \in E \setminus \overline{D(S)} \implies \sup_{s \in G(S)} \left[\langle z - s_1, s_2 \rangle / \|z - s_1\| \right] = \infty$. Then:*

$$\overline{D(S)} = \overline{\operatorname{co} D(S)} = \overline{D(S_\varphi)}.$$

Theorem 27.6(b): The closure of the range. *Let E be a nonzero Banach space, $S\colon E \rightrightarrows E^*$ be monotone, and $z^* \in E^* \setminus \overline{R(S)} \implies \sup_{s \in G(S)} \left[\langle s_1, z^* - s_2 \rangle / \|z^* - s_2\| \right] = \infty$. Then:*

$$\overline{R(S)} = \overline{\operatorname{co} R(S)} = \overline{R(S_\varphi)}.$$

Notation. We write "aff" for "affine hull" and "lin" for "linear hull". If F is a subspace of the nonzero Banach space E, we write

$$F^\perp := \left\{ y^* \in E^* \colon \langle F, y^* \rangle = \{0\} \right\}.$$

Lemma 28.1. *Let E be a nonzero Banach space and $S, T \colon E \rightrightarrows E^*$ be maximally monotone. Then:*

(a) $$\overline{\operatorname{aff}\left[D(S_\varphi) - D(T_\varphi) \right]} = \overline{\operatorname{aff}\left[D(S) - D(T) \right]}.$$

(b) $$\overline{\operatorname{lin}\left[D(S_\varphi) - D(T_\varphi) \right]} = \overline{\operatorname{lin}\left[D(S) - D(T) \right]}.$$

(c) $$\bigcup_{\lambda > 0} \lambda \left[D(S_\varphi) - D(T_\varphi) \right] \subset \overline{\operatorname{lin}\left[D(S) - D(T) \right]}.$$

Corollary 28.2. *Let E be a nonzero Banach space and $S \colon E \rightrightarrows E^*$ be maximally monotone. Then*

$$\overline{\operatorname{aff} D(S_\varphi)} = \overline{\operatorname{aff} D(S)} \quad \text{and} \quad \overline{\operatorname{lin} D(S_\varphi)} = \overline{\operatorname{lin} D(S)}.$$

Theorem 28.6. *Let E be a nonzero Banach space, $S \colon E \rightrightarrows E^*$ be maximally monotone and $D(S)$ be closed and convex. Then*

$$D(S) = \operatorname{co} D(S) = \overline{D(S_\varphi)}.$$

Definition 28.7. *Let F be a subspace of a Banach space E, and $S \colon E \rightrightarrows E^*$ be nontrivial and monotone. We say that S is F-saturated if*

$$x \in D(S) \quad \implies \quad Sx + F^\perp = Sx.$$

Theorem 28.9. *Let E and F be nonzero Banach spaces, $L \colon F \to E$ be continuous and linear with $L(F)$ closed and $L^*(E^*) = F^*$, and $S \colon E \rightrightarrows E^*$ be nontrivial and monotone with $D(S) \subset L(F)$. Then S is maximally monotone $\iff S$ is $L(F)$-saturated and L^*SL is maximally monotone.*

Corollary 28.10: The maximal monotonicity of the restriction to a subspace. *Let F be a nonzero closed subspace of a Banach space E, $S: E \rightrightarrows E^*$ be nontrivial and monotone and $D(S) \subset F$. Define $T: F \rightrightarrows F^*$ by*

$$G(T) := \{(s_1, s_2|_F): s \in G(S)\}.$$

Then S is maximally monotone $\iff$ S is F-saturated and T is maximally monotone.

Monotone multifunctions on reflexive Banach spaces

Theorem 29.2. *Let E be a nonzero reflexive Banach space and $S: E \rightrightarrows E^*$ be nontrivial and monotone. Then S is maximally monotone $\iff$ for all $b \in E \times E^*$, there exists $s \in G(S)$ such that*

$$\|b_1 - s_1\|^2 + \|b_2 - s_2\|^2 + 2q(b - s) = 0.$$

Corollary 29.3: Negative alignment criterion for maximality. *Let E be a nonzero reflexive Banach space and $S: E \rightrightarrows E^*$ be nontrivial and monotone. Then S is maximally monotone $\iff$ for all $b \in E \times E^* \setminus G(S)$, there exists $s \in G(S)$ such that*

$$s_1 \neq b_1, \quad s_2 \neq b_2 \quad \text{and} \quad q(b - s) = -\|b_1 - s_1\|\|b_2 - s_2\|.$$

Theorem 29.5: Rockafellar's surjectivity theorem. *Let E be a nonzero reflexive Banach space and $S: E \rightrightarrows E^*$ be maximally monotone. Then*

$$R(S + J) = E^*.$$

Theorem 29.6: Minimum of the norm of the resolvent. *Let E be a nonzero reflexive Banach space, $S: E \rightrightarrows E^*$ be maximally monotone and*

$$N := \sup_{b \in E \times E^*} \left[\|b\| - \sqrt{2\varphi_S(b) + \|b\|^2} \right] \vee 0.$$

Then

$$\min \left\{ \|x\|: x \in E, \ (S + J)x \ni 0 \right\} = \tfrac{1}{\sqrt{2}} N$$

and

$$\sup \left\{ \|x\|: x \in E, \ (S + J)x \ni 0 \right\} \leq \sqrt{2} \inf_{b \in G(S)} \left[\|b_1\| + \|b_2\| \right].$$

Theorem 29.8: Autoconjugates. *Let E be a nonzero reflexive Banach space and $S: E \rightrightarrows E^*$ be nontrivial and monotone. For all $b \in E \times E^*$, let*

$$h(b) := \inf_{c \in E \times E^*} \left[\tfrac{1}{2}\varphi_S(b + c) + \tfrac{1}{2}\varphi_S{}^@(b - c) + g_0(c) \right].$$

Then:
(a) h is autoconjugate.
(b) $\varphi_S \vee q \leq h \leq \varphi_S{}^@$ on $E \times E^*$ and $\text{pos}\, h$ is a maximally monotone superset of $G(S)$. If S is maximally monotone then $\text{pos}\, h = G(S)$.
(c) $b \in \text{pos}\, h$ if, and only if,

there exists $d \in G(J)$ such that $(b - d, b + d) \in G(\partial \varphi s)$.

Theorem 30.1. Let E be a nonzero reflexive Banach space, $w^* \in E^*$ and $f, g \in \mathcal{PCLSC}(E \times E^*)$ be BC–functions such that $E \times \{w^*\} \subset \operatorname{dom} f$ and $\pi_2 \operatorname{dom} g = E^*$. We define $\rho_2 \colon E \times E^* \to E \times E^*$ by $\rho_2(b) := (b_1, -b_2)$. Then:
(a) $\operatorname{pos} f - \rho_2 \operatorname{pos} g = E \times E^*$ and $\operatorname{pos} f + \rho_2 \operatorname{pos} g = E \times E^*$.
(b) If $x \in E$ then there exist $(y, y^*) \in \operatorname{pos} f$ and $(z, y^*) \in \operatorname{pos} g$ such that $y + z = x$.
(c) If $x^* \in E^*$ then there exist $(y, y^*) \in \operatorname{pos} f$ and $(y, z^*) \in \operatorname{pos} g$ such that $y^* + z^* = x^*$.

Theorem 30.2: Surjectivity theorem. Let E be a nonzero reflexive Banach space, $S, T \colon E \rightrightarrows E^*$ be maximally monotone and $G(S_\varphi) - \rho_2 G(T_\varphi) = E \times E^*$. Then $R(S + T) = E^*$.

Theorem 30.3. Let E be a nonzero reflexive Banach space and $S \colon E \rightrightarrows E^*$ and $T \colon E^* \rightrightarrows E$ be maximally monotone. Suppose that $D(T_\varphi) = E^*$ and $\bigcap_{x \in E} S_\varphi(x) \neq \emptyset$. Then:
(a) If I_E is the identity map on E, $(I_E + TS)(E) = E$.
(b) If I_{E^*} is the identity map on E^*, $(I_{E^*} + ST)(E^*) = E^*$.

Theorem 30.4: Abstract Hammerstein theorem. Let E be a nonzero reflexive Banach space, $S \colon E \rightrightarrows E^*$ and $T \colon E^* \rightrightarrows E$ be maximally monotone. Suppose that **either** $D(T_\varphi) = E^*$ and $\bigcap_{x \in E} S_\varphi(x) \neq \emptyset$ **or** $D(S_\varphi) = E$ and $\bigcap_{x^* \in E^*} T_\varphi(x^*) \neq \emptyset$. Then $(I_E + TS)(E) = E$.

Theorem 31.2. Let E be a nonzero reflexive Banach space and $S \colon E \rightrightarrows E^*$ be maximally monotone. Then

$$\overline{D(S)} = \overline{\operatorname{co} D(S)} = \overline{D(S_\varphi)} \quad \text{and} \quad \overline{R(S)} = \overline{\operatorname{co} R(S)} = \overline{R(S_\varphi)}.$$

Consequently, $\overline{D(S)}$ and $\overline{R(S)}$ are both convex.

Definition 31.5. We say that a monotone multifunction $S \colon E \rightrightarrows E^*$ is rectangular if

$$D(S) \times R(S) \subset G(S_\varphi).$$

Corollary 31.6: Brezis–Haraux theorem. Let E be a nonzero reflexive Banach space, $S, T \colon E \rightrightarrows E^*$ be monotone, and $S + T$ be maximally monotone. If either

$$S \text{ and } T \text{ are both rectangular,}$$

or

$$D(S) \subset D(T) \quad \text{and} \quad T \text{ is rectangular,}$$

then the Brezis–Haraux condition

$$\operatorname{int} R(S + T) = \operatorname{int} \left[R(S) + R(T) \right] \quad \text{and} \quad \overline{R(S + T)} = \overline{R(S) + R(T)}$$

is satisfied.

Theorem 32.2: Sandwiched closed subspace theorem. *Let E be a nonzero reflexive Banach space, $S, T: E \rightrightarrows E^*$ be maximally monotone, and suppose that there exists a closed subspace F of E such that*

$$D(S) - D(T) \subset F \subset \bigcup_{\lambda > 0} \lambda [D(S_\varphi) - D(T_\varphi)].$$

Then $S + T$ is maximally monotone. Furthermore, for all $\varepsilon > 0$,

$$D(S) - D(T) \subset D(S_\varphi) - D(T_\varphi) \subset (1 + \varepsilon)(D(S) - D(T)),$$

(that is to say, $D(S_\varphi) - D(T_\varphi)$ and $D(S) - D(T)$ are almost identical) and

$$\bigcup_{\lambda > 0} \lambda [D(S_\varphi) - D(T_\varphi)] = \bigcup_{\lambda > 0} \lambda [D(S) - D(T)].$$

Corollary 32.3. *Let E be a nonzero reflexive Banach space, $S, T: E \rightrightarrows E^*$ be maximally monotone, and suppose that one of the conditions below is satisfied:*

$$\bigcup_{\lambda > 0} \lambda [D(S) - D(T)] \text{ is a closed subspace of } E.$$

$$\bigcup_{\lambda > 0} \lambda [\operatorname{co} D(S) - \operatorname{co} D(T)] \text{ is a closed subspace of } E.$$

$$\bigcup_{\lambda > 0} \lambda [\pi_1 \operatorname{dom} \varphi_S{}^@ - \pi_1 \operatorname{dom} \varphi_T{}^@] \text{ is a closed subspace of } E.$$

Then $S + T$ is maximally monotone. Furthermore, in any of these cases,

$$\bigcup_{\lambda > 0} \lambda [D(S) - D(T)] = \bigcup_{\lambda > 0} \lambda [\operatorname{co} D(S) - \operatorname{co} D(T)]$$

$$= \bigcup_{\lambda > 0} \lambda [\pi_1 \operatorname{dom} \varphi_S{}^@ - \pi_1 \operatorname{dom} \varphi_T{}^@]$$

$$= \bigcup_{\lambda > 0} \lambda [D(S_\varphi) - D(T_\varphi)].$$

Theorem 33.1: > six set theorem. *Let E be a nonzero reflexive Banach space and the multifunctions $S, T: E \rightrightarrows E^*$ be maximally monotone. Then:*

$$\operatorname{int}[D(S) - D(T)] = \operatorname{int}[\operatorname{co}D(S) - \operatorname{co}D(T)] = \operatorname{int}[D(S_\varphi) - D(T_\varphi)]$$
$$= \operatorname{sur}[D(S) - D(T)] = \operatorname{sur}[\operatorname{co}D(S) - \operatorname{co}D(T)] = \operatorname{sur}[D(S_\varphi) - D(T_\varphi)].$$

Consequently, $\operatorname{int}[D(S) - D(T)]$ is convex.

Theorem 33.2: > nine set theorem. *Let E be a nonzero reflexive Banach space, $S, T: E \rightrightarrows E^*$ be maximally monotone and $\operatorname{sur}[D(S_\varphi) - D(T_\varphi)] \neq \emptyset$. Then:*

$$\overline{D(S) - D(T)} = \overline{\operatorname{co}D(S) - \operatorname{co}D(T)} = \overline{D(S_\varphi) - D(T_\varphi)}$$
$$= \operatorname{int}\overline{[D(S) - D(T)]} = \operatorname{int}\overline{[\operatorname{co}D(S) - \operatorname{co}D(T)]} = \operatorname{int}\overline{[D(S_\varphi) - D(T_\varphi)]}$$
$$= \operatorname{sur}\overline{[D(S) - D(T)]} = \operatorname{sur}\overline{[\operatorname{co}D(S) - \operatorname{co}D(T)]} = \operatorname{sur}\overline{[D(S_\varphi) - D(T_\varphi)]}.$$

Consequently, $\overline{D(S) - D(T)}$ is convex.

Corollary 34.5: Brezis–Crandall–Pazy theorem. *Let E be a nonzero reflexive Banach space, $S\colon E \rightrightarrows E^*$ and $T\colon E \rightrightarrows E^*$ be maximally monotone, $D(S) \subset D(T)$, and suppose that there exist increasing functions $k\colon [0,\infty[\to [0,1[$ and $C\colon [0,\infty[\to [0,\infty[$ such that,*

$$x \in D(S) \implies |Tx| \le k(\|x\|)|Sx| + C(\|x\|).$$

Then $S + T$ is maximally monotone.

Corollary 34.6. *Let E be a nonzero reflexive Banach space, $S\colon E \rightrightarrows E^*$ and $T\colon E \rightrightarrows E^*$ be maximally monotone, $D(S) \subset D(T)$, and suppose that $0 < p < 1$ and there exist increasing functions $k\colon [0,\infty[\to [0,\infty[$ and $C\colon [0,\infty[\to [0,\infty[$ such that,*

$$x \in D(S) \implies |Tx| \le k(\|x\|)|Sx|^p + C(\|x\|).$$

Then $S + T$ is maximally monotone.

Monotone multifunctions on general Banach spaces again

Notation. Let E be a nonzero Banach space and $B := E \times E^*$. In Example 19.2(a), we considered B as a SSD space. We recall from Remark 17.1 that if $b = (b_1, b_2) \in B$ then $q(b) = \langle b_1, b_2 \rangle$. Of course, B is also a normed space under the norm $\|(b_1, b_2)\| = \sqrt{\|b_1\|^2 + \|b_2\|^2}$.

Even though B is a SSD space and a Banach space, it is not a SSDB space if E is not reflexive, since the topology $\mathcal{T}_{\|\ \|}(B)$ is not B-compatible.

Following Notation 16.1, the norm–dual of B is $B^* = E^{**} \times E^*$ under the pairing

$$\langle b, v \rangle := \langle b_1, v_2 \rangle + \langle b_2, v_1 \rangle \quad \big(b = (b_1, b_2) \in B,\ v = (v_1, v_2) \in B^*\big),$$

and the dual norm of B^* is given by $\|(v_1, v_2)\| = \sqrt{\|v_1\|^2 + \|v_2\|^2}$.

B^* is also a SSD space under the bilinear form

$$[u, v] := \langle v_2, u_1 \rangle + \langle u_2, v_1 \rangle \quad \big(u = (u_1, u_2) \in B^*,\ v = (v_1, v_2) \in B^*\big).$$

We define $\widetilde{q}\colon B^* \to \mathbb{R}$ by $\widetilde{q}(v) := \frac{1}{2}[v, v]$. $\widetilde{q}$ is a quadratic form on B^* and, since $\widetilde{q}(v_1, v_2) = \langle v_2, v_1 \rangle$, bilinear on $E^{**} \times E^*$.

We say that $h\colon B \to\,]-\infty, \infty]$ is a $\widetilde{BC}$–function if h is a BC–function and $h^* \ge \widetilde{q}$ on B^*.

As in Definition 21.1, we write $g_0 := \frac{1}{2}\|\cdot\|^2$ on B. Then g_0 is a $\widetilde{BC}$–function.

If $\psi \in \mathcal{PC}(B^*)$ and $\psi \ge \widetilde{q}$ on B^*, let

$$\widetilde{\mathrm{pos}}\, \psi = \big\{v \in B^*\colon \psi(v) = \widetilde{q}(v)\big\}.$$

Lemma 19.8 implies that if $\widetilde{\mathrm{pos}}\, \psi \ne \emptyset$ then $\widetilde{\mathrm{pos}}\, \psi$ is a $\widetilde{q}$–positive subset of B^*.

Lemma 35.1. *Let E be a nonzero Banach space, $k \in \mathcal{PCLSC}(E)$ and $g(c) := k(c_1) + k^*(c_2)$ $\big(c = (c_1, c_2) \in E \times E^*\big)$. Then g is a $\widetilde{BC}$–function.*

Example 35.2: Gossez's skew operator. Let $E := \ell^1$, and $S: \ell^1 \to E^* = \ell^\infty$ be defined by

$$(Sx)_n = -\sum_{k<n} x_k + \sum_{k>n} x_k \quad (x \in \ell^1).$$

Then φ_S is *not* a $\widetilde{BC}$-function.

Notation: The canonical map into the bidual and Gossez's extension. Let E be a nonzero Banach space. We define the linear operator $\widehat{}: E \to E^{**}$ such that $x \in E$ and $x^* \in E^* \implies \langle x^*, \widehat{x} \rangle = \langle x, x^* \rangle$. If $S: E \rightrightarrows E^*$, then $\overline{S}: E^{**} \rightrightarrows E^*$ is defined by:

$$v = (v_1, v_2) \in G(\overline{S}) \iff \inf_{s \in G(S)} \langle s_2 - v_2, \widehat{s_1} - v_1 \rangle \geq 0.$$

Theorem 36.3. *Let E be a nonzero Banach space and $S: E \rightrightarrows E^*$ be maximally monotone.*
(a) *Let S be of type (D). Then S is of type (NI).*
(b) *Let S be of type (NI). Then φ_S is a $\widetilde{BC}$–function, $\overline{\mathrm{pos}}\, \varphi_{S^*} \subset G(\overline{S})$, and there exists $v \in G(\overline{S})$ such that $\|v_1\|^2 = \|v_2\|^2 = -\widetilde{q}(v)$.*

Remark 36.6: The tail operator. Let $E := \ell^1$, and $S: \ell^1 \to E^* = \ell^\infty$ be defined by

$$(Sx)_n = \sum_{k \geq n} x_k \quad (x \in \ell^1).$$

Then S is positive and not of type (FP).

Theorem 37.1. *Let E be a nonzero Banach space, and $S: E \rightrightarrows E^*$ be maximally monotone of type (D). Then S is of type (FP).*

Notation: The full biconjugate. Let E be a nonzero Banach space. We write $\mathcal{PCPC}(E)$ for the set of all those $f \in \mathcal{PC}(E)$ such that $\mathrm{dom}\, f^* \neq \emptyset$. Suppose now that $f \in \mathcal{PCPC}(E)$. Then the *(full) biconjugate*, f^{**}, of f is the function from the bidual, E^{**}, of E into $]-\infty, \infty]$ defined by

$$f^{**}(x^{**}) := (f^*)^*(x^{**}) = \sup_{x^* \in E^*} \left[\langle x^*, x^{**} \rangle - f^*(x^*) \right] \quad (x^{**} \in E^{**}).$$

Definition 38.1. Let E be a nonzero Banach space. We write $\mathcal{CLB}(E)$ for the set of all convex functions $f: E \to \mathbb{R}$ that are Lipschitz on the bounded subsets of E or equivalently, from Theorem 8.7, bounded above on the bounded subsets of E. Consequently

$$f \in \mathcal{CLB}(E) \implies f^{**} \in \mathcal{CLB}(E^{**}).$$

We define the topology $\mathcal{T}_{\mathcal{CLB}}(E^{**})$ on E^{**} to be the coarsest topology on E^{**} making all the functions $h^{**}: E^{**} \to \mathbb{R}$ $\left(h \in \mathcal{CLB}(E)\right)$ continuous. Then $\mathcal{T}_{\mathcal{CLBN}}(E^{**} \times E^*)$ stands for the topology $\mathcal{T}_{\mathcal{CLB}}(E^{**}) \times \mathcal{T}_{\| \cdot \|}(E^*)$ on $E^{**} \times E^*$.

Theorem 38.4. *Let E be a nonzero reflexive Banach space and $S: E \rightrightarrows E^*$ be maximally monotone. Then S is of type (ED).*

Theorem 39.1. *Let E be a nonzero Banach space and $S\colon E \rightrightarrows E^*$ be maximally monotone of type (ED). Then S is of type (FPV).*

Theorem 40.1. *Let E be a nonzero Banach space and $S\colon E \rightrightarrows E^*$ be maximally monotone of type (ED). Then S is strongly maximal.*

Theorem 41.1. *Let E be a nonzero Banach space and $S\colon E \rightrightarrows E^*$ be strongly maximally monotone. Suppose that $y \in E$, $K \geq 0$ and*

$$s \in G(S) \text{ and } \|s_2\| > K \quad\Longrightarrow\quad \langle s_1 - y, s_2 \rangle \geq 0.$$

Then there exists $x^* \in E^*$ such that $\|x^*\| \leq K$ and $(y, x^*) \in G(S)$ and, in particular, $y \in D(S)$.

Corollary 41.2. *Let E be a nonzero Banach space, $S\colon E \rightrightarrows E^*$ be strongly maximally monotone, and $S^{-1}\colon E^* \rightrightarrows E$ be coercive, that is to say $\inf\langle S^{-1}x^*, x^* \rangle / \|x^*\| \to \infty$ as $\|x^*\| \to \infty$. Then $D(S) = E$.*

Theorem 41.3. *Let E be a nonzero reflexive Banach space and $S\colon E \rightrightarrows E^*$ be maximally monotone. Suppose that $y^* \in E^*$, $K \geq 0$ and*

$$s \in G(S) \text{ and } \|s_1\| > K \quad\Longrightarrow\quad \langle s_1, s_2 - y^* \rangle \geq 0.$$

Then there exists $x \in E$ such that $\|x\| \leq K$ and $(x, y^*) \in G(S)$ and, in particular, $y^* \in R(S)$.

Corollary 41.4: Coercivity and surjectivity. *Let E be a nonzero reflexive Banach space and $S\colon E \rightrightarrows E^*$ be maximally monotone and coercive, that is to say $\inf\langle x, Sx \rangle / \|x\| \to \infty$ as $\|x\| \to \infty$. Then $R(S) = E^*$.*

Theorem 42.6. *Let E be a nonzero Banach space, $T\colon E \rightrightarrows E^*$ be maximally monotone of type (ED), $b \in E \times E^*$ and $\alpha, \beta > 0$. Then:*
(b) *If $b \in E \times E^* \setminus G(T)$ then there exists $t \in G(T)$ such that $t_1 \neq b_1$, $t_2 \neq b_2$,*

$$\frac{\|t_1 - b_1\|}{\|t_2 - b_2\|} \text{ is as near as we please to } \frac{\alpha}{\beta}$$

and

$$\frac{q(t - b)}{\|t_1 - b_1\| \|t_2 - b_2\|} \text{ is as near as we please to } -1.$$

In particular, T is of type (ANA) (see Definition 36.11).
(c) *If, further, $\inf_{t \in G(T)} q(t-b) > -\alpha\beta$ then we can take t so that, in addition, $\|t_1 - b_1\| < \alpha$ and $\|t_2 - b_2\| < \beta$. So T is of type (BR).*

Remark 42.7. *Let $S\colon \ell^1 \to \ell^\infty$ be Gossez's skew linear map of Example 35.2. Then S is of type (BR) but not of type (NI).*

Notation. *If $\eta > 0$ then the multifunction $J_\eta\colon E \rightrightarrows E^*$ is defined by declaring that $s \in G(J_\eta)$ when $g_0(s) - q(s) \leq \eta$.*

Theorem 42.8. *Let E be a nonzero Banach space, $T\colon E \rightrightarrows E^*$ be maximally monotone of type (ED), and $\lambda, \eta > 0$. Then $R(T + \lambda J_\eta) = E^*$.*

Theorem 43.1. *Let E be a nonzero Banach space and $S\colon E \rightrightarrows E^*$ be maximally monotone of type (FP). Then*

$$\overline{R(S)} = \overline{\operatorname{co} R(S)} = \overline{R(S_\varphi)}.$$

Theorem 43.4. *Let E be a nonzero Banach space, $S\colon E \rightrightarrows E^*$ be monotone, and suppose that there exists $\eta > 0$ such that, for all $\lambda > 0$, $R(T + \lambda J_\eta) = E^*$. Then*

$$\overline{R(S)} = \overline{\operatorname{co} R(S)} = \overline{R(S_\varphi)}.$$

Notation. Let E be a nonzero Banach space and C be a nonempty closed convex subset of E. Then the *normality multifunction of C*, $N_C\colon E \rightrightarrows E^*$ is defined by

$$(x, x^*) \in G(N_C) \iff x \in C \text{ and } \langle x, x^* \rangle = \max \langle C, x^* \rangle.$$

Theorem 44.1. *Let E be a nonzero Banach space, $S\colon E \rightrightarrows E^*$ be maximally monotone and suppose that if C is a nonempty closed convex subset of E, and $D(S) \cap \operatorname{int} C \neq \emptyset$ then $S + N_C$ is maximally monotone. Then S is necessarily of type (FPV).*

Theorem 44.2. *Let E be a nonzero Banach space and $S\colon E \rightrightarrows E^*$ be maximally monotone of type (FPV). Then*

$$\overline{D(S)} = \overline{\operatorname{co} D(S)} = \overline{D(S_\varphi)}.$$

Definition 45.4. We write $\mathcal{CC}(E)$ for the set of all real convex continuous functions on E.

Corollary 45.5: The biconjugate of a maximum. *Let E be a nonzero Banach space, $h_0 \in \mathcal{PCPC}(E)$ and $h_1, \ldots, h_m \in \mathcal{CC}(E)$. Then*

$$(h_0 \vee \cdots \vee h_m)^{**} = h_0^{**} \vee \cdots \vee h_m^{**} \text{ on } E^{**}.$$

Definition 45.7. Let E be a nonzero Banach space. We define the topology $\mathcal{T}_{\mathcal{CC}}(E^{**})$ on E^{**} to be the coarsest topology on E^{**} such that, for all $f \in \mathcal{CC}(E)$, all of the functions $f^{**}\colon E^{**} \to\,]-\infty, \infty]$ are continuous. Clearly, $\mathcal{T}_{\mathcal{CLB}}(E^{**}) \subset \mathcal{T}_{\mathcal{CC}}(E^{**})$. We write $\mathcal{T}_{\mathcal{CCN}}(E^{**} \times E^*)$ for the topology $\mathcal{T}_{\mathcal{CC}}(E^{**}) \times \mathcal{T}_{\|\ \|}(E^*)$ on $E^{**} \times E^*$.

Lemma 45.9: Density results. *Let E be a nonzero Banach space.*
(a) *Let $f \in \mathcal{PCPC}(E)$ and $x^{**} \in E^{**}$. Then there exists a net $\{x_\gamma\}$ of elements of E such that $\widehat{x_\gamma} \to x^{**}$ in $\mathcal{T}_{\mathcal{CC}}(E^{**})$, $\widehat{x_\gamma} \to x^{**}$ in $\mathcal{T}_{\mathcal{CLB}}(E^{**})$, $f(x_\gamma) \to f^{**}(x^{**})$ and $f^{**}(\widehat{x_\gamma}) \to f^{**}(x^{**})$.*
(b) *$\widehat{E}$ is a dense subset of $(E^{**}, \mathcal{T}_{\mathcal{CC}}(E^{**}))$ and $(E^{**}, \mathcal{T}_{\mathcal{CLB}}(E^{**}))$.*

Corollary 45.10. *Let E be a nonzero Banach space and C be a nonempty convex subset of E. Then the $w(E^{**}, E^*)$–closure of $\widehat{C}$, the $\mathcal{T}_{\mathcal{CLB}}(E^{**})$–closure of $\widehat{C}$ and the $\mathcal{T}_{\mathcal{CC}}(E^{**})$–closure of $\widehat{C}$ are identical.*

Lemma 45.11. *Let E be a nonzero Banach space, and write ι for the linear isometry of $E \times E^*$ into $E^{**} \times E^*$ defined by $\iota(b_1, b_2) := (\widehat{b_1}, b_2)$. Let C be a nonempty convex subset of $E \times E^*$. Then the $\mathcal{T}_{\mathcal{WN}}(E^{**} \times E^*)$–closure of $\iota(C)$, the $\mathcal{T}_{\mathcal{CLBN}}(E^{**} \times E^*)$–closure of $\iota(C)$ and the $\mathcal{T}_{\mathcal{CCN}}(E^{**} \times E^*)$–closure of $\iota(C)$ are identical.*

Remark 45.13. If E is not reflexive then neither of $\big(E^{**}, \mathcal{T}_{\mathcal{CLB}}(E^{**})\big)$ and $\big(E^{**}, \mathcal{T}_{\mathcal{CC}}(E^{**})\big)$ is a topological vector spaces.

Theorem 45.14. *Let E be a nonzero Banach space.*
(a) $\mathcal{T}_{\mathcal{CC}}(E^{**}) = \mathcal{T}_{\| \ \|}(E^{**}) \iff E$ *is reflexive.*
(b) $\mathcal{T}_{\mathcal{CLB}}(E^{**}) = \mathcal{T}_{\| \ \|}(E^{**}) \iff E$ *is reflexive*

Example 45.16. We give an example of a function $f \in \mathcal{CC}(c_0)$ such that f^{**} is not continuous on $c_0^{**} = \ell^\infty$. Define $f \in \mathcal{CC}(c_0)$ by

$$f(x) := \textstyle\sum_{n \geq 1} x_n{}^{2n} \quad (x = \{x_n\}_{n \geq 1} \in c_0).$$

Remark 45.17. Let $f \in \mathcal{CC}(c_0)$ be as in Example 45.16. Then, from the definition of $\mathcal{T}_{\mathcal{CC}}(\ell^\infty)$, f^{**} is $\mathcal{T}_{\mathcal{CC}}(\ell^\infty)$–continuous. Since f^{**} is not $\mathcal{T}_{\| \ \|}(\ell^\infty)$–continuous, it follows that $\mathcal{T}_{\mathcal{CC}}(\ell^\infty) \not\subset \mathcal{T}_{\| \ \|}(\ell^\infty)$.

Theorem 46.1: Operators with convex graph. *Let E be a nonzero Banach space, $S \colon E \rightrightarrows E^*$ be maximally monotone and $G(S)$ be convex. Then:*
(a) S *is strongly maximal.*
(b) S *is of type (FPV).*
(c) *If S is of type (NI) then S is of type (ED) $\big($and hence of type (D) and type (FP)$\big)$.*

Theorem 46.3: Sum theorem for operators with convex graph. *Let E be a nonzero Banach space, $S, T \colon E \rightrightarrows E^*$ be maximally monotone, $G(S)$ and $G(T)$ be convex and $\bigcup_{\lambda > 0} \lambda\big[D(S) - D(T)\big]$ be a closed subspace of E. Then $S + T$ is maximally monotone.*

Remark 47.8. We do not know if S is necessarily of type (ANA) if $D(S)$ is a subspace of E and $S \colon D(S) \to E^*$ is linear and maximally monotone.

Example 47.9. Let $E = \ell^1$, and $S \colon \ell^1 \to \ell^\infty$ be the tail operator defined in Remark 36.6. Then S is of type (ANA) but not of type (BR).

Theorem 48.4: Properties of subdifferentials. *Let E be a nonzero Banach space and $f \in \mathcal{PCLSC}(E)$.*
(b) ∂f *is maximally monotone of type (ED).*
(c) ∂f *is maximally monotone of type (FP).*
(d) ∂f *is maximally monotone of type (FPV).*
(e) ∂f *is strongly maximally monotone.*
(f) ∂f *is maximally monotone of type (ANA).*
(g) ∂f *is maximally monotone of type (BR).*

Theorem 48.6. *Let E be a nonzero Banach space, $f \in \mathcal{PCLSC}(E)$, $b \in E \times E^* \setminus G(\partial f)$ and $\alpha, \beta > 0$. Then:*
(a) *There exists $s \in G(\partial f)$ such that $s_1 \neq b_1$, $s_2 \neq b_2$,*

$$\frac{\|s_1 - b_1\|}{\|s_2 - b_2\|} \ \text{as near as we please to} \ \frac{\alpha}{\beta}$$

and

$$\frac{q(s - b)}{\|s_1 - b_1\| \|s_2 - b_2\|} \ \text{as near as we please to} \ -1.$$

(b) *If, further,* $\inf_{s \in G(\partial f)} q(s - b) > -\alpha\beta$, *then we can take s so that, in addition,* $\|s_1 - b_1\| < \alpha$ *and* $\|s_2 - b_2\| < \beta$.

Corollary 48.8: An almost considerable extension of the Brøndsted–Rockafellar theorem. *Let E be a nonzero Banach space, $f \in \mathcal{PCLSC}(E)$, $b \in E \times E^* \setminus G(\partial f)$, $\alpha, \beta > 0$ and $f(b_1) + f^*(b_2) < q(b) + \alpha\beta$. Then there exists $s \in G(\partial f)$ such that $\|s_1 - b_1\| \in \,]0, \alpha[$, $\|s_2 - b_2\| \in \,]0, \beta[$,*

$$\frac{\|s_1 - b_1\|}{\|s_2 - b_2\|} \ \text{as near as we please to} \ \frac{\alpha}{\beta}$$

and

$$\frac{q(s - b)}{\|s_1 - b_1\| \|s_2 - b_2\|} \ \text{as near as we please to} \ -1.$$

Theorem 49.6. *Let E and F be nonzero Banach spaces, F be reflexive and k be a closed saddle–function on $E \times F$ such that $\operatorname{dom} k \neq \emptyset$. Then the "subdifferential" of k is maximally monotone of type (ED).*

Theorem 51.1: Voisei's theorem. *Let E be a nonzero Banach space and $S, T: E \rightrightarrows E^*$ be maximally monotone. Let $D(S)$ and $D(T)$ be closed and convex and $\bigcup_{\lambda > 0} \lambda[D(S) - D(T)]$ be a closed subspace of E. Then $S + T$ is maximally monotone.*

Corollary 51.2. *Let E be a nonzero Banach space, $S: E \rightrightarrows E^*$ be maximally monotone and $D(S)$ be closed and convex. Then S is maximally monotone of type (FPV).*

Theorem 53.1: Verona–Verona's theorem. *Let E be a nonzero Banach space, $f \in \mathcal{PCLSC}(E)$, $T: E \rightrightarrows E^*$ be maximally monotone and $D(T) = E$. Then the multifunction $\partial f + T$ is maximally monotone.*

References

[1] H. Attouch and H. Brézis, *Duality for the sum of convex funtions in general Banach spaces*, Aspects of Mathematics and its Applications, J. A. Barroso, ed., Elsevier Science Publishers (1986), 125–133.

[2] H. Attouch, H. Riahi and M. Théra, *Somme ponctuelle d'opérateurs maximaux monotones*, Serdica Math. J. **22** (1996), 267–292.

[3] J.-P. Aubin and I. Ekeland, *Applied Nonlinear Analysis*, Wiley, New York – Chichester – Brisbane – Toronto – Singapore (1984).

[4] J.-P. Aubin and H. Frankowska, *Set–Valued Analysis*, Birkhäuser, Boston – Basel – Berlin (1990).

[5] S. Bartz, H. H. Bauschke, J. M. Borwein, S. Reich and X. Wang, *Fitzpatrick functions, cyclic monotonicity and Rockafellar's antiderivative*, Nonlinear Anal. **66** (2007), 1198–1223.

[6] H. H. Bauschke, *Projection algorithms and monotone operators*, PhD thesis, Simon Fraser University, Burnaby BC, Canada, August 1996.

[7] ——, *Fenchel duality, Fitzpatrick functions and the extension of firmly nonexpansive mappings*, Proc. Amer. Math. Soc. **135** (2007), 135–139.

[8] H. H. Bauschke and J. M. Borwein, *Continuous linear monotone operators on Banach spaces*, preprint.

[9] ——, *Maximal monotonicity of dense type, local maximal monotonicity, and monotonicity of the conjugate are all the same for continuous linear operators*, Pacific J. Math. **189** (1999), 1–20.

[10] H. H. Bauschke and P. L. Combettes, *Convex Analysis and Monotone Operator Theory in Hilbert Spaces*, to appear.

[11] H. H. Bauschke and S. Simons, *Stronger maximal monotonicity properties of linear operators*, Bull. Austral. Math. Soc. **60** (1999), 163–174.

[12] H. H. Bauschke, D. A. McLaren and H. S. Sendov, *Fitzpatrick functions: inequalities, examples, and remarks on a problem by S. Fitzpatrick*, J. Convex Anal. **13** (2006), 499–523.

[13] H. H. Bauschke and X. Wang, *The kernel average for two convex functions and its application to the extension and representation of monotone operators*, preprint, May 9, 2007.

[14] G. Beer, *The slice topology: A viable alternative to Mosco convergence in nonreflexive spaces*, Nonlinear Anal. **19** (1992), 271–290.

[15] ——, *Topologies on closed and closed convex sets*, Mathematics and Its Applications **268** (1993), Kluwer Academic Publishers.

[16] J. M. Borwein, *A Lagrange multiplier theorem and a sandwich theorem for convex relations*, Math. Scand. **48** (1981), 198–204.

[17] ——, *Maximal monotonicity via convex analysis*, J. Convex Anal. **13** (2006), 561–586.

[18] ——, *Maximality of sums of two maximal monotone operators in general Banach space*, Proc. Amer. Math. Soc. **135** (2007), 3917–3924.

[19] J. M. Borwein et al., *Simon Fitzpatrick memorial volume*, J. Convex Anal. **13** (2006), 463–476.

[20] J. M. Borwein and A. C. Eberhard, *Maximality of Monotone Operators in General Banach Space*, <http://docserver.cs.dal.ca>.

[21] J. M. Borwein and S. Fitzpatrick, *Local boundedness of monotone operators under minimal hypotheses*, Bull. Austral. Math. Soc. **39** (1988), 439–441.

[22] J. Borwein, S. Fitzpatrick and J. Vanderwerff, *Examples of convex functions and classification of normed spaces*, J. Convex Anal. **1** (1994), 61–73.

[23] J. M. Borwein and A. S. Lewis, *Convex analysis and nonlinear optimization. Theory and examples*. Second edition. CMS Books in Mathematics/Ouvrages de Mathmatiques de la SMC **3**. Springer, New York, 2006.

[24] J. M. Borwein and Q. J. Zhu, *Techniques of variational analysis*. CMS Books in Mathematics/Ouvrages de Mathématiques de la SMC **20**. Springer-Verlag, New York, 2005.

[25] R. I. Boţ, E. R. Csetnek and G. Wanka, *A new condition for maximal monotonicity via representative functions*, Nonlinear Anal. **67** (2007), 2390–2402.

[26] R. I. Boţ, S.–M. Grad and G. Wanka, *Maximal monotonicity for the precomposition with a linear operator*, Siam J. Optim. **17** (2007), 1239–1252.

[27] R. I. Boţ and G. Wanka, *A weaker regularity condition for subdifferential calculus and Fenchel duality in infinite dimensional spaces*, Nonlinear Anal. **64** (2006), 2787–2804.

[28] ——, *The conjugate of the pointwise maximum of two convex functions revisited*, preprint.

[29] H. Brezis, M. G. Crandall and A. Pazy, *Perturbations of nonlinear maximal monotone sets in Banach spaces*, Comm. Pur. Appl. Math. **23** (1970), 123–144.

[30] H. Brezis and A. Haraux, *Image d'une somme d'opérateurs monotone et applications*, Israel J. Math. **23** (1976), 165–186.

[31] A. Brøndsted and R.T. Rockafellar, *On the Subdifferentiability of Convex Functions*, Proc. Amer. Math. Soc. **16** (1965), 605–611.

[32] F. E. Browder, *Nonlinear maximal monotone operators in Banach spaces*, Math. Annalen **175** (1968), 89–113.

[33] R. S. Burachik and B. F. Svaiter, *Maximal monotonicity, conjugation and the duality product*, Proc. Amer. Math. Soc. **131** (2003), 2379–2383.

[34] R. S. Burachik, V. Jeyakumar and Z-Y. Wu, *Necessary and Sufficient Conditions for Stable Conjugate Duality*, Nonlinear Anal. **64** (2006), 1998–2006.

[35] L.-J. Chu, *On the sum of monotone operators*, Michigan Math. J., **43** (1996), 273–289.

[36] M. Coodey, *Examining maximal monotone operators using pictures and convex functions*, Ph. D. dissertation, University of California, Santa Barbara, June 1997.

[37] M. Coodey and S. Simons, *The convex function determined by a multifunction*, Bull. Austral. Math. Soc. **54** (1996), 87–97.

[38] K. Deimling, *Nonlinear Functional Analysis*, Springer–Verlag, New York – Heidelberg – Berlin – Tokyo (1985).

[39] I. Ekeland, *Nonconvex minimization problems*, Bull. Amer. Math. Soc. **1** (1979), 443–474.

[40] K. Fan, *Minimax theorems*, Proc. Nat. Acad. Sci. U.S.A. **39** (1953), 42–47.

[41] K. Fan, I. Glicksberg and A. J. Hoffman, *Systems of inequalities involving convex functions*, Proc. Amer. Math. Soc. **8** (1957), 617–622.

[42] S. Fitzpatrick, *Representing monotone operators by convex functions*, Workshop/Miniconference on Functional Analysis and Optimization (Canberra, 1988), 59–65, Proc. Centre Math. Anal. Austral. Nat. Univ. **20**, Austral. Nat. Univ., Canberra, 1988.

[43] *Simon Fitzpatrick memorial volume*, J. Convex Anal. **13** (2006), No. 3–4.

[44] S. P. Fitzpatrick and R. R. Phelps, *Bounded approximants to monotone operators on Banach spaces*, Ann. Inst. Henri Poincaré, Analyse non linéaire **9** (1992), 573–595.

[45] ——, *Some properties of maximal monotone operators on nonreflexive Banach spaces*, Set–Valued Anal. **3** (1995), 51–69.

[46] S. P. Fitzpatrick and S. Simons, *On the pointwise maximum of convex functions*, Proc. Amer. Math. Soc. **128** (2000), 3553–3561.

[47] N. Ghoussoub, *Maximal monotone operators are selfdual vector fields and vice-versa*, Proc. Amer. Math. Soc., in press.

[48] J.- P. Gossez, *Opérateurs monotones non linéaires dans les espaces de Banach non réflexifs*, J. Math. Anal. Appl. **34** (1971), 371–395.

[49] ——, *On the range of a coercive maximal monotone operator in a nonreflexive Banach space*, Proc. Amer. Math. Soc. **35** (1972), 88–92.

[50] ——, *On a convexity property of the range of a maximal monotone operator*, Proc. Amer. Math. Soc. **55** (1976), 359–360.

[51] R. B. Holmes, *Geometric functional analysis and its applications*, Graduate Texts in Mathematics **24** (1975), Springer–Verlag, New York – Heidelberg.

[52] J. L. Kelley, I. Namioka, and co-authors, *Linear Topological Spaces*, D. Van Nostrand Co., Inc., Princeton – Toronto – London – Melbourne (1963).

[53] H. König, *Über das Von Neumannsche Minimax-Theorem*, Arch. Math. **19** (1968), 482–487.

[54] ——, *On certain applications of the Hahn–Banach and minimax theorems*, Arch. Math. **21** (1970), 583–591.

[55] ——, *Sublineare Funktionale*, Arch. Math. **23** (1972), 500–508.

[56] ——, *Some Basic Theorems in Convex Analysis*, in "Optimization and operations research", edited by B. Korte, North-Holland (1982).

[57] S. Kum, *Maximal monotone operators in the one–dimensional case*, J. Korean Math. Soc. **34** (1997), 371–381.

[58] D. T. Luc, *A resolution of Simons' maximal monotonicity problem*, J. Convex Anal. **3** (1996), 367–370.

[59] D. L. Luenberger, *Optimization by Vector Space Methods*, John Wiley & Sons, Inc, New York – Chichester – Brisbane – Toronto – Singapore (1969).

[60] M. Marques Alves and B. F. Svaiter, *A new proof for maximal monotonicity of subdifferential operators*, IMPA preprint server, A526/2007, <http://www.preprint.impa.br/Shadows/SERIE_A/2007/526.html>.

[61] J-E. Martínez-Legaz and B. F. Svaiter, *Monotone operators representable by l.s.c. convex functions*, Set–Valued Anal. **13** (2005), 21–46.

[62] J.-E. Martínez-Legaz and M. Théra, *ε–Subdifferentials in terms of subdifferentials*, Set–Valued Anal. **4** (1996), 327–332.

[63] ——, *A convex representation of maximal monotone operators*, J. Nonlinear Convex Anal. **2** (2001), 243–247.

[64] J.-J. Moreau, *Fonctionelles convexes*, Séminaire sur les équations aux derivées partielles, Lecture notes, Collège de France, Paris 1966-7.

[65] J.-P. Penot, *Autoconjugate functions and representations of monotone operators*, Bull. Austral. Math. Soc. **67** (2003), 277–284.

[66] ——, *The relevance of convex analysis for the study of monotonicity*, Nonlinear Anal. **58** (2004), 855–871.

[67] J.–P. Penot and C. Zălinescu, *Some problems about the representation of monotone operators by convex functions*, ANZIAM J. **47** (2005), 1–20.

[68] R. R. Phelps, *Convex Functions, Monotone Operators and Differentiability*, Lecture Notes in Mathematics **1364** (1993), Springer–Verlag (Second Edition).

[69] ——, *Lectures on Maximal Monotone Operators*, Extracta Mathematicae **12** (1997), 193–230.

[70] R. R. Phelps and S. Simons, *Unbounded linear monotone operators on nonreflexive Banach spaces*, J. Convex Anal. **5** (1998), 303–328.

[71] J. D. Pryce, *Weak compactness in locally convex spaces*, Proc. Amer. Math. Soc. **17** (1966), 148–155.

[72] S. Reich, *The range of sums of accretive and monotone operators*, J. Math. Anal. Appl. **68** (1979), 310–317.

[73] S. Reich and S. Simons, *Fenchel duality, Fitzpatrick functions and the Kirszbraun-Valentine extension theorem*, Proc. Amer. Math. Soc. **133** (2005), 2657–2660.

[74] J. P. Revalski and M. Théra, *Enlargements and sums of monotone operators*, Nonlinear Anal. **48** (2002), Ser. A, 505–519.

[75] S. M. Robinson, *Regularity and stability for convex multivalued functions*, Math. Oper. Res. **1** (1976), 130–143.

[76] R. T. Rockafellar, *Level sets and continuity of conjugate convex functions*. Trans. Amer. Math. Soc. **123** (1966), 46–63.

[77] ——, *Extension of Fenchel's duality theorem for convex functions*, Duke Math. J. **33** (1966), 81–89.

[78] ——, *Local boundedness of Nonlinear, Monotone Operators*, Michigan Math. J. **16** (1969), 397–407.

[79] ——, *On the Virtual Convexity of the Domain and Range of a Nonlinear Maximal Monotone Operator*, Math. Ann. **185** (1970), 81–90.

[80] ——, *On the Maximality of Sums of Nonlinear Monotone Operators*, Trans. Amer. Math. Soc. **149** (1970), 75–88.

[81] ——, *On the maximal monotonicity of subdifferential mappings*, Pac. J. Math. **33** (1970), 209-216.

[82] ——, *Monotone operators associated with saddle–functions and minimax problems*, Proc. Symp. Pure Math. **18** (1970), 241-250.

[83] R. T. Rockafellar and R. J.-B. Wets, *Variational analysis*. Grundlehren der Mathematischen Wissenschaften/Fundamental Principles of Mathematical Sciences **317**. Springer-Verlag, Berlin, 1998.

[84] W. Rudin, *Functional analysis*, McGraw-Hill, New York (1973).

[85] M. Ruiz Galán and S. Simons, *A new minimax theorem and a perturbed James's theorem*, Bull. Austral. Math. Soc. **66** (2002), 43–56.

[86] E. Saab and P. Saab, *On stability problems of some properties in Banach spaces*, in *Function spaces*, K. Jarosz, ed., Lecture notes in pure and applied mathematics **136** (1992), 367–394, Marcel Dekker, New York.

[87] S. Simons, *Extended and sandwich versions of the Hahn–Banach Theorem*, J. Math. Anal. Appl. **21** (1968), 112–122.

[88] ——, *Minimal sublinear functionals*, Studia Math. **37** (1970), 37–56.

[89] ——, *Formes souslinéaires minimales*, Séminaire Choquet 1970/1971, no. 23, 8 pages.

[90] ——, *Critères de faible compacité en termes du théorème de minimax*, Séminaire Choquet 1970/1971, no. 24, 5 pages.

[91] ——, *The least slope of a convex function and the maximal monotonicity of its subdifferential*, J. of Optimization Theory **71** (1991), 127–136.

[92] ——, *Subdifferentials are locally maximal monotone*, Bull. Austral. Math. Soc. **47** (1993), 465–471.

[93] ——, *Swimming below icebergs*, Set-Valued Anal. **2** (1994), 327–337.

[94] ——, *Subtangents with controlled slope*, Nonlinear Anal. **22** (1994), 1373–1389.

[95] ——, *Minimax theorems and their proofs*, Minimax and applications, Ding-Zhu Du and Panos M. Pardalos eds., Kluwer Academic Publishers, Dordrecht – Boston (1995), 1–23.

[96] ——, *The range of a monotone operator*, J. Math. Anal. Appl. **199** (1996), 176–201.

[97] ——, *Subdifferentials of convex functions*, in "Recent Developments in Optimization Theory and Nonlinear Analysis", Y. Censor and S. Reich eds, American Mathematical Society, Providence, Rhode Island, Contemporary Mathematics **204** (1997), 217–246.

[98] ——, *Pairs of monotone operators*, Trans. Amer. Math. Soc. **350** (1998), 2973–2980.

[99] ——, *Minimax and monotonicity*, Lecture Notes in Mathematics **1693** (1998), Springer–Verlag.

[100] ——, *Maximal monotone multifunctions of Brøndsted–Rockafellar type*, Set-Valued Anal. **7** (1999), 255–294.

[101] ——, *Five kinds of maximal monotonicity*, Set-Valued Anal. **9** (2001), 391–409.

[102] ——, *Excesses, duality gaps and weak compactness*, Proc. Amer. Math. Soc. **130** (2002), 2941–2946.

[103] ——, *A new version of the Hahn–Banach theorem*, Arch. Math. **80** (2003), 630–646.

[104] ——, *Hahn–Banach theorems and maximal monotonicity*, in *Variational analysis and applications*, Nonconvex Optim. Appl. **79**, 1049–1083. Springer, New York, 2005.

[105] ——, *Dualized and scaled Fitzpatrick functions*, Proc. Amer. Math. Soc. **134** (2006), 2983–2987.

[106] ——, *LC–functions and maximal monotonicity*, J. Nonlinear Convex Anal. **7** (2006), 123–138.

[107] ——, *The Fitzpatrick function and nonreflexive spaces*, J. Convex Anal. **13** (2006), 861–881.

[108] ——, *Positive sets and Monotone sets*, J. Convex Anal. **14** (2007), 297–317

[109] S. Simons and C. Zălinescu, *Fenchel duality, Fitzpatrick functions and maximal monotonicity*, J. Nonlinear and Convex Anal. **6** (2005), 1–22.

[110] B. F. Svaiter, *Fixed points in the family of convex representations of a maximal monotone operator*, Proc. Amer. Math. Soc. **131** (2003), 3851–3859.

[111] D. Torralba, *Convergence épigraphique et changements d'échelle en analyse variationnelle et optimisation*, Thesis, Université Montpellier II.

[112] C. Ursescu, *Multifunctions with convex closed graph*, Czechoslovak Math. J. **25** (1975), 438–441.

[113] A. and M. E. Verona, *Remarks on subgradients and ε-subgradients*, Set-Valued Anal. **1** (1993), 261–272.

[114] ——, *Regular maximal monotone operators*, Set-Valued Anal. **6** (1998), 303–312.

[115] ——, *Regular maximal monotone operators and the sum theorem*, J. Convex Anal. **7** (2000), 115–128.

[116] M. D. Voisei, *A maximality theorem for the sum of maximal monotone operators in non–reflexive Banach spaces*, Math. Sci. Res. J. **10** (2006), 36–41.

[117] A. Wilansky, *Modern methods in topological vector spaces*, McGraw–Hill, 1978.

[118] D. Zagrodny, *The maximal monotonicity of the subdifferentials of convex functions: Simons' problem*, Set–Valued Anal. **4** (1996), 301–314.

[119] C. Zălinescu, *Convex analysis in general vector spaces*, (2002), World Scientific.

[120] ——, *A new proof of the maximal monotonicity of the sum using the Fitzpatrick function*, in *Variational analysis and applications*, Nonconvex Optim. Appl. **79** 1159–1172. Springer, New York, 2005.

[121] ——, *A new convexity property for monotone operators*, J. Convex Anal. **13** (2006), 883–887.

[122] E. Zeidler, *Nonlinear Functional Analysis and its Applications*, Vol II/B Nonlinear Monotone Operators, Springer–Verlag, New York–Berlin–Heidelberg (1990).

Subject Index

Absorbing point of a set, 60
Absorbing set, 60
aff, affine hull, 111
Affine function, 20
Asplund decomposition, 13
Asplund's renorming theorem, 118
Attouch–Brezis theorem, 66
Attouch–Riahi–Théra constraint
    qualification, 130
Autoconjugate, 91

Banach–Alaoglu theorem, 26
BC–function, 82
$\overline{BC}$–function, 140
Biconjugate
– restricted, 58
Bidual, 26
Bishop–Phelps theorem, 78
Borwein–Fitzpatrick local boundedness
    theorem, 108
Borwein–Preiss smooth variational
    principle, 75
Brezis–Crandall–Pazy constraint
    qualification, 137
Brøndsted–Rockafellar theorem, full
    version, 76
Brøndsted–Rockafellar theorem, vanilla
    version, 72, 76

Caristi–Kirk fixed–point theorem, 75
Chu's constraint qualification, 130
co, convex hull, 84
Concave function, 20
Conjugate (function), 42, 44, 79
Constraint qualification
– Attouch–Riahi–Théra, 130
– Brezis–Crandall–Pazy, 137
– Chu's, 130
– Rockafellar's, 104
Convex function, 20
Cyclical monotonicity, 13

Decoupling, 46
Deville–Godefroy–Zizler variational
    principle, 75
dom (effective domain of a function),
    20
Dom corollary, 61
Dom lemma, 61
Domain of a multifunction, 71
Drop theorem, 75
Dual space, 25
Duality gap, 28
Duality map, 101

Ekeland's variational principle, 73, 75
Excess, 28
Extended sublinear functional, 17
Extension form of the Hahn–Banach
    theorem, 23

Fan's minimax theorem, 25
Fenchel conjugate, 42, 44, 79
Fenchel duality theorem
– Rockafellar's version, 47
– sharp, 43, 50
Fenchel functional for f and g, 42, 44
Fenchel instability, total, 55
Fenchel–Moreau theorem, 60
Fenchel–Young inequality, 44
Fitpatrification of a multifunction, 100
Fitzpatrick function, 100
Flower–petal theorem, 75

Graph of a multifunction, 71

Hahn–Banach theorem
– extension form, 23
– one–dimensional form, 23
– sublinear form, 16
Hahn–Banach–Lagrange theorem, 21

Indicator function, 54, 73

Inverse graph of a multifunction, 71
Inverse of a multifunction, 71

James's theorem, 27

Karush–Kuhn–Tucker theorem, sharp, 37
KKT functional, 37

Lagrange multiplier, 33, 38
Lagrange multiplier existence theorem, sharp, 33
lin, linear hull, 111
Local boundedness theorem
– Borwein–Fitzpatrick, 108
Locally bounded multifunction, 108
Locally maximally monotone (see "Type (FP)", 149

Maximally monotone multifunction, 72
Maximally monotone subset, 72
Mazur–Orlicz theorem, 19
Mazur–Orlicz–König theorem, 22
Minimax theorem, 25
Minty's theorem (for Hilbert spaces), 118
Monotone multifunction, 72
Monotone subset, 71
Monotone variational inequalities, 13
Monotonically related to, 74
Multifunction, 71
– domain of, 71
– fitzpatrification of, 100
– graph of, 71
– inverse graph of, 71
– inverse of, 71
– locally bounded, 108
– locally maximally monotone, 149
– maximally monotone, 72
– monotone, 72
– range of, 71
– strongly maximally monotone, 151
– type (ANA), 152
– type (BR), 153
– type (D), 148
– type (ED), 156
– type (FP), 149
– type (FPV), 150
– type (NI), 148
– ultramaximally monotone, 164

Negative alignment criterion for maximality, 118

Negative alignment pair, 161
Negative alignment set, 165
Nine set theorem, 109
Nine set theorem for pairs of multifunctions, 132
Normality multifunction, 73

$\ominus$–corollary, 64
$\ominus$–theorem, 63
One–dimensional form of the Hahn–Banach theorem, 23
Open mapping theorem, 64

Positive linear operator, 72

q–negative set, 81
q–positive set, 80

Range of a multifunction, 71
Reflexive Banach space, 26
Rockafellar's
– constraint qualification, 104
– maximal monotonicity theorem (for subdifferentials), 76
– sum theorem, 104
– surjectivity theorem, 118, 119
– version of the Fenchel duality theorem, 47

Saddle–function, 195
Sandwich theorem, 23
Sandwiched closed subspace theorem, 129
Saturated, 113
SBC–function, 87, 99
Seminorm, 15
Sharp Fenchel duality theorem, 43, 50
Sharp Karush–Kuhn–Tucker theorem, 37
Sharp Lagrange multiplier existence theorem, 33
Six set theorem, 109
Six set theorem for pairs of multifunctions, 131
Skew linear operator, 72
Slater condition, 35
SSD space, 79
SSDB space, 89
Strongly maximally monotone, 151
Subdifferential of the sum of convex functions, 74
Sublinear form of the Hahn–Banach theorem, 16

Sublinear functional, 15
Subset
– maximally monotone, 72
– monotone, 71
Sum theorem
– Rockafellar's, 104
sur (a point "surrounded" by a set), 60
Surrounding set, 60
Symmetrically self–dual Banach space, 89
Symmetrically self–dual locally convex space, 79

TBC–function, 83
$\widetilde{\mathrm{TBC}}$–function, 140
Total Fenchel instability, 55
Type (ANA), 152

Type (BR), 153
Type (D), 148
Type (ED), 156
Type (FP), 149
Type (FPV), 150
Type (NI), 148

Ultramaximally monotone multifunction, 164
Uniform boundedness theorem, 62

Voisei's theorem, 197

Weak* topology, 26
Weak topology, 26

Zagrodny's approximate mean value theorem, 77

Symbol index

$^*f^*$, the restricted biconjugate of the function f, 58

$\mathcal{CLB}(E)$, the set of all convex functions that are Lipschitz on the bounded subsets of the Banach space E, 155

$D(S)$, the domain of the multifunction S, 71

$D_{E \setminus N}$, the distance from $E \setminus N$, 32

$\partial k(x)$, the subdifferential of the function k at the point x, 41

$\mathrm{dgap}_f(X, Y)$, duality gap of f over $X \times Y$, 28

D_N, the distance from N, 32

D_Y, the distance from Y, 28

$e(Z, Y)$, excess of Z over Y, 28

E^*, the dual space of the Banach space E, 25

E^{**}, the bidual of the Banach space E, 26

f^*, the conjugate of the function f on a normed space, 42

f^*, the conjugate of the function f with respect to a pairing, 44

$f^{@}$, the conjugate of the function f with respect to a symmetric pairing, 79

$F^\perp$, the subspace of E^* orthogonal to the subspace F of the Banach space E, 111

$G(S)$, the graph of the multifunction S, 71

$G^{-1}(S)$, the inverse graph of the multifunction S, 71

g_0, if $b \in B$ then $g_0(b) = \frac{1}{2}\|b\|^2$, 89

$H(S)$, a set determined by the positive linear map S, 184

$\mathbb{I}_\cdot$, the indicator function of the set $\cdot$, 54, 73

ι, the canonical isometry from $E \times E^*$ into $E^{**} \times E^*$, 139

J, the duality map on the Banach space E, 101

$\mathcal{NAS}$, the negative alignment set of T with respect to b, 165

N_C, the normality multifunction of the set C, 73

$\mathrm{neg}\, g$, the q–negative set determined by a certain convex function g, 81

$\ominus$, 46

$\mathcal{PC}(C)$, the set of proper convex functions on a convex set C, 20

$\mathcal{PCLSC}(E)$, the set of all somewhere finite convex lower semicontinuous functions on the Banach space E, 60

Φ_A, the convex function determined by a q–positive set A, 84

φ_S, the Fitzpatrick function associated with a nontrivial monotone multifunction, S, 100

π_1, π_2, projection maps, 67

π_{12}, π_{34}, projection maps, 133

pos f, the q–positive set determined by a certain convex function f, 81

$R(S)$, the range of the multifunction S, 71

S^{-1}, the inverse of the multifunction S, 71

$\overline{S}$, a multifunction defined by Gossez, 147

S_φ, the fiztpatrification of S, 100

$\mathcal{T}_{CC}(E^{**})$, a certain topology on the bidual E^{**} of the Banach space E, 174

$\mathcal{T}_{CCN}(E^{**} \times E^*)$, the topology $\mathcal{T}_{CC}(E^{**}) \times \mathcal{T}_{\| \; \|}(E^*)$ on $E^{**} \times E^*$, 174

$\mathcal{T}_{CLB}(E^{**})$, a certain topology on the bidual E^{**} of the Banach space E, 155

$\mathcal{T}_{CLBN}(E^{**} \times E^*)$, the topology $\mathcal{T}_{CLB}(E^{**}) \times \mathcal{T}_{\| \; \|}(E^*)$ on $E^{**} \times E^*$, 155

$\mathcal{T}_{CLBN}(E^{**} \times H)$, the topology $\mathcal{T}_{CLB}(E^{**}) \times \mathcal{T}_{\| \; \|}(H)$ on $E^{**} \times H$, 192

$\mathcal{T}_{\| \; \|}$, the norm topology of, 79

$\mathcal{T}_{NW}(B)$, the topology $\mathcal{T}_{\| \; \|}(E) \times w(E^*, E)$ on $E \times E^*$, 79

$\mathcal{T}_{WN}(B^*)$, the topology $w(E^{**}, E^*) \times \mathcal{T}_{\| \; \|}(E^*)$ on $E \times E^*$, 139

$\vee$, 24

$w(E^*, E)$, the weak* topology of the dual, E^*, of the Banach space E, 26

$w(E, E^*)$, the weak topology of the Banach space E, 26

$\wedge$, 24

$\widehat{x}$, the canonical image of x in the bidual, 26

$\lfloor \cdot, \cdot \rfloor$, the symmetric bilinear form on a SSD space, 79

$\lfloor \cdot, \cdot \rfloor$, if $b = (b_1, b_2) \in E \times E^*$ and $c = (c_1, c_2) \in E \times E^*$ then $\lfloor b, c \rfloor :=$ $\langle b_1, c_2 \rangle + \langle c_1, b_2 \rangle$, 93

$\lfloor \cdot, \cdot \rceil$, if $u = (u_1, u_2) \in E^{**} \times E^*$ and $v = (v_1, v_2) \in E^{**} \times E^*$ then $\lfloor u, v \rceil :=$ $\langle v_2, u_1 \rangle + \langle u_2, v_1 \rangle$, 139

Lecture Notes in Mathematics

For information about earlier volumes
please contact your bookseller or Springer
LNM Online archive: springerlink.com

Vol. 1735: D. Yafaev, Scattering Theory: Some Old and New Problems (2000)

Vol. 1736: B. O. Turesson, Nonlinear Potential Theory and Weighted Sobolev Spaces (2000)

Vol. 1737: S. Wakabayashi, Classical Microlocal Analysis in the Space of Hyperfunctions (2000)

Vol. 1738: M. Émery, A. Nemirovski, D. Voiculescu, Lectures on Probability Theory and Statistics (2000)

Vol. 1739: R. Burkard, P. Deuflhard, A. Jameson, J.-L. Lions, G. Strang, Computational Mathematics Driven by Industrial Problems. Martina Franca, 1999. Editors: V. Capasso, H. Engl, J. Periaux (2000)

Vol. 1740: B. Kawohl, O. Pironneau, L. Tartar, J.-P. Zolesio, Optimal Shape Design. Tróia, Portugal 1999. Editors: A. Cellina, A. Ornelas (2000)

Vol. 1741: E. Lombardi, Oscillatory Integrals and Phenomena Beyond all Algebraic Orders (2000)

Vol. 1742: A. Unterberger, Quantization and Non-holomorphic Modular Forms (2000)

Vol. 1743: L. Habermann, Riemannian Metrics of Constant Mass and Moduli Spaces of Conformal Structures (2000)

Vol. 1744: M. Kunze, Non-Smooth Dynamical Systems (2000)

Vol. 1745: V. D. Milman, G. Schechtman (Eds.), Geometric Aspects of Functional Analysis. Israel Seminar 1999-2000 (2000)

Vol. 1746: A. Degtyarev, I. Itenberg, V. Kharlamov, Real Enriques Surfaces (2000)

Vol. 1747: L. W. Christensen, Gorenstein Dimensions (2000)

Vol. 1748: M. Ruzicka, Electrorheological Fluids: Modeling and Mathematical Theory (2001)

Vol. 1749: M. Fuchs, G. Seregin, Variational Methods for Problems from Plasticity Theory and for Generalized Newtonian Fluids (2001)

Vol. 1750: B. Conrad, Grothendieck Duality and Base Change (2001)

Vol. 1751: N. J. Cutland, Loeb Measures in Practice: Recent Advances (2001)

Vol. 1752: Y. V. Nesterenko, P. Philippon, Introduction to Algebraic Independence Theory (2001)

Vol. 1753: A. I. Bobenko, U. Eitner, Painlevé Equations in the Differential Geometry of Surfaces (2001)

Vol. 1754: W. Bertram, The Geometry of Jordan and Lie Structures (2001)

Vol. 1755: J. Azéma, M. Émery, M. Ledoux, M. Yor (Eds.), Séminaire de Probabilités XXXV (2001)

Vol. 1756: P. E. Zhidkov, Korteweg de Vries and Nonlinear Schrödinger Equations: Qualitative Theory (2001)

Vol. 1757: R. R. Phelps, Lectures on Choquet's Theorem (2001)

Vol. 1758: N. Monod, Continuous Bounded Cohomology of Locally Compact Groups (2001)

Vol. 1759: Y. Abe, K. Kopfermann, Toroidal Groups (2001)

Vol. 1760: D. Filipović, Consistency Problems for Heath-Jarrow-Morton Interest Rate Models (2001)

Vol. 1761: C. Adelmann, The Decomposition of Primes in Torsion Point Fields (2001)

Vol. 1762: S. Cerrai, Second Order PDE's in Finite and Infinite Dimension (2001)

Vol. 1763: J.-L. Loday, A. Frabetti, F. Chapoton, F. Goichot, Dialgebras and Related Operads (2001)

Vol. 1764: A. Cannas da Silva, Lectures on Symplectic Geometry (2001)

Vol. 1765: T. Kerler, V. V. Lyubashenko, Non-Semisimple Topological Quantum Field Theories for 3-Manifolds with Corners (2001)

Vol. 1766: H. Hennion, L. Hervé, Limit Theorems for Markov Chains and Stochastic Properties of Dynamical Systems by Quasi-Compactness (2001)

Vol. 1767: J. Xiao, Holomorphic Q Classes (2001)

Vol. 1768: M. J. Pflaum, Analytic and Geometric Study of Stratified Spaces (2001)

Vol. 1769: M. Alberich-Carramiñana, Geometry of the Plane Cremona Maps (2002)

Vol. 1770: H. Gluesing-Luerssen, Linear Delay-Differential Systems with Commensurate Delays: An Algebraic Approach (2002)

Vol. 1771: M. Émery, M. Yor (Eds.), Séminaire de Probabilités 1967-1980. A Selection in Martingale Theory (2002)

Vol. 1772: F. Burstall, D. Ferus, K. Leschke, F. Pedit, U. Pinkall, Conformal Geometry of Surfaces in S^4 (2002)

Vol. 1773: Z. Arad, M. Muzychuk, Standard Integral Table Algebras Generated by a Non-real Element of Small Degree (2002)

Vol. 1774: V. Runde, Lectures on Amenability (2002)

Vol. 1775: W. H. Meeks, A. Ros, H. Rosenberg, The Global Theory of Minimal Surfaces in Flat Spaces. Martina Franca 1999. Editor: G. P. Pirola (2002)

Vol. 1776: K. Behrend, C. Gomez, V. Tarasov, G. Tian, Quantum Comohology. Cetraro 1997. Editors: P. de Bartolomeis, B. Dubrovin, C. Reina (2002)

Vol. 1777: E. García-Río, D. N. Kupeli, R. Vázquez-Lorenzo, Osserman Manifolds in Semi-Riemannian Geometry (2002)

Vol. 1778: H. Kiechle, Theory of K-Loops (2002)

Vol. 1779: I. Chueshov, Monotone Random Systems (2002)

Vol. 1780: J. H. Bruinier, Borcherds Products on O(2,l) and Chern Classes of Heegner Divisors (2002)

Vol. 1781: E. Bolthausen, E. Perkins, A. van der Vaart, Lectures on Probability Theory and Statistics. Ecole d' Eté de Probabilités de Saint-Flour XXIX-1999. Editor: P. Bernard (2002)

Vol. 1782: C.-H. Chu, A. T.-M. Lau, Harmonic Functions on Groups and Fourier Algebras (2002)

Vol. 1783: L. Grüne, Asymptotic Behavior of Dynamical and Control Systems under Perturbation and Discretization (2002)

Vol. 1784: L. H. Eliasson, S. B. Kuksin, S. Marmi, J.-C. Yoccoz, Dynamical Systems and Small Divisors. Cetraro, Italy 1998. Editors: S. Marmi, J.-C. Yoccoz (2002)

Vol. 1785: J. Arias de Reyna, Pointwise Convergence of Fourier Series (2002)

Vol. 1786: S. D. Cutkosky, Monomialization of Morphisms from 3-Folds to Surfaces (2002)

Vol. 1787: S. Caenepeel, G. Militaru, S. Zhu, Frobenius and Separable Functors for Generalized Module Categories and Nonlinear Equations (2002)

Vol. 1788: A. Vasil'ev, Moduli of Families of Curves for Conformal and Quasiconformal Mappings (2002)

Vol. 1789: Y. Sommerhäuser, Yetter-Drinfel'd Hopf algebras over groups of prime order (2002)

Vol. 1790: X. Zhan, Matrix Inequalities (2002)

Vol. 1791: M. Knebusch, D. Zhang, Manis Valuations and Prüfer Extensions I: A new Chapter in Commutative Algebra (2002)

Vol. 1792: D. D. Ang, R. Gorenflo, V. K. Le, D. D. Trong, Moment Theory and Some Inverse Problems in Potential Theory and Heat Conduction (2002)

Vol. 1793: J. Cortés Monforte, Geometric, Control and Numerical Aspects of Nonholonomic Systems (2002)

Vol. 1794: N. Pytheas Fogg, Substitution in Dynamics, Arithmetics and Combinatorics. Editors: V. Berthé, S. Ferenczi, C. Mauduit, A. Siegel (2002)

Vol. 1795: H. Li, Filtered-Graded Transfer in Using Noncommutative Gröbner Bases (2002)

Vol. 1796: J.M. Melenk, hp-Finite Element Methods for Singular Perturbations (2002)

Vol. 1797: B. Schmidt, Characters and Cyclotomic Fields in Finite Geometry (2002)

Vol. 1798: W.M. Oliva, Geometric Mechanics (2002)

Vol. 1799: H. Pajot, Analytic Capacity, Rectifiability, Menger Curvature and the Cauchy Integral (2002)

Vol. 1800: O. Gabber, L. Ramero, Almost Ring Theory (2003)

Vol. 1801: J. Azéma, M. Émery, M. Ledoux, M. Yor (Eds.), Séminaire de Probabilités XXXVI (2003)

Vol. 1802: V. Capasso, E. Merzbach, B. G. Ivanoff, M. Dozzi, R. Dalang, T. Mountford, Topics in Spatial Stochastic Processes. Martina Franca, Italy 2001. Editor: E. Merzbach (2003)

Vol. 1803: G. Dolzmann, Variational Methods for Crystalline Microstructure – Analysis and Computation (2003)

Vol. 1804: I. Cherednik, Ya. Markov, R. Howe, G. Lusztig, Iwahori-Hecke Algebras and their Representation Theory. Martina Franca, Italy 1999. Editors: V. Baldoni, D. Barbasch (2003)

Vol. 1805: F. Cao, Geometric Curve Evolution and Image Processing (2003)

Vol. 1806: H. Broer, I. Hoveijn. G. Lunther, G. Vegter, Bifurcations in Hamiltonian Systems. Computing Singularities by Gröbner Bases (2003)

Vol. 1807: V. D. Milman, G. Schechtman (Eds.), Geometric Aspects of Functional Analysis. Israel Seminar 2000-2002 (2003)

Vol. 1808: W. Schindler, Measures with Symmetry Properties (2003)

Vol. 1809: O. Steinbach, Stability Estimates for Hybrid Coupled Domain Decomposition Methods (2003)

Vol. 1810: J. Wengenroth, Derived Functors in Functional Analysis (2003)

Vol. 1811: J. Stevens, Deformations of Singularities (2003)

Vol. 1812: L. Ambrosio, K. Deckelnick, G. Dziuk, M. Mimura, V. A. Solonnikov, H. M. Soner, Mathematical Aspects of Evolving Interfaces. Madeira, Funchal, Portugal 2000. Editors: P. Colli, J. F. Rodrigues (2003)

Vol. 1813: L. Ambrosio, L. A. Caffarelli, Y. Brenier, G. Buttazzo, C. Villani, Optimal Transportation and its Applications. Martina Franca, Italy 2001. Editors: L. A. Caffarelli, S. Salsa (2003)

Vol. 1814: P. Bank, F. Baudoin, H. Föllmer, L.C.G. Rogers, M. Soner, N. Touzi, Paris-Princeton Lectures on Mathematical Finance 2002 (2003)

Vol. 1815: A. M. Vershik (Ed.), Asymptotic Combinatorics with Applications to Mathematical Physics. St. Petersburg, Russia 2001 (2003)

Vol. 1816: S. Albeverio, W. Schachermayer, M. Talagrand, Lectures on Probability Theory and Statistics. Ecole d'Eté de Probabilités de Saint-Flour XXX-2000. Editor: P. Bernard (2003)

Vol. 1817: E. Koelink, W. Van Assche (Eds.), Orthogonal Polynomials and Special Functions. Leuven 2002 (2003)

Vol. 1818: M. Bildhauer, Convex Variational Problems with Linear, nearly Linear and/or Anisotropic Growth Conditions (2003)

Vol. 1819: D. Masser, Yu. V. Nesterenko, H. P. Schlickewei, W. M. Schmidt, M. Waldschmidt, Diophantine Approximation. Cetraro, Italy 2000. Editors: F. Amoroso, U. Zannier (2003)

Vol. 1820: F. Hiai, H. Kosaki, Means of Hilbert Space Operators (2003)

Vol. 1821: S. Teufel, Adiabatic Perturbation Theory in Quantum Dynamics (2003)

Vol. 1822: S.-N. Chow, R. Conti, R. Johnson, J. Mallet-Paret, R. Nussbaum, Dynamical Systems. Cetraro, Italy 2000. Editors: J. W. Macki, P. Zecca (2003)

Vol. 1823: A. M. Anile, W. Allegretto, C. Ringhofer, Mathematical Problems in Semiconductor Physics. Cetraro, Italy 1998. Editor: A. M. Anile (2003)

Vol. 1824: J. A. Navarro González, J. B. Sancho de Salas, $\mathscr{C}^\infty$ – Differentiable Spaces (2003)

Vol. 1825: J. H. Bramble, A. Cohen, W. Dahmen, Multiscale Problems and Methods in Numerical Simulations, Martina Franca, Italy 2001. Editor: C. Canuto (2003)

Vol. 1826: K. Dohmen, Improved Bonferroni Inequalities via Abstract Tubes. Inequalities and Identities of Inclusion-Exclusion Type. VIII, 113 p, 2003.

Vol. 1827: K. M. Pilgrim, Combinations of Complex Dynamical Systems. IX, 118 p, 2003.

Vol. 1828: D. J. Green, Gröbner Bases and the Computation of Group Cohomology. XII, 138 p, 2003.

Vol. 1829: E. Altman, B. Gaujal, A. Hordijk, Discrete-Event Control of Stochastic Networks: Multimodularity and Regularity. XIV, 313 p, 2003.

Vol. 1830: M. I. Gil', Operator Functions and Localization of Spectra. XIV, 256 p, 2003.

Vol. 1831: A. Connes, J. Cuntz, E. Guentner, N. Higson, J. E. Kaminker, Noncommutative Geometry, Martina Franca, Italy 2002. Editors: S. Doplicher, L. Longo (2004)

Vol. 1832: J. Azéma, M. Émery, M. Ledoux, M. Yor (Eds.), Séminaire de Probabilités XXXVII (2003)

Vol. 1833: D.-Q. Jiang, M. Qian, M.-P. Qian, Mathematical Theory of Nonequilibrium Steady States. On the Frontier of Probability and Dynamical Systems. IX, 280 p, 2004.

Vol. 1834: Yo. Yomdin, G. Comte, Tame Geometry with Application in Smooth Analysis. VIII, 186 p, 2004.

Vol. 1835: O.T. Izhboldin, B. Kahn, N.A. Karpenko, A. Vishik, Geometric Methods in the Algebraic Theory of Quadratic Forms. Summer School, Lens, 2000. Editor: J.-P. Tignol (2004)

Vol. 1836: C. Năstăsescu, F. Van Oystaeyen, Methods of Graded Rings. XIII, 304 p, 2004.

Vol. 1837: S. Tavaré, O. Zeitouni, Lectures on Probability Theory and Statistics. Ecole d'Eté de Probabilités de Saint-Flour XXXI-2001. Editor: J. Picard (2004)

Vol. 1838: A.J. Ganesh, N.W. O'Connell, D.J. Wischik, Big Queues. XII, 254 p, 2004.

Vol. 1839: R. Gohm, Noncommutative Stationary Processes. VIII, 170 p, 2004.

Vol. 1840: B. Tsirelson, W. Werner, Lectures on Probability Theory and Statistics. Ecole d'Eté de Probabilités de Saint-Flour XXXII-2002. Editor: J. Picard (2004)

Vol. 1841: W. Reichel, Uniqueness Theorems for Variational Problems by the Method of Transformation Groups (2004)

Vol. 1842: T. Johnsen, A. L. Knutsen, K_3 Projective Models in Scrolls (2004)

Vol. 1843: B. Jefferies, Spectral Properties of Noncommuting Operators (2004)

Vol. 1844: K.F. Siburg, The Principle of Least Action in Geometry and Dynamics (2004)

Vol. 1845: Min Ho Lee, Mixed Automorphic Forms, Torus Bundles, and Jacobi Forms (2004)

Vol. 1846: H. Ammari, H. Kang, Reconstruction of Small Inhomogeneities from Boundary Measurements (2004)

Vol. 1847: T.R. Bielecki, T. Björk, M. Jeanblanc, M. Rutkowski, J.A. Scheinkman, W. Xiong, Paris-Princeton Lectures on Mathematical Finance 2003 (2004)

Vol. 1848: M. Abate, J. E. Fornaess, X. Huang, J. P. Rosay, A. Tumanov, Real Methods in Complex and CR Geometry, Martina Franca, Italy 2002. Editors: D. Zaitsev, G. Zampieri (2004)

Vol. 1849: Martin L. Brown, Heegner Modules and Elliptic Curves (2004)

Vol. 1850: V. D. Milman, G. Schechtman (Eds.), Geometric Aspects of Functional Analysis. Israel Seminar 2002-2003 (2004)

Vol. 1851: O. Catoni, Statistical Learning Theory and Stochastic Optimization (2004)

Vol. 1852: A.S. Kechris, B.D. Miller, Topics in Orbit Equivalence (2004)

Vol. 1853: Ch. Favre, M. Jonsson, The Valuative Tree (2004)

Vol. 1854: O. Saeki, Topology of Singular Fibers of Differential Maps (2004)

Vol. 1855: G. Da Prato, P.C. Kunstmann, I. Lasiecka, A. Lunardi, R. Schnaubelt, L. Weis, Functional Analytic Methods for Evolution Equations. Editors: M. Iannelli, R. Nagel, S. Piazzera (2004)

Vol. 1856: K. Back, T.R. Bielecki, C. Hipp, S. Peng, W. Schachermayer, Stochastic Methods in Finance, Bressanone/Brixen, Italy, 2003. Editors: M. Fritelli, W. Runggaldier (2004)

Vol. 1857: M. Émery, M. Ledoux, M. Yor (Eds.), Séminaire de Probabilités XXXVIII (2005)

Vol. 1858: A.S. Cherny, H.-J. Engelbert, Singular Stochastic Differential Equations (2005)

Vol. 1859: E. Letellier, Fourier Transforms of Invariant Functions on Finite Reductive Lie Algebras (2005)

Vol. 1860: A. Borisyuk, G.B. Ermentrout, A. Friedman, D. Terman, Tutorials in Mathematical Biosciences I. Mathematical Neurosciences (2005)

Vol. 1861: G. Benettin, J. Henrard, S. Kuksin, Hamiltonian Dynamics – Theory and Applications, Cetraro, Italy, 1999. Editor: A. Giorgilli (2005)

Vol. 1862: B. Helffer, F. Nier, Hypoelliptic Estimates and Spectral Theory for Fokker-Planck Operators and Witten Laplacians (2005)

Vol. 1863: H. Führ, Abstract Harmonic Analysis of Continuous Wavelet Transforms (2005)

Vol. 1864: K. Efstathiou, Metamorphoses of Hamiltonian Systems with Symmetries (2005)

Vol. 1865: D. Applebaum, B.V. R. Bhat, J. Kustermans, J. M. Lindsay, Quantum Independent Increment Processes I. From Classical Probability to Quantum Stochastic Calculus. Editors: M. Schürmann, U. Franz (2005)

Vol. 1866: O.E. Barndorff-Nielsen, U. Franz, R. Gohm, B. Kümmerer, S. Thorbjønsen, Quantum Independent Increment Processes II. Structure of Quantum Lévy Processes, Classical Probability, and Physics. Editors: M. Schürmann, U. Franz, (2005)

Vol. 1867: J. Sneyd (Ed.), Tutorials in Mathematical Biosciences II. Mathematical Modeling of Calcium Dynamics and Signal Transduction. (2005)

Vol. 1868: J. Jorgenson, S. Lang, $Pos_n(R)$ and Eisenstein Series. (2005)

Vol. 1869: A. Dembo, T. Funaki, Lectures on Probability Theory and Statistics. Ecole d'Eté de Probabilités de Saint-Flour XXXIII-2003. Editor: J. Picard (2005)

Vol. 1870: V.I. Gurariy, W. Lusky, Geometry of Müntz Spaces and Related Questions. (2005)

Vol. 1871: P. Constantin, G. Gallavotti, A.V. Kazhikhov, Y. Meyer, S. Ukai, Mathematical Foundation of Turbulent Viscous Flows, Martina Franca, Italy, 2003. Editors: M. Cannone, T. Miyakawa (2006)

Vol. 1872: A. Friedman (Ed.), Tutorials in Mathematical Biosciences III. Cell Cycle, Proliferation, and Cancer (2006)

Vol. 1873: R. Mansuy, M. Yor, Random Times and Enlargements of Filtrations in a Brownian Setting (2006)

Vol. 1874: M. Yor, M. Émery (Eds.), In Memoriam Paul-André Meyer - Séminaire de Probabilités XXXIX (2006)

Vol. 1875: J. Pitman, Combinatorial Stochastic Processes. Ecole d'Eté de Probabilités de Saint-Flour XXXII-2002. Editor: J. Picard (2006)

Vol. 1876: H. Herrlich, Axiom of Choice (2006)

Vol. 1877: J. Steuding, Value Distributions of L-Functions (2007)

Vol. 1878: R. Cerf, The Wulff Crystal in Ising and Percolation Models, Ecole d'Eté de Probabilités de Saint-Flour XXXIV-2004. Editor: Jean Picard (2006)

Vol. 1879: G. Slade, The Lace Expansion and its Applications, Ecole d'Eté de Probabilités de Saint-Flour XXXIV-2004. Editor: Jean Picard (2006)

Vol. 1880: S. Attal, A. Joye, C.-A. Pillet, Open Quantum Systems I, The Hamiltonian Approach (2006)

Vol. 1881: S. Attal, A. Joye, C.-A. Pillet, Open Quantum Systems II, The Markovian Approach (2006)

Vol. 1882: S. Attal, A. Joye, C.-A. Pillet, Open Quantum Systems III, Recent Developments (2006)

Vol. 1883: W. Van Assche, F. Marcellàn (Eds.), Orthogonal Polynomials and Special Functions, Computation and Application (2006)

Vol. 1884: N. Hayashi, E.I. Kaikina, P.I. Naumkin, I.A. Shishmarev, Asymptotics for Dissipative Nonlinear Equations (2006)

Vol. 1885: A. Telcs, The Art of Random Walks (2006)

Vol. 1886: S. Takamura, Splitting Deformations of Degenerations of Complex Curves (2006)

Vol. 1887: K. Habermann, L. Habermann, Introduction to Symplectic Dirac Operators (2006)

Vol. 1888: J. van der Hoeven, Transseries and Real Differential Algebra (2006)

Vol. 1889: G. Osipenko, Dynamical Systems, Graphs, and Algorithms (2006)

Vol. 1890: M. Bunge, J. Funk, Singular Coverings of Toposes (2006)

Vol. 1891: J.B. Friedlander, D.R. Heath-Brown, H. Iwaniec, J. Kaczorowski, Analytic Number Theory, Cetraro, Italy, 2002. Editors: A. Perelli, C. Viola (2006)

Vol. 1892: A. Baddeley, I. Bárány, R. Schneider, W. Weil, Stochastic Geometry, Martina Franca, Italy, 2004. Editor: W. Weil (2007)

Vol. 1893: H. Hanßmann, Local and Semi-Local Bifurcations in Hamiltonian Dynamical Systems, Results and Examples (2007)

Vol. 1894: C.W. Groetsch, Stable Approximate Evaluation of Unbounded Operators (2007)

Vol. 1895: L. Molnár, Selected Preserver Problems on Algebraic Structures of Linear Operators and on Function Spaces (2007)

Vol. 1896: P. Massart, Concentration Inequalities and Model Selection, Ecole d'Été de Probabilités de Saint-Flour XXXIII-2003. Editor: J. Picard (2007)

Vol. 1897: R. Doney, Fluctuation Theory for Lévy Processes, Ecole d'Été de Probabilités de Saint-Flour XXXV-2005. Editor: J. Picard (2007)

Vol. 1898: H.R. Beyer, Beyond Partial Differential Equations, On linear and Quasi-Linear Abstract Hyperbolic Evolution Equations (2007)

Vol. 1899: Séminaire de Probabilités XL. Editors: C. Donati-Martin, M. Émery, A. Rouault, C. Stricker (2007)

Vol. 1900: E. Bolthausen, A. Bovier (Eds.), Spin Glasses (2007)

Vol. 1901: O. Wittenberg, Intersections de deux quadriques et pinceaux de courbes de genre 1, Intersections of Two Quadrics and Pencils of Curves of Genus 1 (2007)

Vol. 1902: A. Isaev, Lectures on the Automorphism Groups of Kobayashi-Hyperbolic Manifolds (2007)

Vol. 1903: G. Kresin, V. Maz'ya, Sharp Real-Part Theorems (2007)

Vol. 1904: P. Giesl, Construction of Global Lyapunov Functions Using Radial Basis Functions (2007)

Vol. 1905: C. Prévôt, M. Röckner, A Concise Course on Stochastic Partial Differential Equations (2007)

Vol. 1906: T. Schuster, The Method of Approximate Inverse: Theory and Applications (2007)

Vol. 1907: M. Rasmussen, Attractivity and Bifurcation for Nonautonomous Dynamical Systems (2007)

Vol. 1908: T.J. Lyons, M. Caruana, T. Lévy, Differential Equations Driven by Rough Paths, Ecole d'Été de Probabilités de Saint-Flour XXXIV-2004 (2007)

Vol. 1909: H. Akiyoshi, M. Sakuma, M. Wada, Y. Yamashita, Punctured Torus Groups and 2-Bridge Knot Groups (I) (2007)

Vol. 1910: V.D. Milman, G. Schechtman (Eds.), Geometric Aspects of Functional Analysis. Israel Seminar 2004-2005 (2007)

Vol. 1911: A. Bressan, D. Serre, M. Williams, K. Zumbrun, Hyperbolic Systems of Balance Laws. Lectures given at the C.I.M.E. Summer School held in Cetraro, Italy, July 14–21, 2003. Editor: P. Marcati (2007)

Vol. 1912: V. Berinde, Iterative Approximation of Fixed Points (2007)

Vol. 1913: J.E. Marsden, G. Misiołek, J.-P. Ortega, M. Perlmutter, T.S. Ratiu, Hamiltonian Reduction by Stages (2007)

Vol. 1914: G. Kutyniok, Affine Density in Wavelet Analysis (2007)

Vol. 1915: T. Bıyıkoğlu, J. Leydold, P.F. Stadler, Laplacian Eigenvectors of Graphs. Perron-Frobenius and Faber-Krahn Type Theorems (2007)

Vol. 1916: C. Villani, F. Rezakhanlou, Entropy Methods for the Boltzmann Equation. Editors: F. Golse, S. Olla (2008)

Vol. 1917: I. Veselić, Existence and Regularity Properties of the Integrated Density of States of Random Schrödinger (2008)

Vol. 1918: B. Roberts, R. Schmidt, Local Newforms for GSp(4) (2007)

Vol. 1919: R.A. Carmona, I. Ekeland, A. Kohatsu-Higa, J.-M. Lasry, P.-L. Lions, H. Pham, E. Taflin, Paris-Princeton Lectures on Mathematical Finance 2004. Editors: R.A. Carmona, E. Çinlar, I. Ekeland, E. Jouini, J.A. Scheinkman, N. Touzi (2007)

Vol. 1920: S.N. Evans, Probability and Real Trees. Ecole d'Été de Probabilités de Saint-Flour XXXV-2005 (2008)

Vol. 1921: J.P. Tian, Evolution Algebras and their Applications (2008)

Vol. 1922: A. Friedman (Ed.), Tutorials in Mathematical BioSciences IV. Evolution and Ecology (2008)

Vol. 1923: J.P.N. Bishwal, Parameter Estimation in Stochastic Differential Equations (2008)

Vol. 1924: M. Wilson, Littlewood-Paley Theory and Exponential-Square Integrability (2008)

Vol. 1925: M. du Sautoy, L. Woodward, Zeta Functions of Groups and Rings (2008)

Vol. 1926: L. Barreira, V. Claudia, Stability of Nonautonomous Differential Equations (2008)

Vol. 1927: L. Ambrosio, L. Caffarelli, M.G. Crandall, L.C. Evans, N. Fusco, Calculus of Variations and Non-Linear Partial Differential Equations. Lectures given at the C.I.M.E. Summer School held in Cetraro, Italy, June 27–July 2, 2005. Editors: B. Dacorogna, P. Marcellini (2008)

Vol. 1928: J. Jonsson, Simplicial Complexes of Graphs (2008)

Vol. 1929: Y. Mishura, Stochastic Calculus for Fractional Brownian Motion and Related Processes (2008)

Recent Reprints and New Editions

Vol. 1618: G. Pisier, Similarity Problems and Completely Bounded Maps. 1995 – 2nd exp. edition (2001)

Vol. 1629: J.D. Moore, Lectures on Seiberg-Witten Invariants. 1997 – 2nd edition (2001)

Vol. 1638: P. Vanhaecke, Integrable Systems in the realm of Algebraic Geometry. 1996 – 2nd edition (2001)

Vol. 1702: J. Ma, J. Yong, Forward-Backward Stochastic Differential Equations and their Applications. 1999 – Corr. 3rd printing (2007)

Vol. 830: J.A. Green, Polynomial Representations of GL_n, with an Appendix on Schensted Correspondence and Littelmann Paths by K. Erdmann, J.A. Green and M. Schocker 1980 – 2nd corr. and augmented edition (2007)

Vol. 1693: S. Simons, From Hahn-Banach to Monotonicity (Minimax and Monotonicity 1998) – 2nd exp. edition (2008)